RCM3™: Risk-Based Reliability Centered Maintenance

Marius Basson

INDUSTRIAL PRESS, INC.

Industrial Press, Inc.

32 Haviland Street, Suite 3
South Norwalk, Connecticut 06854
Phone: 203-956-5593
Toll-Free in USA: 888-528-7852
Email: info@industrialpress.com

Author(s): Marius Basson
RCM3: Risk-Based Reliability Centered Maintenance
Library of Congress Control Number: 2018959514

© by Industrial Press.
All rights reserved. Published in 2018.
Printed in the United States of America.

ISBN (print):   978-0-8311-3632-1
ISBN (ePUB):   978-0-8311-9486-4
ISBN (eMOBI):   978-0-8311-9487-1
ISBN (ePDF):   978-0-8311-9485-7

Editorial Director: Judy Bass
Copy Editor: Judy Duguid
Compositor: Paradigm Data Services (P) Ltd., Chandigarh
Cover Designer: Janet Romano-Murray

books.industrialpress.com
ebooks.industrialpress.com

10 9 8 7 6 5 4 3 2 1

For Werner and Conrad

Acknowledgments

It has only been possible to write this book with the knowledge and assistance of the many members of the Aladon Network and the people that influenced my career. In particular, the training and mentorship I received from John Moubray followed by the many relationships I built over the years along this career path. The first person who impacted my decision to join The Aladon Network, is Theuns Koekemoer. Thank you for the personal friendship, for introducing me to RCM2 and Aladon. and for the many hours, days, weeks, months and years we worked together on building the Aladon brand and developing the risk-based RCM3 methodology. Theuns laid the foundation for RCM3 built on the original ideas of John Moubray. Theuns spent many unpaid and relentless hours assisting me with the development and testing of the methodology and material. Theuns is also the main developer of the RCM3 software and many other ideas, too many to mention. I will always be grateful.

John Moubray obviously had a huge impact on my career and how I view the world of asset care. Another person I wish to mention is Chris Fynn who gave me a job with New Dimensions Solutions (NDS later acquired by PWC) and relocated me to the USA in 2002. I worked for NDS and PWC collectively for a period of eight years and established myself as a reliability expert working with some of the best organizations in the world. I am truly grateful for the opportunities and for the many clients that helped form and develop me.

Scott Haskins and the leadership at CH2M (now Jacobs Engineering) were also very influential in my career and the decision to acquire Aladon in 2013. All of these folks are friends and custodians of asset management.

To all my colleagues and friends in The Aladon Network, working with you is both fulfilling and challenging and I trust this book will be as much yours as it is mine. Thank you for the feedback and the advice, it also is much appreciated.

I also want to thank Edith Moubray for her support and friendship and for allowing me to write this book—John's journey is alive and continuing.

Finally, a special thank you to Gretha for allowing me to do what I enjoy . . . . .

Contents

Preface   xv

CHAPTER 1
RCM3 Background   1
    About This Book   4

CHAPTER 2
Introduction to RCM   7
  2.1 The Changing World of Maintenance   7
    Safety and Reliability Are Directly Related   10
    The First-Generation Maintenance   11
    The Second-Generation Maintenance   11
    The Third-Generation Maintenance   12
    The Fourth-Generation Maintenance   13
    New Expectations and Reality   23
    New Techniques   24
    The New Developments   24
    The Challenges Facing Maintenance   26
  2.2 Maintenance and RCM   28
  2.3 RCM3: The Eight Questions   29
    Operating Context   30
    Functions and Performance Standards   31
    Failed States (Functional Failures)   32
    Failure Mode   32
    Failure Effects and Consequence Severity   33
    Inherent Risk   34
    Risk Management   34
    Proactive Risk Management Strategies   37

Default Actions for Tolerable Risk Decisions   42
The RCM Strategy Selection Process   43
2.4 Applying the RCM Process   45
2.5 The RCM Risk Mitigation Strategy Selection Process 46
Asset Criticality and Asset Prioritization (ACAP) Process 46
What ACAP Should Deliver   47
2.6 Asset Registry and Verification   48
Asset Verification Process   48
What Asset Verification Should Deliver   49
RCM Planning   50
Setting Clear Performance Standards and
    How These Will be Measured   50
Review Groups   52
Facilitators   53
The Outcomes of an RCM3 Analysis   53
Auditing and Implementation   54
2.7 What RCM Achieves   55
2.8 International Standards   58

CHAPTER 3

Operating Context   **59**
3.1 Documenting the Operating Context   70

CHAPTER 4

Functions   **77**
4.1 Describing Functions   78
4.2 Performance Standards   79
Multiple Performance Standards   82
Quantitative Performance Standards   82
Qualitative Standards   82
Absolute Performance Standards   83
Variable Performance Standards   83
Upper and Lower Limits   84
4.3 Different Types of Functions   86
Primary Functions   87
Secondary Functions   90
4.4 How Functions Should Be Listed   99

CHAPTER 5

Failed States   **101**
5.1 Failures   102

5.2 Failed States (Functional Failures) 103
    Performance Standards and Failure 103
    Who Should Set the Standard? 111
    How Failed States Should Be Recorded 113

CHAPTER 6

Failure Modes and Effects Analysis   115

6.1 What Is a Failure Mode? 115
6.2 Why Analyze Failure Modes 118
6.3 Categories of Failure Modes 122
    Falling Capability 123
    Increase in Desired Performance (or Increase in
        Applied Stress) 126
    Initial Incapability 131
6.4 How Much Detail 132
    Causation 133
    Failure Causes—Root Cause of Failure 134
    Failure Mechanisms 138
    Normal Deterioration Mechanisms 139
    Other Mechanisms 140
    Identifying the Root Cause 140
    Latent Causes 143
    Behavior-Based Safety 143
    Probability 144
    Consequences 145
    What Is Meant by Reasonably Likely? 147
    Cause Versus Effect 148
    Failure Modes and the Operating Context 149
6.5 Failure Effects and Consequence Severity 149
6.6 Sources of Information about Failure Modes
        and Effects 158
    The Manufacturer or Vendor of the Equipment 158
    Generic Lists of Failure Modes 160
    Other Users of the Same Equipment 161
    The People Who Operate and Maintain the Equipment 161
    Resnikoff Conundrum 162
6.7 Levels of the Analysis and the Information Worksheet 164
    Level of Analysis 164
    How Failure Modes and Effects Should Be Recorded 171
    How to Record Failure Effects 174
6.8 A Completed Information Worksheet 181

CHAPTER 7

Failure Consequences and Risk   **183**

7.1 Risk Management for Physical Risks   184
     How Do We Manage Risk?   184
7.2 Defining Risk and Developing a Risk Matrix   195
     What Methodology Is Acceptable?   196
7.3 Technically Feasible and Worth Doing   204
7.4. Hidden and Evident Functions   207
     Categories of Evident Failures   209
7.5 Safety and Environmental Consequences (Physical Risks)   210
     Safety First   210
     Safety and Proactive Maintenance   211
     RCM, Safety Legislation, and Management Standards   213
7.6 Operational Consequences (Economic Risks)   214
     How Failures Affect Operations   216
     Avoiding Operational Consequences   217
7.7 Tolerable Risk (Also Referred to as Non-operational
       Consequences)   222
     Further Points Concerning Tolerable Risks   224
7.8 Risks Associated with Hidden Failures   226
     Hidden Failures and Protective Devices   228
     The Required Availability of Hidden Functions   231
     Calculating the Probability of a Multiple Failure   233
     Routine Maintenance and Hidden Functions Posing a
       Physical Risk   237
     Routine Maintenance and Hidden Functions Posing
       Operational Risk   241
     Hidden Functions: The Decision Process   242
     Further Points about Hidden Functions   243
     Layers of Protection   248
7.9 Conclusion   249

CHAPTER 8

Risk Management Strategies—Technical Feasibility   **253**

8.1 Proactive Strategies and Technical Feasibility   253
8.2 Treating Intolerable Risks—Proactive Failure
       Management Strategies   254
8.3 Age and Deterioration   256
     Age-Related Failures   257
8.4 Age-Related Failures and Preventive Maintenance   260

8.5 Scheduled Restoration and Scheduled Discard Tasks 261
     The Frequency of Scheduled Restoration and
          Scheduled Discard Tasks 263
     The Technical Feasibility of Scheduled Restoration 267
     The Technical Feasibility of Scheduled Discard Tasks 265
     The Effectiveness of Scheduled Restoration and
          Scheduled Discard Tasks 267
8.6 Failures That Are Not Age-Related 269
     Variable Stress 269
     Complexity 272
     Patterns D, E, and F 273
8.7 Potential Failures and On-Condition Maintenance 274
8.8 The P-F Interval 276
     The Net P-F Interval 278
     P-F Interval Consistency 280
8.9 Technical Feasibility of On-Condition Tasks 281
8.10 Categories of On-Condition Techniques 282
     Condition Monitoring 282
     Product Quality Variation 283
     Primary Effects Monitoring 285
     The Human Senses 286
     Selecting the Right Category 287
8.11 On-Condition Tasks: Some of the Pitfalls 289
     Potential and Functional Failures 289
     The P-F Interval and Operating Age 290
8.12 Nonlinear and Linear P-F Curves 291
     The Final Stages of Deterioration 291
     Linear P-F Curves 294
8.13 How to Determine the P-F Interval 298
     Continuous Observation 299
     Start with a Short Interval and Gradually Extend It 299
     Arbitrary Intervals 300
     Research 300
     A Rational Approach 300
8.14 When On-Condition Tasks Are Worth Doing 302
8.15 Selecting Proactive Risk Management Strategies
          (Proactive Tasks) 303
     On-Condition Tasks 305
     Scheduled Restoration Tasks 305
8.16 Combinations of Tasks 306
     The Task Selection Process 307

8.17 Optimization of Protective Devices 313
8.18 Failure-Finding Tasks 314
    Multiple Failures and Failure-Finding 314
    Technical Aspects of Failure-Finding 315
8.19 Failure-Finding Task Intervals 317
    Failure-Finding Intervals, Availability, and Reliability 317
8.20 Rigorous Methods for Calculating FFI 322
    Sources of Data for FFI Calculations 327
    An Informal Approach to Setting Failure-Finding Intervals 328
    Other Methods of Calculating Failure-Finding Intervals 329
    The Practicality of Task Intervals 329
8.21 The Technical Feasibility of Failure-Finding 330
    Failure-Finding Is a Proactive Risk Management Strategy! 331
    What If Failure-Finding Is Not Suitable? 332
8.22 One-Time Change—Engineering Solution
    Applied Proactively 333
    Redesign and Modification 334
    Design and Maintenance 336
    One-Time Changes Applied Proactively 338
    Reliability Centered Design 347

CHAPTER 9

Managing Tolerable Risks—Why Care? 349

9.1 Strategies Dealing with Tolerable Risks 350
    Economic Risks—Operational 350
    Physical Risk—Safety and the Environment 350
9.2 No Scheduled Maintenance 351
9.3 Walk-Around Checks 352
9.4 Spare Parts Optimization and Logistics 353

CHAPTER 10

The RCM3 Decision Process 355

10.1 Developing Risk Mitigation Strategies 355
10.2 The RCM3 Decision Process 357
    Failure Consequences 360
    Proactive Risk Management Strategies 361
    Proactive Maintenance—Predictive and Preventive
        Maintenance Tasks 363
    Proactive Detection—Failure-Finding Tasks and
        Functional Checks 364
    Optimize Existing Protective Devices 366

Combination of Tasks 367
One-Time Change 367
No Scheduled Maintenance 369
Proposed Risk Management Strategy 370
Initial Task Interval 370
Unit of Measure 371
Who Must Do The PM? 371
Quantity of Tradespeople Required to Do the Task 372
Duration of the Task 372
One-Time Change Recommendations 373
Revised Risk Ranking 373
10.3 Completing the Decision Worksheet 374
10.4 Software and RCM 376
Software Functionality 377

CHAPTER 11

Implementing RCM Recommendations   379

11.1 Implementation—the Key Steps 379
11.2 The RCM Audit 380
Who Should Do the Audit 381
When the Audit Should Be Done 381
What the Audit Entails 383
11.3 Risk Management Strategy (Task) Descriptions 386
Basic Information 388
ISO Standards and RCM 388
11.4 Implementing One-Off Changes 389
Changes to the Physical Asset 390
Changes to the Way in Which the Plant
Is Operated 391
Changes to the Capability of People 391
11.5 Work Packages 391
Standard Operating Procedures 391
Maintenance Schedules 392
11.6 Maintenance Planning and Control Systems 394
High- and Low-Frequency Maintenance Schedules 394
Schedules Done by Operators 395
Schedules and Quality Checks 396
High-Frequency Schedules Done by Maintenance 396
Low-Frequency Schedules Done by Maintenance 400
11.7 Reporting Defects 403
11.8 Eliminating Defects 404

CHAPTER 12

Actuarial Analysis and Failure Data   **405**

12.1 The Six Failure Patterns   405
    Failure Pattern B   406
    Failure Pattern E   409
    Failure Pattern C   414
    Failure Pattern D   418
    Failure Pattern F   418
    Failure Pattern A   421
12.2 Technical History Data   422
    The Role of Actuarial Analysis in Establishing
       Maintenance Policies   422
    Specific Uses of Data in Formulating Maintenance Policies   429

CHAPTER 13

Applying the RCM Process   **437**

13.1 Who Knows?   437
13.2 RCM Review Groups   443
13.3 RCM3 Facilitators   447
    Performing Asset Criticality Review and Asset Prioritization   448
    Developing a Risk Framework and Tolerability Thresholds
       (Developing a Risk Matrix)   450
    Planning the RCM Project (Scope and Analysis Boundary)   450
    Applying the RCM Logic   453
    Managing the Analysis   453
    Conducting the Meetings   455
    Time Management and Implementation Planning   457
    Interacting with People Outside the Review Group—
       Administration, Logistics, and Managing Upward   459
    Who Should Facilitate?   462
13.4 Implementation Strategies   463
    The Task Force Approach   463
    The Selective Approach   464
    The Comprehensive Approach   468
    Deciding Which Approach to Use   469
13.5 RCM in Perpetuity—a Living Program   471
13.6 How RCM Should *Not* Be Applied   473
    The Analysis Is Performed at Too Low a Level   473
    Too Hurried or Too Superficial an Application   473
    Too Much Emphasis on Failure Data   473
    Asking a Single Individual to Apply the Process   474

Using the Maintenance Department on Its Own to
    Apply RCM   475
Asking Manufacturers or Equipment Vendors to
    Apply RCM on Their Own   476
Using Outsiders to Apply RCM   477
Using Computers to Drive the Process   478
Conclusion   479
13.7   Building Skills in RCM3   479

CHAPTER 14

What RCM Achieves   481

14.1   Maintenance Effectiveness   482
     Different Ways of Measuring Maintenance Effectiveness   482
     Different Expectations   485
     Different Functions   487
     Conclusion   499
14.2   Maintenance Efficiency   499
     Maintenance Costs   499
     Labor   500
     Spares and Materials   501
     Planning and Control   502
14.3   What RCM Achieves   503
     Greater Safety and Environmental Integrity   503
     Higher Plant Availability and Reliability   505
     Improved Product Quality   507
     Greater Maintenance Efficiency (Cost-Effectiveness)   508

CHAPTER 15

A Brief History of RCM   515

15.1   The Experience of the Airlines   515
15.2   The Evolution of RCM3   519

APPENDIX I

Comparison Between RCM2 and RCM3   521

APPENDIX II

Asset Hierarchies and Functional Block Diagrams   527

Plant Registers and Asset Hierarchies   523
Functional Hierarchies and Functional Block Diagrams   525
System Boundaries   529

APPENDIX III

Human Error   **535**

Principal Categories of Human Error   531
     Anthropometric Factors   532
     Human Sensory Factors   532
     Physiological Factors   533
     Psychological Factors   534
Conclusion   540

APPENDIX IV

A Continuum of Risk   **545**

INDEX   553

Preface

When I first learned about RCM in the 90s, I never thought that one day my career would center around it, nor that I would never stop learning about RCM and its powerful philosophy. A philosophy is a way of thinking and works by asking very basic questions about all things and the connections between them. RCM is more than a methodology or a process, it is a philosophy. When you study RCM, you realize how things *really* work and how they are connected.

The power of RCM and what it can achieve is often underestimated. In the forty years since RCM was first applied in the airline industry, it has been applied to almost all types of assets. As RCM practitioners, we have learned much about how and where to apply it. Over the years, true RCM has drawn criticism for taking too long and tying up too many resources. For me, a properly executed RCM process is the *only* way to ensure an asset will continue to do what its users want it to do (intended function) in its present operating context.

The SAE JA 1011 standard defines the criteria that any process must possess to be called RCM. For many years the Aladon RCM2™ methodology has been recognized as the gold standard for RCM processes. The RCMII book by John Moubray was a key reference in the standard, and has sold more than 100,000 copies. RCM2 has been applied globally on more sites than any other RCM process and The Aladon Network trained more people in RCM than any other organization. That collective knowledge and collaboration of the Aladon Network led to the development of the RCM3™ methodology.

RCM3 is a risk-based approach, and profoundly different from the RCM process defined by the SAE JA 1011 standard. RCM3 is not only more advanced and aligned with the international standards for physical asset management and risk management, but it also allows users to fully understand and quantify the risks associated with owning and operating assets. The RCM3 methodology is based on the initial work introduced by John Moubray in 2003 and is a continuation of many years of rigorous development and testing. In the process of improving the hugely successful RCM2 methodology (RCM2), we have come to realize the absolute brilliance and pioneering work of John Moubray. People who know RCMII will recognize the terminologies and process. The RCM3 process, however. changes the way we look at the importance of the operating context, how protective systems are managed and more importantly, how risk is quantified and mitigated.

The major difference between the two processes is the treatment of failures. In RCM, the major distinction was between hidden and evident failures. Proactive treatment of evident failures (routine maintenance) were considered first before the default actions (run to failure or redesign) were selected. In RCM3, the focus is still on the distinction between hidden and evident failures, but RCM3 now differentiates clearly between intolerable and tolerable risks. This has a profound impact on how decisions are made— the proactive risk management strategies are now more comprehensive, and more decisions are made during the analysis. The time it takes to perform an RCM3 analysis is impacted by the treatment of identified risks. The RCM3 facilitator now focuses on the mitigation of intolerable risks (as defined by the organization's risk framework) during the RCM workshop and treats tolerable risks (if it can be done in a cost-effective way) outside the RCM analysis meetings, using the expertise of individuals rather than the whole review group, thus saving time and money.

This work is intended for everyone who wants to learn more about the risk and reliability associated with operating and maintaining physical assets. The book further provides an overview of the RCM3 process and its benefits, how to apply it and how to build a sustainable reliability program.

Enjoy your RCM3 journey!

Marius Basson

Wilmington
North Carolina
November 2018

RCM3 Background

Since the release of the Nowlan and Heap report in 1978 when the process was first called reliability-centered maintenance (RCM), RCM became very popular, but it also became very distorted through many variations and derivatives that followed since. RCM has been in the minds and on the lips of many people throughout the industrial world for more than 37 years. A number of attempts by different people and organizations have been made to industrialize RCM. There have been spectacular successes, the most notable of which was the development of RCM2™ by John Moubray and the application thereof by the worldwide Aladon Network since 1991. Subsequent significant developments include the development and release of the standards SAE JA1011, "Evaluation Criteria for RCM Processes," and SAE JA1012, "A Guide to the RCM Standard." Many RCM service providers started up during the same time, but very few maintained the rigor and intent that was produced through the years of research and development. A number of other works on the subject of RCM (formal and informal) have been published around the world. This simply illustrates that RCM has become a familiar name in the industry.

Over the years, John Moubray and Marius Basson (the author of this edition) realized that the RCM process, if applied correctly, would change not just the way we do things (in maintenance), but also the way we think.

Marius Basson was trained by John Moubray in RCM2 and has been implementing RCM2 for over 20 years on a full-time basis. After his acquisition of Aladon, he realized it was time for a change and approached a longtime friend and colleague, Theuns Koekemoer, who had left the Aladon Network a decade earlier to develop his own version of RCM, a risk-based approach. This risk-based approach was started by John Moubray more than 15 years ago. John unfortunately passed away before he could finalize and launch his risk-based process.

John struggled with two issues at the time when he started to revise the popular RCM2 approach: the first centered on commercial acceptance (RCM2 was and still is a very popular methodology), and the second was the confusion it may have created in the industry. Theuns Koekemoer took John's ideas and ground-breaking work and developed a revised RCM process, calling it "risk-based RCM". Then Marius and Theuns came together and polished the process and rebranded it as the Aladon RCM3 process. Since the start of their work together in 2014, it soon became evident that the new process had some holes in it, and it needed to be tested and implemented before it could be fully qualified and released. This took 4 years of hard work and many revisions, mostly around the decision logic. To most people familiar with RCM, the changes in RCM3 may seem to be minor (even cosmetic), but it is a dramatic departure from the RCM process described in SAE Standard JA1011, having profound benefits and a shift in the way it is performed and implemented.

RCM3 adds a new dimension to how maintenance and risk management strategies are defined. RCM3 complies fully with SAE Standard JA1011 and surpasses it by extending the functionality to align with the newer ISO standards for risk and asset management while incorporating all the valid RCM methodology steps.

Reliability management has become highly specialized, and with the introduction of new standards and technology, RCM3 places reliability mainstream with organization management systems. RCM3 moves closer to directly influencing and contributing to other business processes.

RCM3 fully integrates with other risk-based approaches such as risk-based inspection (RBI) and root cause failure analysis (RCFA). RBI is mainly focused on vessels under pressure, which are subject to statutory prescriptive inspections. Traditional statutory inspections have some shortcomings, and RBI attempts to address these shortcomings through identifying risks associated with equipment failures. The mechanical deterioration mechanisms are identified through inspections. RBI, like RCM3, takes into consideration the condition of equipment, the risks associated with possible failures, and the specific operating context. RCM3 is applicable to all equipment and plant types. The scope of RCM3 is not limited to any specific type of equipment, plant, or processes. The Aladon RCFA process fully integrates with RCM3 and provides the continuous improvement cycle.

The combination of a failure, the effect of the failure, and the associated consequence poses a specific risk. RCM3 is focused on first identifying the risks involved with possible failures, then quantifying the risks, and then determining the most effective way to deal with such risks in the most appropriate way. When considering physical risk (safety and environmental) and economic risk, the RCM3 process deals with the risks altogether, eliminating or mitigating the risk of equipment failure to a level that will be tolerable to the organization.

Every organization has very specific responsibilities to the organization's owners and investors, to its employees, and to society in general. Therefore, the risks associated with plant, equipment (physical assets), and processes should always be assessed within the context of the organization as a whole.

Risk management has always been an inherent characteristic of RCM. To the experienced practitioner, risk management is, in fact, intuitive, and integrating risk management into RCM is considered a necessary evolution.

RCM3 sets out to highlight and formalize the identification, categorization, and management of risk as part of developing the failure management and maintenance management plan.

About This Book

This book is an updated version of the popular book by John Moubray, called *RCMII*, which has been translated in multiple languages with over 100,000 copies sold. This book uses many of the same concepts and terminology, but includes the updated methodology also known as "risk-based RCM." The book describes the RCM3 process and covers the requirements and the process as well as the implementation strategies.

While Appendix I traces the progression from RCM2 to RCM3 in detail, the following summarizes the highlights, presents the eight questions that characterize the RCM3 process, and describes how RCM3 fully aligns with ISO 31000 and ISO 55000:

- The development of the operating context is not only referred to as an important step; in RCM3 it is now a definite requirement, and it also is the first question of the RCM3 process: "What are the operating conditions (how the equipment or system is being used)?" Together with the functional requirements and associated performance standards, it provides the context for risk management associated with physical assets. The second question in the RCM3 process asks, "What are the functions and associated performance standards of the asset in its present operating context?"

- ISO 31000 defines risk as the combination of the severity of the consequence and the probability that the consequence will happen. The events that pose a risk to the organization (negative deviation from what is expected) can be compared with failed states or functional failures in RCM, which leads to the third question: "In what ways does it fail to fulfill its functions (failed states)?" Each failed state poses a risk, and the failure modes are the events that cause the failed states.

- RCM3 makes a clear distinction between the causes of failure and the mechanisms of failure, therefore minimizing mistakes and ambiguities. The fourth question deals with failure modes and asks, "What causes each failed state (failure modes)?"

- Both the failure characteristics associated with each failure mode and the inherent reliability of the asset or component under consideration will determine the likelihood or probability of failure. RCM3 categorizes risk in two categories—*physical* and *economic risks*—and it uses the severity of the consequence and the probability to quantify the inherent risk for each likely failure mode. This information is captured as failure effects, which is the fifth question in the process: "What happens when each failure occurs (failure effects and consequence severity)?"

- RCM3 focuses on the risk associated with each failure mode. To determine whether the failure matters or not—in other words, whether the failure poses an intolerable risk—the sixth question asks, "What are the risks associated with each failure (inherent risk *quantified*)?"

- All intolerable risks must be mitigated. The seventh question of the RCM3 process asks, "What *must* be done to reduce intolerable risks to a tolerable level (using proactive risk management strategies)?" Where risks are tolerable, the RCM review group doesn't have to consider any further action, saving time, money, and valuable resources. The eighth and last question asks, "What *can* be done to reduce or manage tolerable risks in a cost-effective way?" When risks are tolerable, further optimization can be achieved provided it is done in a cost-effective manner.

- RCM ensures that the minimum safe amount of *maintenance and engineering solutions* will be considered to reduce or manage the risks to tolerable levels as specified by the organization's risk management framework.

CHAPTER 2

Introduction to RCM

2.1 The Changing World of Maintenance

Over the past 50 years, maintenance has changed, and continues to change perhaps more so than any other management discipline. The changes are due to a huge increase in the number and variety of physical assets (plant, equipment, and buildings) that must be maintained throughout the world, much more complex designs, new maintenance techniques, and changing views on maintenance organization and responsibilities.

Maintenance is also responding to changing expectations. These include a rapidly growing awareness of the extent to which equipment failure affects safety and the environment, a growing awareness of the connection between maintenance and product quality, and increasing pressure to achieve high plant availability and to contain costs. The developments and awareness of renewable energy, carbon trading, greenhouse gas (GHG) effects, and climate change brought renewed focus on equipment reliability and the carbon footprint these assets possess. Attempting to comply with governments' and society's environmental expectations, while at the same time maintaining profitable production facilities, tends to create conflict and work in opposite directions.

The changes are testing attitudes and skills in all branches of industry to the limit. Maintenance people are having to adopt completely new ways of thinking and acting, as engineers and as managers. At the same time the limitations of maintenance systems are becoming increasingly apparent, no matter how much they are computerized.

In the face of this avalanche of change, managers everywhere continue to look for the newest approach to maintenance. They want to avoid the false starts and dead ends that always accompany major upheavals. Instead they seek a strategic framework that synthesizes the new developments into a coherent pattern, so that they can evaluate the developments sensibly and apply those likely to be of most value to them and their companies.

RCM3 is a philosophy that provides just such a framework, and if applied correctly within the maintenance and operations organizations, RCM3 will provide the asset integrity and reliability to fulfill the business needs set by the managers and stakeholders.

If it is applied correctly, RCM transforms the relationships between the undertakings that use it, their existing physical assets, and the people who operate and maintain those assets. It also enables new assets to be put into effective service with great speed, confidence, and precision.

This chapter provides a brief introduction to RCM, starting with a look at how maintenance has evolved over the past 80 years.

Since the 1930s, the evolution of maintenance can be traced through four generations. RCM quickly became the cornerstone of the third generation, and the current generation (the fourth generation) can only be viewed in the light of the previous generations. During three previous generations the focus was very much on availability, reliability, safety, and environmental integrity, and even with the strong focus on these aspects, companies still failed to deliver to society's expectations. If we consider all the major industrial accidents caused by equipment failure since the turn of the century (Chernobyl, Piper Alpha, Bhopal, Texas City explosion, Deep Water Horizon, etc.), it is very much proof that a lot must still be done even though the number of deaths caused by industrial accidents is declining. Figure 2.1 illustrates the decline in fatal injuries over the past 20 years.

Data from multiple sources reflect the large decreases in work-related deaths from the high rates and numbers of deaths among workers during the early twentieth century. The earliest systematic survey of workplace fatalities in the United States in the twentieth century covered Allegheny County, Pennsylvania, from July 1906 through June 1907; that year in the

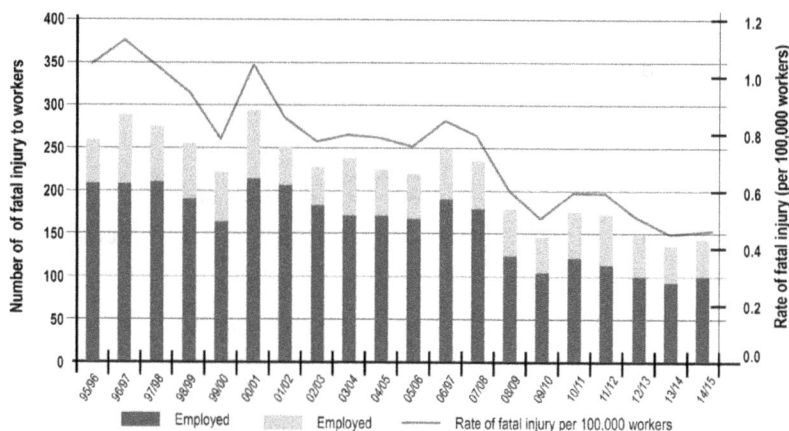

FIGURE 2.1 Rate of fatalities

one county, 526 workers died in *work accidents*; 195 of these were steel-workers. In contrast, in 1997, 17 steelworker fatalities occurred nationwide. The National Safety Council estimated that in 1912, 18,000–21,000 workers died from work-related injuries. And in 1913, the Bureau of Labor Statistics documented approximately 23,000 industrial deaths among a workforce of 38 million, equivalent to a rate of 61 deaths per 100,000 workers.

Under a different reporting system, data from the National Safety Council from 1933 through 1997 indicate that deaths from uninten-tional work-related injuries declined 90%, from 37 per 100,000 workers to 4 per 100,000. The corresponding annual number of deaths decreased from 14,500 to 5,100; during this same period, the workforce more than tripled, from 39 million to approximately 130 million.

More recent and probably more complete data from death certifi-cates were compiled from the CDC's National Institute for Occupational Safety and Health (NIOSH) National Traumatic Occupational Fatalities (NTOF) surveillance system. These data indicate that the annual number of deaths declined 28%, from 7,405 in 1980 to 5,314 in 1995 (the most recent year for which complete NTOF data are available). The average rate of deaths from occupational injuries decreased 43% during the same time, from 7.5 to 4.3 per 100,000 workers. Industries with the highest average rates for fatal occupational injury during 1980–1995 included mining (30.3 deaths per 100,000 workers), agriculture/forestry/fishing (20.1), construction (15.2), and transportation/communications/public utilities (13.4) as illustrated in Figure 2.2. Leading causes of fatal occupational injury during the period include motor vehicle–related injuries, workplace homicides, and machine-related injuries.

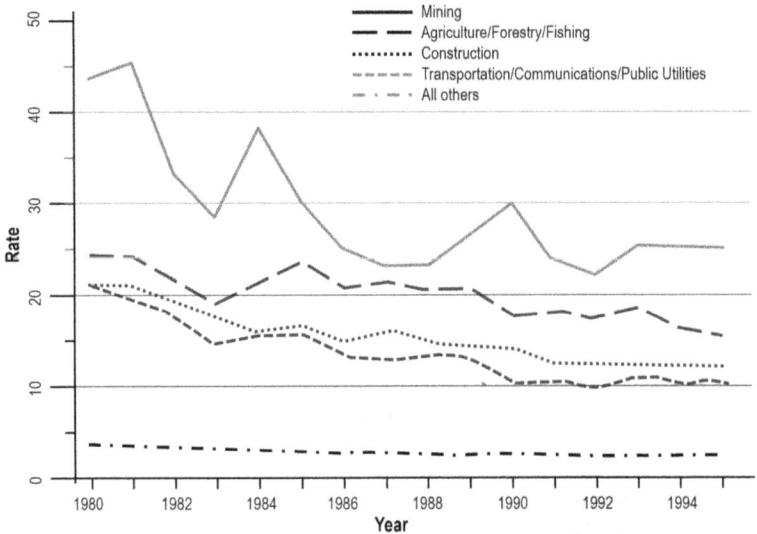

FIGURE 2.2   Occupational injury death rates by industry division, United States, 1980–1995

Companies became preoccupied with safety, and it became top management's performance indicator.

Safety and Reliability Are Directly Related

One reason for the decline in fatalities is the fact that operations people are no longer working next to or close to the equipment, as was the case 50 years ago. However, because of automatization and modernization, incidents and accidents became fewer in numbers but much more severe. Safety is now related to the state of equipment and assets. Safety and reliability are therefore directly related; however, more maintenance does not mean safer operations, which brings us to realize that there is no direct relationship between the amount of maintenance and safety and thus the amount of maintenance and reliability.

More maintenance ≠ more reliable

The First-Generation Maintenance

It is important to understand how industry moved maintenance and the changes in maintenance along. The First Industrial Revolution was a fundamental change in the way goods were produced, from human labor to machines and more efficient means of production and subsequent higher levels of production, triggering far-reaching changes in industrialized societies. The First Industrial Revolution started largely in the UK during the last quarter of the eighteenth century; achievements included the harnessing of steam power, the mechanization of the textile industry, developments in transportation (trains and trams), advances in communication (telegraph, telephone, and radio), and the birth of the modern factory. Machines started to replace humans in agriculture and manufacturing.

The First-Generation Maintenance covers the period up to World War II. In those days, industry was not very highly mechanized, so downtime did not matter much. This meant that the prevention of equipment failure was not a very high priority in the minds of most managers. At the same time, most equipment was simple and much of it was overdesigned. This made it reliable and easy to repair. As a result, there was no need for systematic maintenance of any sort beyond simple cleaning, servicing, and lubrication routines. The need for skills was also lower than it is today.

The Second Industrial Revolution, also known as the Technological Revolution, was a phase of the larger Industrial Revolution corresponding to the latter half of the nineteenth century until World War I. It is considered to have begun with the development of the Bessemer process for making inexpensive steel in the 1860s and culminated in mass production and the production line, improved workflow, and scientific management. The Second Industrial Revolution was driven by electricity, a cluster of inventions, internal combustion engines, airplanes, and moving pictures. Increased mechanization of industry and improvements in worker efficiency increased the productivity of factories while undercutting the need for skilled labor.

The Second-Generation Maintenance

Things changed dramatically during World War II. Wartime pressures increased the demand for goods of all kinds while the supply of industrial labor dropped sharply. This led to increased mechanization. By the 1950s, machines of all types were more numerous and more complex. Industry was beginning to depend on them.

As this dependence grew, downtime came into sharper focus. This led to the idea that equipment failures could and should be prevented, which led, in turn, to the concept of preventive maintenance. In the 1960s, this consisted mainly of equipment overhauls done at fixed intervals.

The cost of maintenance also started to rise sharply relative to other operating costs. This led to the growth of maintenance planning and control systems. These systems have helped greatly to bring the cost of maintenance under control and are now an established part of the practice of maintenance.

Finally, the amount of capital tied up in fixed assets, together with a sharp increase in the cost of that capital, led people to start seeking ways in which they could maximize the life of the assets.

The Third-Generation Maintenance

The first two Industrial Revolutions made people richer and more urban. The Third Industrial Revolution made life and manufacturing go digital. The Third Industrial Revolution included digital technology, personal computing, the internet, and mass customization. According to the *Economist* the Third Industrial Revolution is known for a number of remarkable technologies that converged: clever software, novel materials, more dexterous robots, new processes (notably three-dimensional printing), and a whole range of web-based services. The factory of the past was based on producing identical products: Ford famously said that car buyers could have any color they liked, as long as it was black. Factories now focus on mass customization, using new and lighter materials that are stronger and more durable than the old ones. New techniques shape engineering, and the internet allows ever more designers to collaborate on new products.

Since the mid-seventies, the process of change in industry has gathered even greater momentum. The changes can be classified under the headings of new expectations, new research, and new techniques

Expectations changed from having high availability to wanting reliability also. It placed a great emphasis on maintenance.

Digital technology changed the media and retailing industries. Factories are no longer full of grimy machines manned by workers in oily overalls. Many factories are squeaky clean and almost deserted. Most jobs are no longer on the factory floor but in the offices nearby, full of designers, engineers, IT specialists, logistics experts, marketing staff, and other professionals. Manufacturing jobs require more skills, and dull repetitive tasks have become almost obsolete.

The Third Industrial Revolution affected not only how things are made today, but also where. As a result of the Third Industrial Revolution, labor costs are growing less and less important: Offshore production is increasingly moving back to rich countries, not because Chinese wages are rising, but because companies now want to be closer to their customers so that they can respond more quickly to changes in demand and customization.

The lines between manufacturing and services are blurring. Rolls-Royce no longer sells jet engines; it sells the hours that each engine is actually thrusting an airplane through the sky. OEMs (original equipment manufacturers) would rather sell equipment capability while continuing to own and to maintain the assets. Another example is a large shovel manufacturer that sells tons moved but keeps ownership of the shovels. The operating company pays for the tons and has guaranteed uptime, while the equipment maintenance and repairs remain the responsibility of the OEM. This places an even greater emphasis on inherent reliability and maintenance efficiency.

The Fourth-Generation Maintenance

The Fourth Industrial Revolution is upon us. The following tweet by an IBM global entrepreneur captures the essence of the technological upheavals of the Fourth Industrial Revolution that is currently sweeping through the global economy (this was at the time of writing):

- World's largest taxi company owns no taxis (Uber).
- Largest accommodation provider owns no real estate (Airbnb).

- Largest phone companies own no telecom infrastructure (Skype, WeChat).
- World's most valuable retailers keep no inventory (Amazon and Alibaba).
- Most popular media owner creates no content (Facebook).
- Fastest-growing bank has no actual money (SocietyOne).
- World's largest movie distributor owns no cinemas (Netflix).
- Largest software vendors don't write the apps (Apple and Google).

The Fourth Industrial Revolution, also known as the Second Machine Age, is fundamentally changing each and every aspect of our life and is very different from the previous ones!

Professor Klaus Schwab, founder and executive chairman of the World Economic Forum and author of the recently published book *The Fourth Industrial Revolution*, mentions that the lines between the physical, digital, and biological spheres are getting blurred in the Fourth Industrial Revolution. The revolution is disrupting almost every industry in every country. The "breadth and depth of these changes herald the transformation of entire systems of production, management, and governance."

Not only does the Fourth Industrial Revolution change what we are doing, but it also changes us. We need new economic models and a value shift. It changes the way we collaborate on every level of society and civilization. It will fundamentally alter the way we live, work, and relate to one another, how we generate, supply, and move energy around and interact with machines. Millions of traditional jobs may be lost to technology and robots, but with new education and innovation, millions more will be created. The Fourth Industrial Revolution (and how we respond to it) will lead to better productivity and improved safety, reliability, and quality.

Digitalist Magazine summarizes five key factors that are changing modern businesses:

1. Hyperconnected products that wirelessly collect, store, and send data through the Internet of Things
2. Supercomputing analytical tools that provide, store, and interpret Big Data

3. Cloud computing platforms that collect and store large sets of data
4. Smart technology like wearables, robotics, machine learning, artificial intelligence, and 3D printing
5. Cybersecurity solutions that protect data and soothe privacy concerns from varying physical, human, and virtual threats

For manufacturers this means real-time factory and enterprise-level insights, zooming in on granular levels of the supply chain with high-tech sensors. These sensors securely enable virtual tracking of assets, processes, resources, and products to optimize and automate supply and demand. As more manufacturers employ smart processes into workflows, the amount of waste, energy, and unplanned downtime is forecasted to decrease. None of this will be possible without the correct maintenance response.

Since the early 2000s, more emphasis has been placed on controlling the way organizations apply asset management. Although it is still a long way from being legislated, and very much left to the organizations to demonstrate due diligence toward safety and environmental responsibility, more organizations are being sued and more individuals have had to stand in front of a court explaining why their asset management programs were flawed.

RCM3 provides the rigor and a robust methodology to face the challenges of the Fourth Industrial Revolution. This is our Fourth-Generation Maintenance approach. John Moubray wrote about the Third-Generation Maintenance in his book *Reliability-Centered Maintenance* (*RCMII, Second Edition*), where he explained the *growing expectations of maintenance, the changing views of equipment failure, and changing maintenance techniques.* Similar to how the Fourth Industrial Revolution is building on the Third, Aladon took the foundation of the Third-Generation Maintenance and lessons learned over 30 years in industry and developed the Fourth-Generation Maintenance methodology (RCM3™), which recognizes the shift in demographics, even more changing expectations (outcome based), asset performance monitoring and predictive analytics (Industrial Internet of Things, or IIoT), mobility (World in Motion), and defect elimination (Reliability Centered Design™), all to meet the challenges the Fourth Industrial Revolution brings.

According to Marius Basson from Aladon, the Fourth-Generation Maintenance will bring about the same change in how industry views maintenance when compared with industry's response after Nowlan and Heap released their report "Reliability-Centered Maintenance" in 1978. However, the velocity, scope, and system impact of the Fourth-Generation Maintenance are exponentially faster than the generations passed. See Figure 2.3.

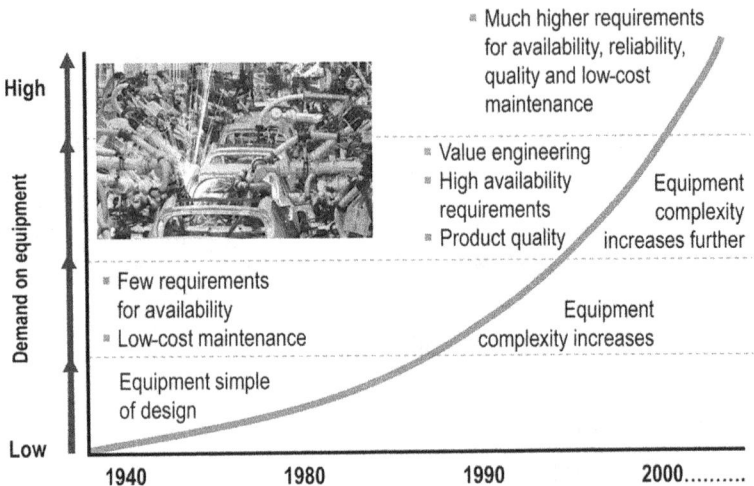

FIGURE 2.3 Increasing demand on equipment

It is the view of the author that management of critical physical assets will be regulated and could soon be legislated, similar to bookkeeping. Companies will no longer be allowed to get away with ignoring their responsibility to maintain assets and the infrastructure better. The risk associated with owning and operating critical assets is just becoming too far-reaching and important.

For example, owners of data centers are now regulated, and compliance is essential. Keeping medical records and financial data is regulated and overseen by government agencies. Companies responsible for keeping data are liable for the data customers put on to those servers. The following is from an article written on the compliance issues faced by data center designers and operators.

Health care is a major area where compliance is critical to offering data center services. Here, the primary concern is maintaining privacy of patient records and information. Laws protect the data that is kept on these servers and data centers that store, process or transmit electronic protected health information must comply with these legislative standards. Again, since health care is such a tremendous portion of the U.S. economy, neglecting compliance in this area cuts data center companies off from access to a very large base of potential customers.

Another big concern is dealing with financial customers. Here, financial regulations such as Sarbanes-Oxley are the focus.

A broader area of compliance is the Payment Card Industry (PCI) Data Security Standard (DSS), which applies to companies handling credit card information, for example. In this case, the scope of clients for whom PCI DSS compliance is critical goes beyond market segments like health care or finance. This multidimensional security standard includes requirements for security management, policies, procedures, network architecture, software design and other critical protective measures to ensure a controlled and secure environment for processing the sensitive information.

It is almost certain that any of the above (if not all) impact our lives directly. The Fourth-Generation Maintenance is required to go beyond traditional methods of dealing with symptoms, most of the requirements are now based on information and how it is collected and distributed. See Figure 2.4.

				4th Generation
			3rd Generation	• Managing physical and economic risks
		2nd Generation	• Higher availability, reliability and throughput	• Standardization and adopting standards (i.e. ISO 55000, ISO 31000)
			• Greater cost-effectiveness	• Globalization
			• Greater safety	• Stewardship and social responsibility
	1st Generation	• Higher availability	• Better product quality	• Renewable strategies
		• Lower costs	• No damage to the environment	• Defect elimination
	• Fix it when it breaks	• Longer asset life	• Longer asset life	• Innovation
1930	1940	1950	1960	1970	1980	1990	2000	2010	2020.....

FIGURE 2.4 Changing expectations

Downtime has always affected the productive capability of physical assets by reducing output, increasing operating costs, and interfering with customer service. By the 1960s and 1970s this was already a major concern in the mining, manufacturing, and transport sectors. In manufacturing, the effects of downtime are being aggravated by the worldwide move towards just-in-time systems, where reduced stocks of work-in-progress mean that quite small breakdowns are now much more likely to stop a whole plant. In recent times, the growth of mechanization and automation has meant that reliability and availability have now also become key issues in sectors as diverse as health care, data processing and warehousing, telecommunications, space exploration, global internet and shared networks (the cloud), and building management.

Greater automation also means that more and more failures affect our ability to sustain satisfactory quality standards. This applies as much to standards of service as it does to product quality. For instance, equipment failures can affect data integrity, financial security, stability of governments, climate control in buildings and the punctuality of transport networks as much as they can interfere with the consistent achievement of specified tolerances in manufacturing.

More and more failures have serious safety or environmental consequences at a time when standards in these areas are rising rapidly. In most parts of the world, the point is approaching where organizations either conform to society's safety and environmental expectations or they cease to operate. This adds an order of magnitude to our dependence on the integrity of our physical assets, one which goes beyond cost and which becomes a simple matter of organizational survival.

At the same time as our dependence on physical assets is growing, so too is their cost to operate and to own. To secure the maximum return on the investment which they represent, they must be kept working efficiently for as long as we want them to.

Finally, the cost of maintenance itself is still rising, in absolute terms and as a proportion of total expenditure. In some industries, it is now the second highest or even the highest element of operating costs. As a result, in only 50 years it has moved from almost nowhere to the top of the league as a cost control priority.

Managing Physical and Economic Risks. Management has always been chartered with maximizing stakeholder and investor share price while trying to avoid disruptions and production losses. This creates opposing priorities. On the one hand, managers are trying to save cost (reducing maintenance expenditure), while on the other hand, they are trying to increase revenue (extending the equipment beyond its capability). Management also takes a view of risk avoidance rather than risk management.

An avoidance policy forces management to *avoid* risks by adopting a zero-tolerance approach (which is unaffordable and unsustainable) or by ignoring the risks and doing nothing (taking zero action) to manage them appropriately. Management either takes a paranoiac view (*avoid it at all costs*) or takes a view of feeling lucky (*this will not happen to us*). Neither of these approaches is sustainable; the first is not achievable, and the second is no longer acceptable. With the release of the international asset management and risk management standards (ISO 55000 and ISO 31000), a renewed focus is placed on managing physical and economic risks rather than *avoiding* or *ignoring* them.

Companies may acknowledge the criticality of their operations and assets but do nothing or little about it unless they recognize it is a risk to attaining business objectives. Once organizations fully understand the risks to their operations, they will act ("How can we reduce our risk?"). Most industrial incidents and accidents are proof of how companies ignored the risk of very critical assets and operations.

RCM3 is such a philosophy that brings reliability and risk management mainstream with other important business processes. RCM3 cuts through and eliminates any differences between industries, cultures, and societies and allows a pragmatic view of risk management.

Example: A Canadian-based coal company acquired a coal mine in Wales, UK. When the Canadian managers showed up at the mine in Wales for the first time, they were shocked to find out that the underground workers used the incline conveyor to "ride out of the mine." At every shift change, production would stop and the workers would jump on the belt at the bottom, lie down to avoid being bumped off, and jump off at the top. It was a very safe and acceptable practice that had been performed at the mine for many years. According to the Canadian Health and Safety Laws, this practice was unacceptable, and the company's management banned this practice immediately. Miners had to walk out of the mine now. This created a big uproar with the Welsh, and management was under pressure to find alternative ways for transporting the miners quickly; there was no rail system in place, and the mine was not set up for using vehicles for personnel transport. After some review and the review of the incident records, management determined that riding the belt was actually a safe practice and reinstated it. The culture, laws, and regulations between the two countries were vastly different, and it became quite the experience when Canadians went to visit. This was not the only difference that had to be overcome.

A pragmatic view of the risk associated with this practice allowed the Canadians to be more open to the idea. The risk associated with this practice was well understood and managed and did not have to be avoided.

Standardization and Adopting Standards. More emphasis is being placed on standardization, not only in equipment selection but also in the way we do things. In order to measure and compare performance across an organization, the need for establishing rules and metrics becomes crucial. Companies want to measure and compare between divisions or locations (where they operate in multiple locations), different regions, market sectors, and industries. Standardization leads to adoption of *best practices* and the creation of *centers of excellence* (COEs). This further leads to *standardized* work management practices and standardized approaches for operations and maintenance.

In our example of the Canadian company and the Welsh mine, standardization and *templating* turned out to be problematic, especially when considering diversity in cultures, regulations and regulatory requirements, governments, and government agencies. What was needed was better risk management strategies where everyone was involved.

Globalization. With the popularity of the internet and internet buying, the market has opened up to everyone. People are buying and getting exposed to information and equipment they never have seen or heard of before. It allows organizations to buy at competitive and lower overall costs. The challenges this brings also have never been experienced before: Controls and program logic are programmed in foreign languages, and so are the maintenance and operating manuals that come with the equipment. As well, equipment reliability (the prevention of failure) has become much more important, because engineers and maintainers are no longer dealing with the supplier or agent two towns over; they are now dealing with OEMs on the opposite side of the world, most times not even speaking the same language. This emphasizes the need for a *common language* and *common set of values* to ensure reliability and maintainability.

Responsible Custodians and Environmental Responsibility. Society as a whole is placing pressure on organizations to be responsible and accountable. The four components of corporate social responsibility (CSR)—economic, legal, ethical, and altruistic duties—should be recognized and supported by management. The different perspectives and proper role of business in society, from profit making to community service provider, are required to be successful.

Much of the confusion and controversy over CSR stems from a failure to distinguish among ethical, altruistic, and strategic forms of

CSR. On the basis of a thorough examination, altruistic CSR is not a legitimate role of business. The ethical CSR, grounded in the concept of ethical duties and responsibilities, is mandatory, and strategic CSR is good for business and society.

This responsibility places enormous pressure on the maintenance function. Failures causing health problems (and injuries or deaths), damage to the environment, and harm to society's valuables (drinking water, wild life, air quality, etc.) are no longer acceptable. Managers and asset owners are being tried and jailed for ignoring their responsibilities.

Renewable Strategies. Renewable strategies are created for sustainable development and a sustainable environment. These strategies typically involve three major technological changes: savings on the demand side (more effective utilization and preservation), efficiency improvements on the production side (increased reliability and availability), and the replacement of current sources (e.g., oil and coal). Consequently, large-scale renewable strategies must include strategies for integrating renewable sources in coherent systems influenced by energy savings and efficiency measures.

These strategies introduced new equipment and thinking to the industry. It also changed the way maintenance is done. Through preservation, recycling, renewal, and conservation, societies are more sustainable compared with traditional thinking of obsolescence and replacement. Companies are incentivized through buying renewable energy, and soon this incentive will spread to other essential resources.

Defect Elimination. Defect elimination is a *proactive* maintenance strategy with seemingly obvious value. Defects are also referred to as "failed states." This is so whether the machine has a defect and still works but produces unacceptable quality or whether the machine is unable to produce anything at all. Maybe the machine still works, but production is slow and the number produced over time is less. The value of the product is reduced in all three cases, and all defects are undesirable.

Defects cost money. If end users candidly investigate, they may see that they are investing money in keeping their defects. How? Money is lost because the products are being sold at a reduced price or because the maximum number of products was not produced. Slowed production or downtime is money invested in keeping defects.

According to *Webster's*, "eliminate" means to "put an end to or get rid of." An asset that is free of defects is an asset that can be optimally profitable.

Innovation. Organizations are constantly looking for innovative ideas. More millionaires are produced through people selling innovative ideas than millionaires producing things. Consider the twenty-first-century ideas and the products we are getting used to more and more: for example, the innovation in energy (wind, solar, wave, and biomass), electric cars and other modes of transport, drones, unmanned vehicles, autonomous mining, etc. The list goes on. For all these new ideas and assets, we have to come up with new maintenance techniques and operating procedures. Equipment is monitored by other equipment, operations are centralized and remote (sometimes in other countries), "driving" is done via satellite and GPS, and maintenance has to use the diagnostics rather than the firsthand feedback from an "operator" at the end of a mission or shift. Maintenance must produce innovative ideas for maintaining new types of equipment also.

New Expectations and Reality

Quite apart from greater expectations and new technologies, research changed many of our most basic beliefs about age and failure. In particular, it is apparent that there is less and less connection between the operating age of most assets and the likelihood they will fail. It is even truer in modern facilities and installations.

Figure 2.5 shows that the earliest view of failure was simply that as things got older, they were more likely to fail. A growing awareness of "start-up failure" led to a widespread second-generation belief in the "bathtub" curve.

However, third-generation research has revealed that not one or two, but six failure patterns actually occur in practice. This is discussed in detail later, but it too had a profound effect on maintenance.

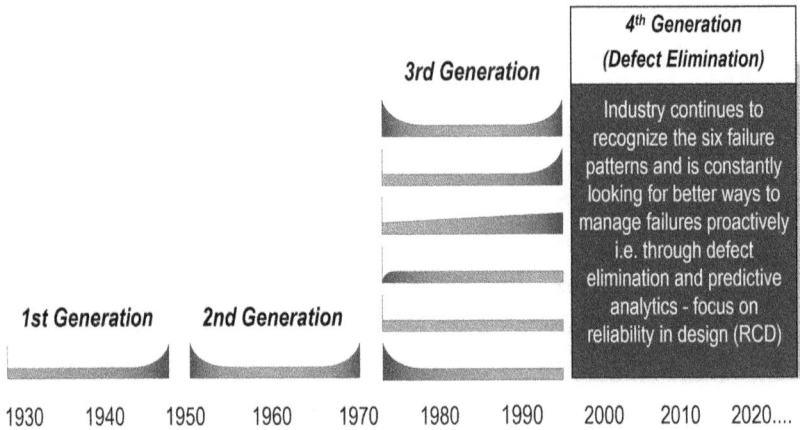

FIGURE 2.5   Changing world of maintenance

In the fourth generation the focus is still on recognizing the six failure patterns, but more so on eliminating failures altogether—a tough ask and even more pressure on *design integrity, production assurance practices, process safety,* and *maintenance.* Failures can no longer be tolerated, and equipment has to be performing at acceptable standards right from the start. Understanding the failure characteristics is essential in the development of risk management strategies, especially predictive and preventive maintenance strategies. The impact this has on the development of cost-effective maintenance strategies is still underestimated and misunderstood.

New Techniques

There has been explosive growth in new maintenance concepts and techniques. Hundreds have been developed over the past 30 years, and more are emerging every week. Figure 2.6 shows how the classical emphasis on overhauls and administrative systems has grown to include many new developments in a number of different fields.

The New Developments

With the development of new tools such as decision support, hazard studies, failure modes and effects analysis, expert systems, and condition

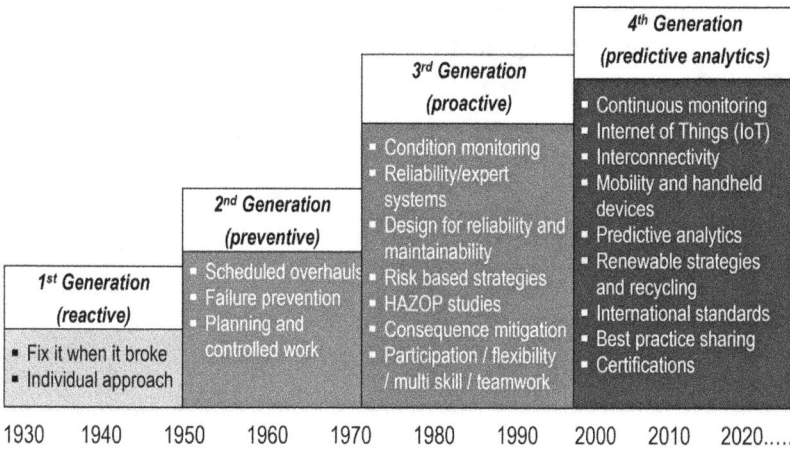

1st Generation (reactive)	2nd Generation (preventive)	3rd Generation (proactive)	4th Generation (predictive analytics)
• Fix it when it broke • Individual approach	• Scheduled overhauls • Failure prevention • Planning and controlled work	• Condition monitoring • Reliability/expert systems • Design for reliability and maintainability • Risk based strategies • HAZOP studies • Consequence mitigation • Participation / flexibility / multi skill / teamwork	• Continuous monitoring • Internet of Things (IoT) • Interconnectivity • Mobility and handheld devices • Predictive analytics • Renewable strategies and recycling • International standards • Best practice sharing • Certifications

1930   1940   1950   1960   1970   1980   1990   2000   2010   2020.....

FIGURE 2.6   New maintenance techniques

monitoring, and with a much greater emphasis on reliability and maintainability when designing equipment, as well as a shift in organizational thinking toward participation, teamwork, and flexibility (internal collaboration), the industry has seen a movement toward communities of practice and knowledge sharing and the establishment of standards and certifications. More reliance is placed on collective experience and proven best practices (external collaboration).

A major challenge for maintenance people continues to be learning and understanding what these new techniques are and to decide which are worthwhile and which are not in their own organizations. Figure 2.7 illustrates how the world of maintenance evolved from a simplistic few of fix it when it broke mentality to more comprehensive maintenance management of planning and scheduling. Figure 2.7 further illustrates that modern thinking involves much more than just maintenance management, encompassing a holistic view of physical asset management. They have to deal not only with these new techniques but also with new equipment they have never experienced before. If we make the right choices, it is possible to improve asset performance and at the same time contain and even reduce the cost of maintenance. If we make the wrong choices, new problems are created while existing problems only get worse.

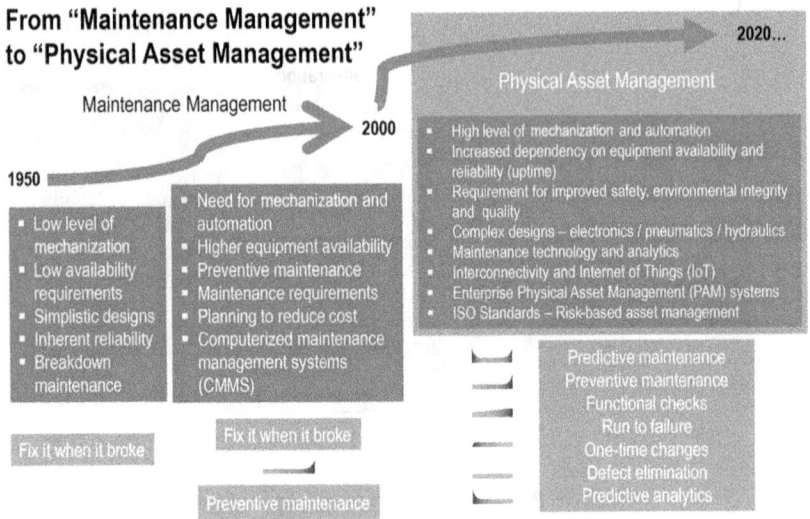

FIGURE 2.7 Physical asset management

The Challenges Facing Maintenance

In a nutshell, the key challenges facing modern maintenance managers can be summarized as follows:

- To select the most appropriate techniques
- To deal with each type of failure process
- In order to fulfill all the expectations of the owners of the assets, the users of the assets, and of society as a whole
- In the most cost-effective and enduring fashion
- With the active support and cooperation of all the people involved
- To demonstrate corporate and social responsibility while achieving production targets
- To understand the challenges that globalization brings
- To understand how data and its use could improve decision-making
- To deal with changing demographics, and
- Build succession around diminishing skills and experience

How we respond to these challenges in an ever-changing world of maintenance can be best compared to building a house:

- **Lay the foundation:** Using RCM3, determine the maintenance requirements of each physical asset in its operating context in order to meet the growing organizational and social expectations while managing physical and economic risks.
- **Build the walls:** Obtain the resources to meet the maintenance requirements. Encourage the use of new reliability techniques in the development of reliability and asset care strategies.
- **Put up the roof:** Meet the requirements as described in the newly adopted international standards and asset management systems (SAE JA 1011, ISO 55000, ISO 31000, etc.). Implement an enterprise asset management (EAM) system and maintenance management systems for effective resource management to ensure compliance.

Many organizations have the roofs up before the walls are in place or even the foundation is laid. This causes ineffective maintenance practices where the equipment schedules the work and not the planners and schedulers.

RCM3 provides a framework that enables users to respond to these challenges, quickly and simply. It does so because it never loses sight of the fact that maintenance is not just about physical assets but also about the people who interact with the assets. If these assets did not exist, the maintenance function itself would not exist, and the maintenance people would not be around. So RCM starts with a comprehensive, *zero-based* review of the inherent risk posed by owning, operating, and maintaining assets, the maintenance and engineering requirements (risk mitigation strategies) of each asset in its operating context utilizing the knowledge and experience of the people who know the equipment best.

All too often, these requirements are taken for granted. This results in the development of organizational structures, the deployment of resources, and the implementation of systems on the basis of incomplete or incorrect assumptions about the real needs of the assets. On the other hand, if these requirements are defined correctly in the light of modern thinking, it is possible to achieve quite remarkable step changes in maintenance efficiency and effectiveness.

The rest of this chapter introduces RCM and RCM3 in more detail. It begins by exploring the meaning of "maintenance" itself. Then it goes on to define RCM and to describe the eight key steps involved in applying the RCM3 process.

2.2 Maintenance and RCM

From the engineering viewpoint, there are two elements in the management of any physical asset. The asset must be maintained, and from time to time it may also need to be modified.

The major dictionaries define "maintain" as "cause to continue" (*Oxford*) or "keep in an existing state" (*Webster's*). This suggests that maintenance means preserving something. On the other hand, both dictionaries agree that to "modify" something means to change it in some way. This distinction between "maintain" and "modify" has profound implications, which are discussed at length in later chapters. However, we focus on maintenance at this point.

When we set out to maintain something, what is it that we wish to cause to continue? What is the existing state that we wish to preserve?

The answer to these questions can be found in the fact that every physical asset is put into service because someone wants it to do something. In other words, someone expects it to fulfill a specific function or functions. So, it follows that when we maintain an asset, the state we wish to preserve must be one in which it continues to do whatever its users want it to do.

Maintenance: Ensuring that physical assets continue to do what their users want them to do.

What the users want will depend on exactly where and how the asset is being used (the operating context). This leads to the following formal definition of reliability-centered maintenance:

Reliability-centered maintenance: A process used to determine the maintenance requirements of any physical asset in its operating context.

In the light of the earlier definition of maintenance, a fuller definition of RCM3 could be:

A process used to define the minimum required safe amount of maintenance, engineering, and other risk management strategies to ensure the minimum tolerable level of safety and environmental integrity and cost-effective operational capability as specified in the organization's asset management system.

2.3 RCM3: The Eight Questions

The RCM3 process requires eight questions to be answered for each asset or system under review:

1. What are the operating conditions (how the equipment or system is being used)?
2. What are the functions and associated performance standards of the asset in its present operating context?
3. In what ways does it fail to fulfill its functions (failed states)?
4. What causes each failed state (failure modes)?
5. What happens when each failure occurs (failure effects and consequence severity)?
6. What are the risks associated with each failure (inherent risk quantified)?
7. What *must* be done to reduce intolerable risks to a tolerable level (using proactive risk management strategies)?
8. What *can* be done to reduce or manage tolerable risks in a cost-effective way?

These questions are touched on briefly in the following paragraphs and are considered in detail in Chapters 3 to 9.

Operating Context

The operating context is defined in SAE JA1012 as "the circumstances in which a physical asset or system is expected to operate." Technically, identical equipment will perform differently if the operating context is different. Therefore, the maintenance program for technically identical equipment can be radically different if the operating contexts are different.

It is of utmost importance that we define the operating context of a physical asset before we attempt to establish what maintenance we should be doing. The following typical operating parameters and conditions should be considered. These will be discussed in detail in Chapter 3.

- Batch process or flow process
- Physical conditions and operating environment
- Product or service quality standards
- Environment and environmental standards
- Safety standards, regulations, and regulatory requirements
- Shift arrangements
- Standby capacity or redundancy
- Work in progress
- Utilization
- Spares policies and logistics
- Current asset condition
- Market demand and raw material supply
- Available skills and technology
- Repurposing, reuse, or recycling

The operating context is the basis on which the risk management strategies and maintenance program are developed. When a decision is made to use RCM3 to develop risk management strategies for physical assets, the first step is to develop the operating context. The operating context is described in more detail in Chapter 3.

Functions and Performance Standards

Before it is possible to apply a process used to determine what must be done to ensure that any physical asset continues to do whatever its users want it to do in its present operating context, we need to do two things:

- Determine what its users want it to do.
- Ensure that it is capable of doing what its users want to start with.

This is why the second step in the RCM3 process is to define the functions of each asset in its operating context, together with the associated desired standards of performance.

What users expect assets to be able to do can be split into two categories: primary functions and secondary functions:

- Primary functions summarize why the asset was acquired in the first place. This category of functions covers issues such as speed, output, carrying or storage capacity, product, quality, and customer service.
- Secondary functions recognize that every asset is expected to do more than simply fulfill its primary functions. Users also have expectations in areas such as safety, control, containment, comfort, cleanliness, structural integrity, economy, protection, efficiency of operation, compliance with environmental regulations, future requirements for recycling, reuse, or repurposing, and even the appearance of the asset.

The users of the assets are usually in the best position by far to know exactly what contribution each asset makes to the physical and financial well-being of the organization as a whole, so it is essential that they are involved in the RCM process from the outset.

Done properly, this step alone usually causes the team doing the analysis to learn a remarkable amount, often a frightening amount, about how the equipment actually works.

Functions are explored in more detail in Chapter 4.

Failed States (Functional Failures)

The objectives of maintenance are defined by the functions and associated performance expectations of the asset under consideration. But how does maintenance achieve these objectives?

The only occurrence that is likely to stop any asset performing to the standard required by its users is some kind of failure. This suggests that maintenance achieves its objectives by adopting a suitable approach to the management of failure. However, before we can apply a suitable blend of failure management tools, we need to identify what failures can occur. The RCM3 process does this at two levels:

- First by identifying what circumstances amount to a failed state
- Then by asking what events can cause the asset to get into a failed state

In the world of RCM, failed states are known as "functional failures" because they occur when an asset is unable to fulfill a function to a standard of performance that is acceptable to the user.

In addition to the total inability to function, this definition encompasses partial failures, where the asset still functions but at an unacceptable level of performance (including situations where the asset cannot sustain acceptable levels of quality or accuracy). Clearly these can only be identified after the functions and performance standards of the asset have been defined.

Functional failures are discussed at greater length in Chapter 5.

Failure Mode

As mentioned above, once each failed state has been identified, the next step is to try to identify all the events that are reasonably likely to cause each failed state. These events are known as "failure modes." "Reasonably likely" failure modes include those that have occurred on the same or similar equipment operating in the same context, failures that are currently being prevented by existing maintenance regimes, and failures that have not happened yet but that are considered to be real possibilities in the context in question. "Unlikely" failure modes (ones that have severe consequences) should also be listed so that the consequences can be assessed early on.

Most traditional lists of failure modes incorporate failures caused by deterioration or normal wear and tear. However, the list should include failures caused by human errors (on the part of operators and maintainers) and design flaws, so that all reasonably likely causes of equipment failure can be identified and dealt with appropriately. It is also important to identify the cause of each failure in enough detail to ensure that time and effort are not wasted trying to treat symptoms instead of causes. On the other hand, it is equally important to ensure that time is not wasted on the analysis itself by going into too much detail.

Failure modes consist of a cause and mechanism (mechanism leading to the failure). Failure modes are discussed at greater length in Chapter 6.

Failure Effects and Consequence Severity

The fifth step in the RCM3 process entails listing failure effects, which describe what happens when each failure mode occurs. These descriptions should include all the information needed to support the evaluation of the consequences of the failure, such as:

- When is the failure most likely to occur (start-up, normal operation, after maintenance)?
- How often would it happen if no attempt is made to prevent it (probability)?
- What evidence (if any) is there that the failure has occurred?
- In what ways (if any) does it pose a threat to safety or the environment?
- In what ways (if any) does it affect production or operations?
- What physical damage (if any) is caused by the failure?
- What must be done to repair the failure?
- Does it cause any secondary damage?
- What is the revenue loss (if any)?

Failure effects are discussed at greater length in Chapter 6.

The process of identifying functions, functional failures, failure modes, and failure effects quantifies the risk and yields surprising and often very exciting opportunities for managing or reducing the risk, improving performance and safety, and also for eliminating waste.

Inherent Risk

A detailed analysis of an average industrial undertaking is likely to yield between 3,000 and 10,000 possible failure modes. Each of these failures affects the organization in some way, but in each case, the effects are different. They may affect operations. They may also affect product quality, customer service, safety, or the environment. They will all take time and cost money to repair.

It is these consequences that most strongly influence the extent to which we try to prevent each failure. In other words, if a failure has serious consequences, we are likely to go to great lengths to try to avoid it. On the other hand, if it has little or no effect, then we may decide to do no routine maintenance beyond basic cleaning and lubrication.

A great strength of RCM3 is that it recognizes that the risk associated with failures are far more important than their technical characteristics. In fact, it recognizes that the only reason for doing any kind of proactive maintenance is not to avoid failures per se, but to avoid or at least to reduce the risk associated with failure. Defining risk and tolerability, developing risk matrixes and performing risk assessments are discussed in Chapter 7.

Risk Management

Risk is generally defined as:

A probability or threat of damage, injury, liability, loss, or any other negative occurrence that is caused by external or internal vulnerabilities, and that may be avoided through preemptive action

According to ISO 31000, "risk" is the "effect of uncertainty on objectives," and an "effect" is a positive or negative deviation from what is expected.

Risk is the combination of the effect (severity) of the event and the likelihood (probability) of the event happening.

$$\text{Risk} = \text{consequence severity} \times \text{probability of consequence happening}$$

RCM3 goes further to distinguish between risks that impact safety and/or the environment (physical risk) and risks with operational and financial impact (economic risk).

The RCM3 process classifies the risks associated with failures into five groups:

Hidden physical risk. Hidden failures have no direct impact, but they expose the organization to multiple failures with serious, often catastrophic, consequences. (Most of these failures are associated with protective devices that are not fail-safe.) In cases where the multiple failures could affect safety or the environment, RCM3 also refers to them as failures with hidden physical risks.

Hidden economic risk. Not all protective devices are installed to protect humans; some protect our equipment from damage and any consequential damage. The failure or bridging of protective devices will increase the risk of the failure of the equipment that it is meant to protect, and it may have very expensive consequences. RCM3 refers to these failures as failures with hidden economic risks.

Evident physical risk (safety and environmental). A failure has safety consequences if it could hurt or kill someone. It has environmental consequences if it could lead to a breach of any corporate, regional, national, or international environmental standard or regulation.

Evident economic risk (operational). A failure has operational consequences if it affects production (output, product quality, customer service, or operating costs in addition to the direct cost of repair).

Evident tolerable risk (nonoperational). Evident failures which fall into this category do not pose an intolerable risk to safety, the environment or operations. These risks are tolerable and for the ones that have operational consequences, the direct cost of repair (and any secondary damage) may be the only effect. *Physical risks* that are considered tolerable according to the organization or society's standard, should not be ignored but be reviewed and managed appropriately. Physical risk with severe consequences should never be tolerated, unless the likelihood of occurrence is extremely low (preferably zero). Tolerable risks are discussed in detail in Chapter 10.

We will see later how the RCM3 process uses these risk categories as the basis of a strategic framework for maintenance decision making. Together with the likelihood of failure (probability), the risk is "quantified." By forcing a structured review of the risk associated with each failure mode in terms of the above categories, RCM3 integrates the operational, environmental, and safety objectives of the maintenance function. This helps to bring safety and the environment into the mainstream of physical asset management.

The risk evaluation process also shifts emphasis away from the idea that all failures are bad and must be prevented. In so doing, it focuses attention on the maintenance activities and other engineering practices that have the most effect on the performance of the organization, and it diverts energy away from those that have little or no effect. It also encourages us to think more broadly about different ways of managing failure and the risk associated with it, rather than to concentrate only on failure prevention. Risk management techniques are applied in two ways, according to the following two categories:

Intolerable risks. These are *physical and economic risks* that are not tolerable to the organization, and all attempts must be made to reduce the risk to a tolerable level. Risk management techniques are applied proactively before a failure occurs or in order to prevent the item from getting into a failed state. They embrace what is traditionally known as "predictive" and "preventive" maintenance, although we will see later that RCM also addresses one-time changes and optimization of protective devices (where relevant) as techniques to reduce the risk proactively. RCM3 uses the terms "scheduled restoration," "scheduled discard," and "on-condition maintenance" for proactive maintenance tasks.

Tolerable risks. These are risks associated with failures that do not impact safety and/or the environment *(although some risks associated with safety or environmental consequences may actually be tolerable if their occurrence is rare, i.e., airline crash rate, failures of nuclear power plants)*, and the economic impact is tolerable (as defined by the organization's risk tolerance). The failed state where the risk is tolerable could be ignored for any further action. Optimization of tolerable risks is, however, possible provided the effort is worth doing (reducing the risk does not cost more than the cost of the failure plus the consequential damage that it is meant to prevent). The default action is run-to-failure while other strategies will be considered to reduce overall downtime or increase inherent reliability *(keeping spare parts or performing a modification)*.

The consequence evaluation and risk mitigation processes are discussed again briefly later in this chapter and then in much more detail in Chapter 6. The next section of this chapter looks at proactive tasks in more detail.

Proactive Risk Management Strategies

Many people still believe that the best way to optimize plant availability is to do some kind of proactive maintenance on a routine basis. Second-generation wisdom suggested that this should consist of overhauls or component replacements at fixed intervals. Figure 2.8 illustrates the fixed-interval view of failure.

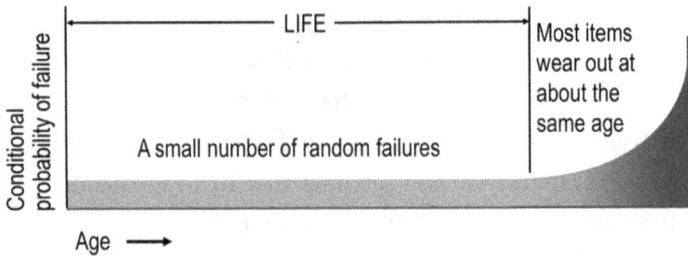

FIGURE 2.8 Traditional view of equipment failure

Figure 2.8 is based on the assumption that most items operate reliably for a period X and then wear out. Classical thinking suggests that extensive records about failure will enable us to determine this life and so make plans to take preventive action shortly before the item is due to fail in the future.

This model is true for certain types of simple equipment and for some complex items with dominant failure modes. In particular, wear-out characteristics are often found where equipment comes into direct contact with the product. Age-related failures are also often associated with fatigue, corrosion, abrasion, and evaporation.

However, equipment in general is far more complex than it was 40 years ago. This has led to startling changes in the patterns of failure, as shown in Figure 2.9. The graphs show conditional probability of failure against operating age for a variety of electrical and mechanical items.

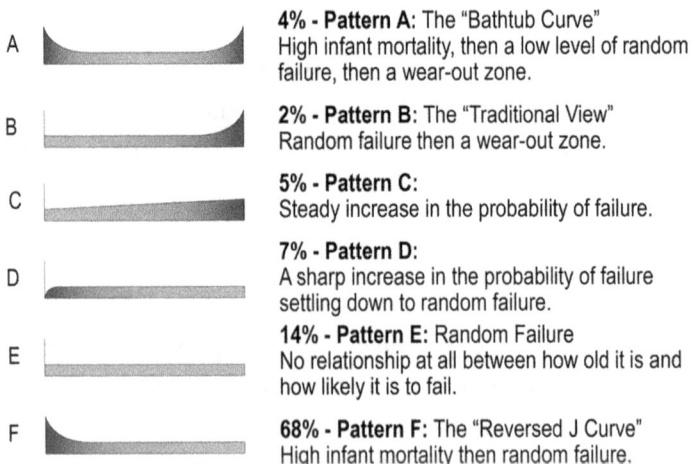

4% - Pattern A: The "Bathtub Curve"
High infant mortality, then a low level of random failure, then a wear-out zone.

2% - Pattern B: The "Traditional View"
Random failure then a wear-out zone.

5% - Pattern C:
Steady increase in the probability of failure.

7% - Pattern D:
A sharp increase in the probability of failure settling down to random failure.

14% - Pattern E: Random Failure
No relationship at all between how old it is and how likely it is to fail.

68% - Pattern F: The "Reversed J Curve"
High infant mortality then random failure.

FIGURE 2.9 Six failure patterns

Pattern A is the well-known bathtub curve. It begins with a high incidence of failure (known as start-up failure), followed by a constant or gradually increasing conditional probability of failure, followed by a wear-out zone. Pattern B shows a constant or slowly increasing conditional probability of failure, ending in a wear-out zone (the same as Figure 2.8).

Pattern C shows a slowly increasing conditional probability of failure, but there is no identifiable wear-out age. Pattern D shows a low conditional probability of failure when the item is new or just out of the shop, then a rapid increase to a constant level, while pattern E shows a constant conditional probability of failure at all ages (random failure). Pattern F starts with a high start-up failure, which drops eventually to a constant or very slowly increasing conditional probability of failure.

Studies done on civil aircraft showed that 4% of the items conformed to pattern A, 2% to B, 5% to C, 7% to D, 14% to E, and no fewer than 68% to pattern F. (The number of times these patterns occur in aircraft is not necessarily the same as in industry. But there is no doubt that as assets become more complex, we see more and more of patterns E and F.)

These findings contradict the belief that there is always a connection between reliability and operating age. This belief led to the idea that the more often an item is overhauled, the less likely it is to fail. Nowadays, this is seldom true. Unless there is a dominant age-related failure mode, age limits do little or nothing to improve the reliability of complex items. In fact, scheduled overhauls can actually increase overall failure rates by introducing start-up failure into otherwise stable systems. It is the authors' experience that more and more equipment installed in modern-day plants and facilities falls into the category of non-age-related failures. It is also true, based on the previous statement, that calendar or time-based maintenance programs contribute very little toward integrity and reliability; they may even be detrimental to the asset's health. It is further true that for many failures no effective form of proactive maintenance exists.

An awareness of these facts has led some organizations to abandon the idea of proactive maintenance altogether. In fact, this can be the right thing to do for failures with minor consequences. But when the failure consequences are significant, something must be done to prevent or predict the failures, or at least to reduce the consequences.

This brings us back to the question of proactive tasks, or proactive risk management strategies as defined in RCM3. RCM3 divides proactive risk management strategies into six categories:

- Scheduled on-condition tasks
- Scheduled restoration tasks
- Scheduled discard tasks
- Combination of tasks (for failures with safety and environmental consequences)
- Failure-finding tasks and functional checks (optimization of protective systems)
- One-time changes (compulsory when no proactive strategy can be found for failures posing a physical risk and recommended for failures posing an economic risk)

On-Condition Tasks. The continuing need to prevent certain types of failure, and the growing inability of classical techniques to do so, is behind the growth of new types of failure management. The majority of these techniques rely on the fact that most failures give some warning that they are about to occur. These warnings are known as potential failures and are defined as identifiable physical conditions that indicate that a functional failure is about to occur or is in the process of occurring.

The new techniques are used to detect potential failures so that action can be taken to avoid the consequences that could occur if the potential failures degenerate into functional failures. They are called "on-condition tasks" because items are left in service on the condition that they continue to meet desired performance standards. (On-condition maintenance includes predictive maintenance, condition-based maintenance, and condition monitoring.)

Used appropriately, on-condition tasks are a very good way of managing failures, but they can also be an expensive waste of time. RCM enables decisions in this area to be made with particular confidence. Many condition monitoring or asset health monitoring devices come standard with a variety of self-diagnostic and monitoring devices. The "asset health indicators" are monitored and are available for download (as in the case of some engine monitoring) or live streaming. Real-time diagnostics are electronically transmitted to centers where people or data analyzers monitor the performance of other equipment. We rely more and more on equipment to monitor the health of yet other equipment, and although very reliable, the consequences of failure are much more severe. Anomalies or deviations generate alarms or notifications for people to investigate. The further we remove the people from the equipment, the more important it becomes to understand the failure characteristics and how to manage them.

Scheduled Restoration and Scheduled Discard Tasks. Scheduled restoration entails remanufacturing a component or overhauling an assembly at or before a specified age limit, regardless of its condition at the time. Similarly, scheduled discard entails discarding an item at or before a specified life limit, regardless of its condition at the time.

Collectively, these two types of tasks are now generally known as preventive maintenance. They used to be by far the most widely used form of proactive maintenance. However, for the reasons discussed above, they are much less widely used than they were 20 years ago.

Combination of Tasks. RCM3 treats a combination of tasks as also a "proactive risk management strategy." A combination of tasks will only be considered where the failure or a multiple failure has physical risk implications (health and safety or damage to the environment). In both cases, RCM3 seeks to proactively reduce the associated risk if a single task cannot be found that would reduce the risk to a tolerable level.

Failure-Finding Tasks and Functional Checks. Failure-finding tasks entail checking hidden functions periodically to determine whether they have failed (whereas condition-based tasks entail checking if *something is failing*). Failure-finding is no longer a default action. Increasing the reliability of a protected function may not be enough to reduce the risk associated with its failure to a tolerable level, but increasing the availability of the protective device (where the function is protected by one or more protective devices) may do so. Furthermore, a one-time change must be considered where the risk of the multiple is intolerable.

Functional checks entail checking the operation of protective devices that fail evident and where the associated risk of the failure is intolerable. The task reduces the risk of the unanticipated failure to a tolerable level, but it does not necessarily prevent the failure altogether.

One-Time Changes. One-time changes entail making any one-off change to the built-in capability of an asset or system. This includes modifications to the hardware and covers one-off changes to procedures and training of operations and maintenance personnel. Where the risk is intolerable, and no proactive strategy would result in reducing the risk to a tolerable level, a one-time change will be necessary. One-time changes can be applied proactively in order to reduce the severity of the consequence.

Default Actions for Tolerable Risk Decisions

RCM recognizes the following default actions:

> **No scheduled maintenance.** As the name implies, this default entails making no effort to anticipate or prevent failure modes to which it is applied, and so those failures are simply allowed to occur and then be repaired. This default is also called "run-to-failure."
>
> **Spare part optimization.** Corrective maintenance done quickly and effectively reduces downtime and the economic risk associated with failures. Spare part policies are considered for risk mitigation in both the tolerable and intolerable risk categories, but they are especially useful in mitigating risks associated with unanticipated failures—where no proactive maintenance task exists.
>
> **Other.** Risk mitigation for tolerable risks is considered only if the mitigation can be applied in a cost-effective way. This does not

exclude the strategies associated with proactive maintenance and one-time changes. These may also be considered provided they are cost-effective.

The RCM Strategy Selection Process

A great strength of RCM is the way it provides simple, precise, and easily understood criteria for deciding which (if any) of the proactive risk management strategies are technically feasible in any context, and if so, for deciding how often they should be done and who should do them. These criteria are discussed in more detail in Chapter 8.

Whether or not a proactive task is technically feasible is governed by the technical characteristics of the task and of the failure that it is meant to prevent. Whether it is worth doing is governed by how well it deals with the risk associated with the failure. If a proactive task cannot be found that is both technically feasible and worth doing, or a combination of tasks cannot be found for a failure or a multiple failure with safety and environmental consequences, then a suitable one-time change must be implemented. The essence of the task selection process is as follows:

- For hidden failures posing a safety or environmental risk (physical risk), a proactive task is worth doing if it reduces the risk of the multiple failure associated with that function to a tolerable low level. If such a task cannot be found, then a scheduled failure-finding task must be performed. If a suitable failure-finding task cannot be found, then a compulsory one-time change must be implemented. The item must be redesigned, or the process has to be changed, to reduce the probability of a multiple failure to a tolerable level.
- For hidden failures posing an economic risk, a proactive task is worth doing if it reduces the probability of the multiple failure associated with that function to a tolerable low level. If such a task cannot be found, then a scheduled failure-finding task must be performed (at intervals where the sum of the cost of doing

the failure-finding task and the associated cost of the multiple failure is a minimum). If a suitable failure-finding task cannot be found, then a one-time change must be considered to reduce the overall risk. The item must be redesigned, or the process has to be changed, to reduce the probability of a multiple failure to a tolerable level.

- For failures posing an evident safety or environmental risk (physical risk), a proactive failure management strategy is only worth doing if it reduces the risk of that failure on its own to a very low level indeed, if it does not eliminate it altogether. If a single task cannot be found that reduces the risk of the failure to an acceptably low level, a combination of tasks will be considered; and in the event that the combination of tasks is not feasible, the optimization of the protective device (where functions are protected by one or more protective devices) must be considered. In the event that the risk is still intolerable, a one-time change is compulsory (the item must be redesigned, or the process must be changed, in order to make it tolerably safe).

- If the failure poses an evident operational risk (economic risk), a proactive task is only worth doing if the total cost of doing it over a period of time is less than the cost of the operational consequences and the cost of repair over the same period. In other words, the task must be justified on economic grounds. If this occurs and the operational consequences are still unacceptable (intolerable), then the optimization of protective systems will be considered for functions that are protected by one or more protective devices. If the risk associated with failure is still intolerable, a one-time change will be considered next in order to reduce the economic risk. If it is not justified, no scheduled maintenance will be selected as a default task.

- If a failure poses an evident nonoperational or tolerable risk, a proactive task is only worth doing if the cost of the task over a period of time is less than the cost of repair over the same period. So these tasks must also be justified on economic grounds. If they are not justified, the initial default decision is again no scheduled maintenance, and if the repair costs are too high, the secondary default decision is a one-time change.

This approach means that proactive strategies are only specified for failures that really need them, which, in turn, leads to substantial reductions in routine workloads and savings. Less routine work also means that the remaining tasks are more likely to be done properly. This together with the elimination of counterproductive tasks leads to more effective maintenance.

Compare this with the traditional approach to the development of maintenance policies. Traditionally, the maintenance requirements of each asset are assessed in terms of its real or assumed technical characteristics, without considering the consequences of failure. The resulting schedules are used for all similar assets, again without considering that different consequences apply in different operating contexts. This results in large numbers of schedules that are wasted, not because they are wrong in the technical sense, but because they achieve nothing.

Note also that the RCM3 process considers the maintenance requirements of each asset before asking whether it is necessary to reconsider the design. This is simply because the maintenance engineer who is on duty today has to maintain the equipment as it exists today, not what should be there or what might be there at some stage in the future.

2.4 Applying the RCM Process

Before setting out to analyze the maintenance requirements of the assets in any organization, we need to know what these assets are and to decide which of them are to be subjected to the RCM review process. This means that a plant register must be prepared if one does not exist already. In fact, the vast majority of industrial organizations nowadays already possess plant registers that are adequate for this purpose, so this book does not cover this issue. Of note, however, is how to determine which assets should be covered by RCM. This is again discussed at length in Chapter 15, and is an important consideration prior to the start of the RCM program.

2.5 The RCM Risk Mitigation Strategy Selection Process

Asset criticality and asset prioritization (ACAP) is a methodology that sets the basis for developing the required business case to support reliability improvement initiatives. ACAP could lead to additional benefits when determining asset strategies and subsequent maintenance management.

Asset Criticality and Asset Prioritization (ACAP) Process

The ACAP process includes both the review of the data and information normally contained in the Work Management System (WMS) as well as field observation. Based on the outcome of the ACAP process, it is essential to determine a tolerable level of risk for each asset system. The ACAP process considers redundancy *(standby systems)* and protective systems and follows basic principles to determine the relative risk each asset *(in its operating context)* poses to the organization. These are:

- What is the worst-case, reasonably likely functional failure (the asset no longer does what the users want it to do)?
- If a hidden protective device is present, can the failure consequences be reduced or eliminated? If so, analyze the protective device separately and ensure that the device is present in the physical asset hierarchy.
- Do all the subsystems share the same consequences of failure? If not, start at a lower level.
- What are the consequences of this failure by category (Safety, Environmental, Operations)?
- How often does this failure occur (likelihood)?
- What is a tolerable level of risk?

Following the asset criticality assessment and asset prioritization, the organization should have an overview of the assets organized by the asset hierarchy and categorized by relative importance and the impact they have on business objectives. The ACAP outcomes are used to prioritize the RCM process (where to start) and what to include.

> Maintenance frequencies are not determined by asset criticality, but asset criticality determines which systems to consider for maintenance.

What ACAP Should Deliver

The ACAP process delivers the following benefits:

- Focused on the organization's objectives; it is developed and applied by the stakeholders, which include operations and maintenance
- A fast and low-resource intensive methodology that provides solid, repeatable results
- A sensible and defensible methodology that follows a well-documented consistent structured approach
- Considers safety, environment, and operational consequences as well as any other category (i.e., regulatory requirements)
- Provides a measure of the impact of failure and the impact of non-reliability; two very different measures with unique applications
- Can easily be adapted to the organization's operating context
- Well defined and maintained protective devices and redundant systems
- Work order prioritization based on asset criticality—focus resources where they make the biggest impact

Once completed, the ACAP process would, based on asset criticality, determine the assets that should be considered for a *zero-base* RCM3 analysis. The ACAP process should also reveal where the process of analysis process should start. For assets that are below the criticality threshold, the analysis could be pushed out to a later stage in the reliability improvement program or it may even be possible to apply a less rigorous process such as Maintenance Task Analysis (MTA) to determine or optimize the maintenance program for these assets. The MTA methodology is not covered in this book but can be compared to a planned maintenance optimization process (PMO), which is not a *zero-base* process and should not be considered for critical assets.

The author, however, does not recommend that no analyses are performed on assets that are not deemed critical. It is typical that between 15% and 20% of the asset install-base fall in the *critical category*. If the remaining 80% and 85% *non-critical assets* perform badly, the improvement benefits of the *critical assets* may be lost, or the loss may even exceed the gains made by the RCM program. A comprehensive review of all the assets are required to improve performance and add to the bottom line; using RCM on all of them is not required although some users decide to do so.

2.6 Asset Registry and Verification

Asset registers and hierarchies are rarely complete or correct. This may be for many reasons ranging from incorrect data entry or asset modifications and/or replacements. Critical assets are normally recorded more accurately but it pays to verify that the asset registry contained in the Work Management System and the install-base are true reflections of one another. The asset verification process should consist of a table-top review and field verification. The RCM Facilitator or RCM project team should review and compare the information and data contained in the WMS with the equipment installed through the verification process and notify the stakeholders of any anomalies or discrepancies.

Asset Verification Process

The AV process includes both a table-top review of the data and information contained in the MS as well as field observation and verification. The project team should verify the assets exist and that there are no duplicate, missing, or obsolete assets. The AV process includes a qualitative condition assessment where obvious and relevant findings should be shared with the project team. These may include evidence of leaks, corrosion, vibration, exposed wiring, hazards, loose connections and fittings, obsolescence, missing warning signs and safety equipment, etc.

An AV process follows the following steps:

- What equipment is included? (asset identification and location)
- Asset classification and priority as given by failure category
- Equipment attributes and information (make, model, manufacturer, serial number, specific asset information)
- Standalone or redundant system (primary or secondary system)
- Protective devices and safety systems

Following the Asset Verification, the RCM project team should have an accurate record of the critical assets in each location. Each asset registry verification will produce a list with anomalies (if any) and urgent findings (if any).

What Asset Verification Should Deliver

- Single source of information kept in a single database (WMS)
- RCM analysis focused on the correct equipment information and configuration
- No duplication or omissions of assets
- Obsolescence identified early on in the process (no need for RCM review and analysis)
- Asset performance criteria verified (important for defining functions in the RCM process)
- Secondary functions identified and recorded (important for defining secondary functions in the RCM process)
- Qualitative condition assessment leads to proactive failure management strategy, even before the RCM process recommendations are available

RCM Planning

If it is correctly applied, RCM leads to remarkable improvements in maintenance effectiveness, and it often does so surprisingly quickly. However, the successful application of RCM depends on meticulous planning and preparation. The key elements of the planning process are as follows:

- Decide which assets are most likely to benefit from the RCM process and exactly how they will benefit. Our recommendations are to use a formal approach for determining which assets and systems will be analyzed using RCM and also the priority in which they will be analyzed. The asset criticality is still the best method for determining which assets pose the biggest threat for achieving business objectives. The level of effort required to analyze the assets should also be considered during the planning phase. Both the criticality assessment and level of effort will be discussed in more detail in Chapter 11.
- Assess the resources required to apply the process to the selected assets.
- In cases where the likely benefits justify the investment, decide in detail who is to perform and who is to audit each analysis, and when and where, and arrange for them to receive appropriate training.
- Develop the operating context and ensure that the operating context of the asset is clearly understood.

Setting Clear Performance Standards and How These Will be Measured

Organizations consider RCM for different reasons ranging from low equipment reliability and availability, one-time events causing catastrophic failure, excessive downtime caused by unplanned stoppages, regulations and regulatory requirements, excessive maintenance being performed or compliance to international standards for physical asset management (ISO 55000). The reason for doing RCM should not be: "others do it and it sounds like a good idea." Selecting RCM for these reasons will lead to failure. Before any RCM-based reliability improvement program starts, the organization should have a very clear understanding of what is to be achieved and how it will be measured. The success of the program can only be evaluated if these standards are set at the start and the to-be state can be compared to the as-is state. There are many metrics and performance indicators that can be used for measuring the success of the program and many books and whitepapers have been written on the topic. What is worthy of noting in this book is how availability and reliability are determined and how they are used throughout this book.

Availability = Uptime + Downtime

Downtime = unplanned stoppages + planned maintenance downtime

Reliability is defined as: "the probability that any physical asset will perform its function to satisfactory performance standards for a given period of time under specified operating conditions."

Reliability is often expressed as the Mean Time Between Failures or Mean Time To Failures. Figure 2.10 illustrates how these are determined.

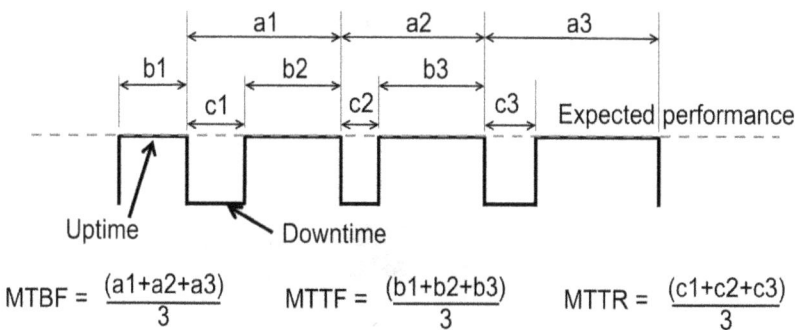

$$MTBF = \frac{(a1+a2+a3)}{3} \qquad MTTF = \frac{(b1+b2+b3)}{3} \qquad MTTR = \frac{(c1+c2+c3)}{3}$$

MTTR is the "Mean Time to Repair"

MTTR in this example is also the "Mean Down Time" (MDT)

FIGURE 2.10   Mean Time Between Failure, Mean Time To Failure, and Mean Time To Repair

It is worth mentioning at this point that MTBF or MTTF is not used for determining any maintenance intervals. It is a useful metric for determining inherent reliability and overall performance and used when budgets and spare parts are determined. No maintenance schedule should be based on the MTBF of an asset or asset system (other than the failure-finding intervals for protective systems). Similarly, the criticality of an asset does not have any bearing on how often it should be maintained.

Review Groups

We have seen how the RCM3 process embodies eight basic questions. In practice, maintenance people simply cannot answer all these questions on their own. This is because many (if not most) of the answers can only be supplied by engineering, production, or operations people. This applies especially to questions concerning functions, desired performance, failure effects, and failure consequences.

For this reason, a review of the maintenance requirements of any asset should be done by small teams that include at least one person from the engineering and maintenance function and one from the operations function. The seniority of the group members is less important than the fact that they should have a thorough knowledge of the asset under review. Each group member should also have been trained in RCM. The makeup of a typical RCM review group is shown in Figure 2.11.

Facilitator

| Operations Supervisor | RCM3 Review Group | Maintenance Supervisor |
| Operator | | Maintainer |

Specialist

FIGURE 2.11 Typical RCM review group

Not only do these groups enable management to gain access to the knowledge and expertise of each member of the group on a systematic basis, but the members themselves gain a greatly enhanced understanding of the asset in its operating context. The people who operate and maintain the equipment participate.

Facilitators

RCM review groups work under the guidance of highly trained specialists in RCM, known as facilitators. The facilitators are the most important people in the RCM review process. Their role is to ensure that:

- The RCM analysis is carried out at the right level, the system boundaries are clearly defined, no important items are overlooked, and the results of the analysis are properly recorded.
- RCM is correctly understood and applied by the group members.
- The group reaches consensus in a brisk and orderly fashion while retaining the enthusiasm and commitment of individual members.
- The analysis progresses reasonably quickly and finishes on time.
- Facilitators also work with RCM project managers or sponsors to ensure that each analysis is properly planned and receives appropriate managerial and logistic support.

Facilitators and RCM review groups are discussed in more detail in Chapter 11.

The Outcomes of an RCM3 Analysis

If it is applied in the manner suggested above, an RCM3 analysis results in six tangible outcomes:

- **Operating context** (especially useful for ensuring perpetuity of the RCM analysis)
- **Inherent risk** associated with operating and maintaining the asset *defined* and *quantified*

- **Maintenance schedules** (risk management strategies) to be done by the maintenance department (including functional checks)
- **Revised operating and maintenance procedures** for the operators and maintainers of the asset
- **Revised or residual risk** after implementation of the risk management strategies
- **One-time changes** for the safe continued operation of the assets (change the way it matters) made to the design of the asset or the way in which it is operated to deal with situations where the asset cannot deliver the desired performance in its current configuration and where proactive risk management strategies associated with proactive maintenance alone will not reduce the risk to a tolerable low level

Two less tangible outcomes are that participants in the process learn a great deal about how the asset works, and also tend to function better as teams.

Auditing and Implementation

Immediately after the review has been completed for each asset, senior managers with overall responsibility for the equipment must satisfy themselves that decisions made by the group are sensible and defensible.

After each review is approved, the recommendations are implemented by incorporating maintenance schedules into maintenance planning and control systems, by incorporating operating procedure changes into the standard operating procedures for the asset, and by handing recommendations for design changes to the appropriate design authority. Key aspects of auditing and implementation are discussed in Chapter 11.

The RCM3 process determines the inherent risk associated with failures (while nothing is done to prevent the failures from happening) and the revised risk based on the risk mitigation strategy selected—implementation is key to the actual risk management since the risk will not be revised unless the strategy is fully implemented. Organizations can no longer afford not to implement an RCM recommendation.

2.7 What RCM Achieves

Desirable as they are, the outcomes listed above should only be seen as a means to an end. Specifically, they should enable the maintenance function to fulfill all the expectations listed in Figure 2.4 at the beginning of this chapter. Third-generation expectations are still valid and how the maintenance function fulfills these and the expectations of the fourth generation is summarized in the following paragraphs, and discussed again in more detail in Chapter 14.

- **Greater safety and environmental integrity**. RCM considers the safety and environmental implications of every failure mode before considering its effect on operations. This means that steps are taken to minimize all identifiable equipment-related safety and environmental hazards, if not eliminate them altogether. By integrating safety into the mainstream of maintenance decision making, RCM also improves attitudes to safety. It is a known fact by now that safety and reliability are directly related.

- **Improved operating performance (output, product quality, and customer service)**. RCM recognizes that all types of maintenance have some value, and provides rules for deciding which is most suitable in every situation. By doing so, it helps ensure that only the most effective forms of maintenance are chosen for each asset and that suitable action is taken in cases where maintenance cannot help. This much more tightly focused maintenance effort leads to quantum jumps in the performance of existing assets where these are sought.

- **Reduced risk.** The biggest benefit achieved from the RCM3 process is reduced risk. Risk is "quantified" throughout the process, and intolerable risk levels are addressed proactively, first, through the maintenance policies (predictive and preventive maintenance) changing the likelihood of failure and, second, through failure-finding tasks and on-time changes changing the consequences of failure (although one-time changes may also change the probability of failure).

RCM was developed to help airlines draw up maintenance programs for new types of aircraft before they enter service. As a result, it is an ideal way to develop such programs for new assets, especially complex equipment for which no historical information is available. This saves much of the trial and error that is so often part of the development of new maintenance programs—trial, which is time-consuming and frustrating, and error, which can be very costly.

• **Greater maintenance cost-effectiveness.** RCM continually focuses attention on the maintenance activities that have the most effect on the performance of the plant. This helps to ensure that everything spent on maintenance is spent where it will do the best.

In addition, if RCM is correctly applied to existing maintenance systems, it reduces the amount of routine work (in other words, maintenance tasks to be undertaken on a cyclic basis) issued in each period, usually by 40% to 70%. On the other hand, if RCM is used to develop a new maintenance program, the resulting scheduled workload is much lower than if the program is developed by traditional methods.

• **Longer useful life of expensive items.** This is due to a carefully focused emphasis on the use of on-condition maintenance techniques.

• **A comprehensive database.** An RCM review ends with a comprehensive and fully documented record of the maintenance requirements of all the significant assets used by the organization. This makes it possible to adapt to changing circumstances (such as changing shift patterns or new technology) without having to reconsider all maintenance policies from scratch. It also enables equipment users to demonstrate that their maintenance programs are built on rational foundations (the audit trail required by more and more regulators). Finally, the information stored on RCM worksheets reduces the effects of staff turnover with its attendant loss of experience and expertise.

An RCM review of the maintenance requirements of each asset also provides a much clearer view both of the skills required to maintain each asset and of what spares should be held in stock. A valuable by-product is the improved drawings and manuals.

- **Greater motivation of individuals.** especially people who are involved in the review process. This leads to a greatly improved general understanding of the equipment in its operating context, together with wider ownership of maintenance problems and their solutions. It also means that solutions are more likely to endure.
- **Better teamwork.** RCM provides a common, easily understood technical language for everyone who has anything to do with maintenance. This gives maintenance and operations people a better understanding of what maintenance can (and cannot) achieve and what must be done to achieve it.
- **More people learn more about the equipment.** The process is especially useful to capture and retain the knowledge and experience of the workforce.
- **Standardization.** When applied on multiple assets across multiple sites, RCM provides a robust framework for ensuring consistent, repeatable maintenance strategies with uniform performance standards.

All these issues are part of the mainstream of physical asset management, and many are already the target of improvement programs. A major feature of RCM is that it provides an effective step-by-step framework for tackling all of them at once, and for involving everyone who has anything to do with the equipment in the process.

RCM yields results very quickly. In fact, if the RCM reviews are correctly focused and correctly applied, they can pay for themselves in a matter of months and sometimes even a matter of weeks, as discussed in Chapter 12. The reviews transform both the perceived maintenance requirements of the physical assets used by the organization and the way in which the maintenance function as a whole is perceived. The result is more cost-effective, more harmonious, and much more successful maintenance.

2.8 International Standards

Standards are not new to industry, and for many years, organizations adopted standards, e.g., ISO 9001, ISO 14001, etc., to improve their business and demonstrate compliance. Companies used standards not only to improve their business but also to achieve a competitive advantage. The new international standards for asset management, ISO 55000 and risk, ISO 31000, will have the same effect—once the standards are adopted and more widely implemented, organizations will feel compelled to implement the standard. Standards are guidance documents about what businesses should do, but standards never prescribe how businesses should go about it. RCM is not a requirement, but the process underpins the requirements of the standard, and companies that do RCM will find it easier to comply with the ISO standard than those that do not. This book does not cover the ISO standards in more detail.

RCM3 fully complies with SAE JA1011 and SAE JA1012 standards for reliability-centered maintenance. This SAE JA 1011 standard for reliability-centered maintenance (RCM) is intended for use by any organization that has or makes use of physical assets or systems that it wishes to manage responsibly.

RCM is a specific process used to identify the policies that must be implemented to manage the failure modes that could cause the functional failure of any physical asset in a given operating context. The SAE standard is intended to be used to evaluate any process that purports to be an RCM process, in order to determine whether it is a true RCM process. This book supports such an evaluation by specifying the minimum characteristics that a process must have in order to be an RCM process.

CHAPTER 3

Operating Context

A requirement to do a proper RCM analysis is the development of the operating context. This is also the first step of RCM3 and probably the most undervalued step in the whole process.

Operating context can be defined as:

The circumstances in which a physical asset or system is expected to operate

The operating context not only provides the context in which we operate, but also lets us determine how functions, failed states, failure modes, failure effects, failure consequences and associated risks, and risk management strategies for identical equipment may differ if the operating context is different. SAE Standard JA1011 refers to the operating context and describes how important it is. Because of its importance, RCM3 makes the operating context the definite first question of the process.

In Chapter 2, RCM was defined as "a process used to determine the maintenance requirements of any physical asset in its operating context." This context pervades the entire maintenance strategy formulation process, starting with the definition of functions.

For example, consider a situation where a maintenance program is being developed for electric shovels used by a mining company with distributed assets. The shovels were bought from the same OEM and are used to load overburden and commodities (coal and copper), respectively. The mining company has mines operating in Canada and Chile. Before the functions and associated performance standards of these shovels can be defined, the people developing the program need to ensure that they thoroughly understand the operating context. Many of the functions and performance standards for identical assets and asset systems may be similar, but we will see that the operating context determines specifics for consideration.

For instance, how abrasive is the material being handled? What are the weather and ambient conditions at different locations? What are the typical worst-case weather and adverse conditions that may be encountered (e.g., earthquakes)? What dust is generated at each operation? What is the availability of spare parts and skilled labor at each location? What environmental and regulatory constraints apply to the operations? How stable is the power supply at each location? What are the shift arrangements and utilization at each operation? What is the demand, and what deviations can be expected? What technology is being used, and are there any constraints?

The operating context also profoundly influences the requirements for secondary functions. In the case of the shovel being used in Chile, the climate may demand air conditioning, while the shovel being used in Canada may require heating; regulations may demand special lighting and special dust filters; the remoteness of the mine site may demand that special spares be carried; and so on.

We discussed earlier how the operating context provides the context for how functions, failed states, failure modes, failure effects, and failure consequences for every asset or system should be listed, and it also provides the context for how similar assets perform differently from one to the other if the operating context is different. The following example illustrates how similar equipment could perform completely differently if the operating context is different.

From the above we can see that the mining company will have great difficulty in comparing availability (uptime) and performance for the shovels across the fleet. In order to achieve the required performance, the correct maintenance program is essential. In its attempt to standardize the maintenance and spares program across its fleet, the company quickly realizes that the shovels operating in the mines in Chile and the shovels operating in Canada do not perform the same. The operating environments and operations are completely different. Some of the criteria the company has to consider to *tailor* the maintenance programs are:

- Ambient conditions (temperature and rainfall/snowfall)
- Dust (texture and fineness)
- Material being loaded (overburden versus coal, copper, density, texture)
- Blasting patterns (rock size)
- Operator and maintainer skills
- Market demand
- Utilization (shift arrangements)
- Logistics

Not only does the context drastically affect functions and performance expectations, but it also affects the nature of the failure modes that could occur, their effects and consequences, the frequency with which they happen, and ways to manage them.

For instance, the primary function of the pump in Figure 3.1 would be listed as:

To pump water from tank X to tank Y at not less than 300 gallons per minute

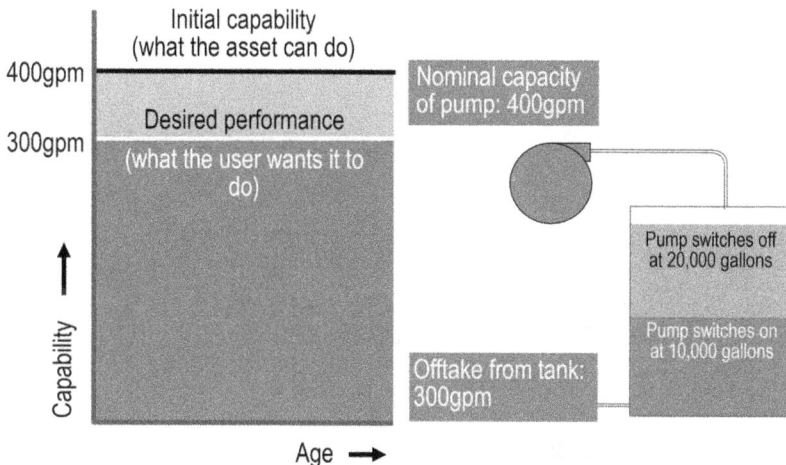

FIGURE 3.1   Initial capability versus desired performance

If it were moved to a location where it pumps mildly abrasive slurry (instead of water) into a tank B from which the slurry is being drawn at a rate of 300 gallons per minute, the primary function would be:

To pump slurry into tank B at not less than 300 gallons per minute

This is a higher performance standard than in the previous location, so the standard to which it has to be maintained rises accordingly. Because it is now pumping slurry instead of water, the nature, frequency, and severity of the failure modes also change. As a result, although the pump itself is unchanged, it is likely to end up with a completely different maintenance program in the new context.

What all this means is that individuals setting out to apply RCM to any asset or process must ensure that they have a crystal-clear understanding of the operating context before they start. Some of the most important factors that need to be considered are as follows:

- Whether the asset is part of a batch or a flow process
- The presence of a redundant or standby plant
- Utilization and loading
- The quality standards that apply to the finished product
- Safety standards, regulations, and regulatory requirements
- The operating environment (ambient conditions) and environmental standards
- Safety hazards
- Shift arrangements
- Volumes of work in progress and finished goods stocks
- Repair time
- Spares stockholding policies
- Skills available and logistics
- Recycling, reuse, or repurposing
- Trends and fluctuations in market demand
- Raw material supply

Each of these will be described in more detail in the following paragraphs

Batch and Flow Processes. In manufacturing plants, the most important feature of the operating context is the type of process. This ranges from flow process operations where nearly all the equipment is interconnected, to jobbing operations where most of the machines are independent.

In flow processes, the failure of a single asset can either stop the entire plant or significantly reduce output, unless surge capacity (buffer stock) or a standby plant is available. On the other hand, in batch or jobbing facilities, most failures only curtail the output of a single machine or line. The consequences of such failures are determined mainly by the duration of the stoppage and the amount of work in progress queuing in front of the machine that is down or in front of subsequent operations.

These differences mean that the maintenance strategy applied to an asset in a flow process is often radically different from the strategy applied to an identical asset in a batch environment.

Redundancy. The presence of redundancy—or alternative means of production—is a feature of the operating context that must be considered in detail when defining the functions of any asset.

The importance of redundancy is illustrated by the three identical pumps shown in Figure 3.2. Pump B has a standby, while pump A does not. This means that the primary function of pump A is to transfer liquid *on its own*, and that of pump B is to do it *in the presence of a standby*. Traditional thinking suggests that identical pumps should have identical (e.g., generic) maintenance programs (and spare parts). This difference in operating context means that the maintenance requirements of these pumps will be different (just how different we see later), even though the pumps are identical.

Utilization and Loading. Some equipment may have to operate all the time, whereas other equipment is only used as the need arises. Also, some equipment operates very close to design capability, whereas other equipment will have a healthy margin between capability and the way it is operated. The difference in the utilization and margin may have

Stand alone	Duty	Stand-by
If A fails, production is affected.	If B fails, switch to C and repair B.	No one knows if C fails on its own.
Operational risk	Tolerable risk	Hidden (Operational)
Preventive/predictive maintenance.	No scheduled maintenance?	Check if C has failed.

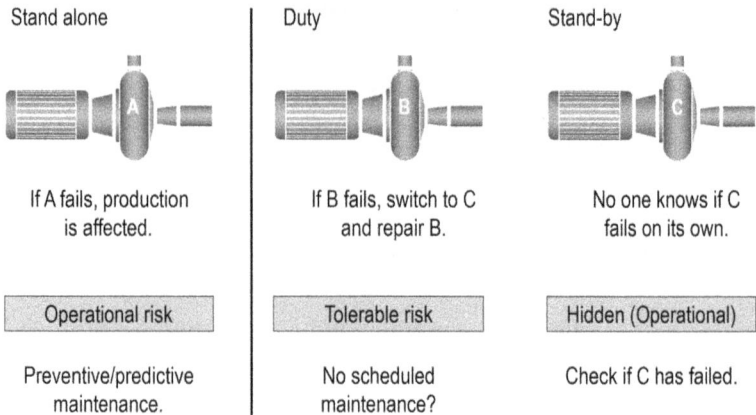

FIGURE 3.2 Different operating contexts

significant impact on how the equipment behaves, how it fails, and what kind of maintenance is required. Furthermore, the maintenance intervals may also be significantly different and may depend on the way the equipment is operated.

Quality Standards. Quality standards and standards of customer service are two more aspects of the operating context which can lead to differences between the descriptions of the functions of otherwise identical machines.

For example: Identical water treatment units used for softening and filtration may be of the exact same design and deliver water for different usage. The one is used for supplying treated water to pump seals and heat exchangers in the process plant, while the other is used to supply treated water to the eyewash station and safety shower. The quality requirements of the water supplied by the two units may differ significantly and impact the subsequent maintenance requirements.

The Operating Environment and Environmental Standards. The environment that the equipment operates in plays a large role in how the equipment behaves and fails. For example, a pump and switchgear operating outside in a humid corrosive environment will behave completely different than will the same equipment operating in a dry and controlled environment (inside an HVAC-controlled building).

Further, an increasingly important aspect of the operating context of any asset is the impact that it has (or could have) on the environment. Growing worldwide interest in environmental issues means that when we maintain any asset, we have to satisfy two sets of *users*. The first is the people who own and operate the asset, and the second is society as a whole, which wants both the asset and the process of which it forms part not to cause undue harm to the environment.

What society wants is expressed in the form of increasingly stringent environmental standards and regulations. These are international, national, regional, municipal, or even corporate standards. They cover an extraordinarily wide range of issues, from the biodegradability of detergents to the content of exhaust gases. In the case of processes, they tend to concentrate on unwanted liquid, solid, and gaseous by-products. Most industries are responding to society's environmental expectations by ensuring that equipment is designed to comply with the associated standards. However, it is not enough to ensure that the plant and processes are environmentally sound at the moment of commissioning, but steps have to be taken to ensure that it remains in compliance throughout its life. More focus is placed on renewables and recycling, considerations that place additional responsibilities on designers and maintainers. Taking the right steps is becoming a matter of urgency, because all over the world, more and more incidents that seriously affect the environment are occurring because some physical asset did not behave as it should—in other words, because something failed. The associated penalties are becoming very harsh indeed, so long-term environmental integrity is now a particularly important issue for maintenance people.

Care needs to be taken to establish whether and how any of these standards affect any asset that is to be analyzed using RCM.

Safety Standards, Regulations, and Regulatory Requirements. An increasing number of organizations either have developed their own standards or subscribe to formal standards concerning tolerable levels of risk. In some cases, these apply at the corporate level, in others to individual sites, and in yet others to individual processes or assets. Clearly, wherever such standards exist, they are an important part of the operating context. The RCM3 process uses the risk management standards to guide the RCM review group in the development of maintenance strategies, and therefore the level of tolerable risk must be defined prior to the start of any RCM3 analysis.

The regulations and regulatory requirements should also be identified and listed in the operating context. Applying RCM3 will develop a defensible risk and reliability-based maintenance program, while some organizations prefer to fall back on the regulations that apply; however, these may not be enough. The author refers to this as a liability-based maintenance program and not necessarily a reliability-based program.

Safety Hazards. Safety hazards associated with specific operations and/or material must be identified. Safety hazards may also require special care, storage, identification, warning signs, and protective systems. All of these would determine whether adequate precaution is taken or whether additional containment structures, fire suppression, suppression media, material selection, ventilation, and restrictions may apply and these would impact how failures matter.

Shift Arrangements. Shift arrangements profoundly affect the operating context. Some undertakings operate for 8 hours per day, 5 days a week (and even less in bad times). Others operate continuously 7 days a week, and yet others somewhere in between.

In a single-shift plant, production lost due to failures can usually be made up by working overtime. This overtime leads to increased production cost, so maintenance strategies are evaluated in the light of these costs. It is, however, possible to schedule regular maintenance activities in times when the operational shift workers are not on-site.

On the other hand, if a plant is operating 24 hours per day, 7 days per week, it is seldom possible to make up lost time, and so downtime causes lost production and lost sales and revenue. This costs a great deal more than overtime, so it is worth trying much harder to prevent failures under these circumstances. However, it is also much harder to make equipment available for maintenance in a fully loaded plant, so maintenance strategies need to be formulated with the greatest care.

As products move through their life cycles or as economic conditions change, organizations can move from one end of the spectrum to the other surprisingly quickly. For this reason, it is wise to review maintenance strategies every time this aspect of the operating context changes.

Work in Progress and Buffer Stock. *Work in progress* refers to any material that has not yet been through all the steps of a manufacturing process. The material might be stored in tanks, in bins, in hoppers, on pallets, on conveyors, or in special stores. The consequences of the failure of any machine are greatly influenced by the amount of this work in process between it and the next machines in the process.

Consider an example where the volume of work in the queue is sufficient to keep the next operation working for 6 hours and it only takes 4 hours to repair the failure mode under consideration. In this case, the failure would be unlikely to affect overall output. Conversely, if it took 8 hours to repair, it could affect overall output because the next operation would come to a halt. The severity of these consequences, in turn, depends on:

- The amount of work in progress between that operation and the next and so on down the line
- The extent to which any of the operations affected is a bottleneck operation (in other words, an operation that governs the output of the whole line)

Although plant stoppages cost money, it also costs money to hold stocks of work in progress. Nowadays the costs of holding stock of any kind are so high that reducing them to an absolute minimum is a top priority. This is a major objective of just-in-time systems and their derivatives.

These systems reduce work-in-process stocks, and so the cushion that the stocks provided against failure is rapidly disappearing. This is a vicious circle, because the pressure on maintenance departments to reduce failures in order to make it possible to do without the cushion is also increasing.

So, from the maintenance viewpoint, a balance has to be struck between the economic implications of operational failures and:

- The cost of holding work-in-process stocks in order to mitigate the effects of those failures

Or:

- The cost of doing proactive maintenance tasks with a view to anticipating or preventing the failures

To strike this balance successfully, this aspect of the operating context must be particularly clearly understood in manufacturing operations.

Logistics and Repair Times.   Repair times are influenced by the speed of response to the failure, which is a function of failure reporting systems and staffing levels, and the speed of the repair itself, which is a function of the availability of spares and appropriate tools and of the capability of the person doing the repairs. The logistics (where spare parts are kept) and the distance from the facility (how long it takes for craftspeople to travel to the plant) further influence the downtime and repair times. These factors heavily influence the effects and the consequences of failures, and they vary widely from one organization to another. As a result, this aspect of the operating context also needs to be clearly understood.

Spares and Stocking Policies.   It is possible to use a derivative of the RCM process to optimize spares stocks and the associated risk management strategies. This derivative is based on the fact that the only reason for keeping a stock of spare parts is to avoid or reduce the risks associated with the failure by reducing the mean time to repair (MTTR).

The relationship between spares and risk hinges on the time it takes to procure spares from suppliers. If it could be done instantly, there would be no need to stock any spares at all. But in the real world, procuring spares takes time. This is known as the lead time, and it ranges from a matter of minutes to several months or years. If the spare is not a stock item, the lead time often dictates how long it takes to repair the failure, and hence the severity of its consequences. On the other hand, holding spares in stock also costs money, so a balance needs to be struck, on a case-by-case basis, between the cost of holding a spare in stock and the total cost of not holding it. In some cases, the weight and/ or dimensions of the spares also need to be taken into account because of load and space restrictions, especially in facilities like oil platforms and ships.

This spares optimization process is beyond the scope of this book. However, when applying RCM to an existing facility, one has to start somewhere. In most cases, the best way to deal with spares is as follows:

- Use RCM to develop a maintenance strategy based on existing spares-holding policies.
- Review the failure modes associated with key spares on an exception basis by establishing what impact (if any) a change in the present stockholding policy would have on the initial maintenance strategy and then by picking the most cost-effective maintenance strategy/ spares-holding policy.

If this approach is adopted, then the existing spares-holding policy can be seen as part of the (initial) operating context.

Skills Availability and Logistics. The skills available on-site (or in the area) are important for determining how failures will be managed. It may be necessary to contract work that requires a higher skill. Furthermore, some repairs may have to be done off-site, which will extend downtime and increase the cost of repairs. The changing work-force and changing demographics must be considered when developing an operating context, the impact of this may have to be considered throughout the living program.

Recycling, Reuse, or Repurposing. The intentions to recycle, reuse or repurpose certain equipment at the end of the equipment's *useful life* will have an impact on how equipment is being used and maintained. If this is known at the time of the RCM analysis, the information should be captured in the operating context.

Market Demand and Seasonal and Daily Fluctuations. The operating context sometimes features cyclic variations in demand for the products or services provided by the organization.

Take soft drink companies, for example. They experience greater demand for their products in summer than in winter. Another example is urban transport companies, which experience peak demand during rush hours.

Still another example may be taken from the water and wastewater utility industry where demands vary daily. More people use water during the morning and evening hours when people use showers, washing machines, and dishwashers. The daily demand may vary significantly, and seasonal demand may also vary in areas where utilities must deal with melting snow and storm events (hurricanes).

In cases like these, the operational consequences of failure are much more serious at the times of peak demand, so in this type of industry, this aspect of the operating context needs to be especially clearly understood when defining functions and assessing failure consequences.

Raw Material Supply. Sometimes the operating context is influenced by cyclic fluctuations in the supply of raw materials. Food manufacturers often experience periods of intense activity during harvest times and periods of little or no activity at other times. This applies especially to fruit processors and sugar mills. During peak periods, operational failures not only affect output, but can lead to the loss of large quantities of raw materials if these cannot be processed before they deteriorate.

3.1 Documenting the Operating Context

For all the above reasons, it is essential to ensure that everyone involved in the development of a maintenance program for any asset fully understands the operating context of that asset. The best way to do so is to document the operating context, if necessary up to and including the overall mission statement of the entire organization, as part of the RCM process.

Figure 3.3 shows a hypothetical operating context statement for a crankshaft grinding machine. The crankshaft is used in a type of engine installed in motor car Model X.

Make Car Model X Corresponding asset: Model X Division	Model X Division employs 4,000 people to produce 220,000 cars this year. Sales forecasts indicate that this could rise to 320,000 per year within 3 years. We are now number 18 in national customer satisfaction rankings, and we intend to reach 15th place next year and 10th place the following year. The target for lost-time injuries throughout the division is 1 per 500,000 paid hours. The probability of a fatality occurring anywhere in the division should be less than 1 in 50 years. The division plans to conform to all known environmental standards.
Make engines Corresponding asset: Motown Engine Plant	The Motown Engine Plant produces all the engines for Model X cars. 140,000 Type 1 and 80,000 Type 2 engines are produced per year. In order to achieve the customer satisfaction targets for the entire vehicle, warranty claims for engines must drop from the present level of 20 per 1,000 to 5 per 1,000. The plant suffered 3 reportable environmental excursions last year—our target is not more than 1 in the next 3 years. The plant shuts down for 2 weeks per year to allow production workers to take their main annual vacations.
Make Type 2 engines Corresponding asset: Type 2 engine line	The Type 2 engine line presently works 110 hours per week (two 10-hour shifts 5 days per week and one 10-hour shift on Saturdays). The assembly line could produce 140,000 engines per year in these hours if it ran continuously with no defects, but the overall output of engines is limited by the speed of the crankshaft manufacturing line. The company would like as much maintenance as possible to be done during normal hours without interfering with production.

FIGURE 3.3 Example of an operating context

Machine crankshafts Corresponding asset: Crankshaft machining line 2	The crankshaft line consists of 25 operations and is nominally able to produce 20 crankshafts per hour (2,200 per week, 110,000 per 50-week year). It currently sometimes fails to produce the requirement of 1,600 per week in normal time. When this happens, the line has to work overtime at an additional cost of $800 per hour. (Since most of the forecast growth will be for Type 2 engines, stoppages on this line could eventually lead to lost sales of Model X cars unless the performance is improved.) There should be no crankshafts stored between the end of the crankshaft line and the engine assembly line, but operations, in fact, keep a pallet of about 60 crankshafts to provide some "insurance" against stoppages. This enables the crankshaft line to stop for up to 3 hours without stopping assembly. Crankshaft machining defects have not caused any warranty claims, but the scrap rate on this line is 4%. The initial target is 1.5%.
Finish grind crankshaft main and big-end journals Corresponding asset: Ajax Mark 5 grinding machine	The finish grinding machine grinds 5 main and 4 big-end journals. It is the bottleneck operation on the crankshaft line, and the cycle time is 3.0 minutes. The finished diameter of the main journals is 75 mm ± 0.1 mm, and of the big ends 53 mm ± 0.1 mm. Both journals have a surface finish of Ra 0.2. The grinding wheels are dressed every cycle, a process that takes 0.3 minute out of each 3-minute cycle. The wheels need to be replaced after 3,500 crankshafts, and replacement takes 1.8 hours. There are usually about 10 crankshafts on the conveyor between this machine and the next operation, so a stoppage of 25 minutes can be tolerated without interfering with the next operation. Total buffer stocks on the conveyors between this machine and the end of the line mean that this machine can stop for about 45 minutes before the line as a whole stops. Finish grinding contributes 0.4% to the present overall scrap rate.

FIGURE 3.3   Example of an operating context (*continued*)

The hierarchy starts with the division of the corporation that produces this model, but it could have gone up one level further to include the entire corporation. Note also that a context statement at any level should apply to all the assets below it in the hierarchy, not just the asset under review.

The context statements at the higher levels in this hierarchy are simply broad function statements. Performance standards at the highest levels quantify expectations from the viewpoint of the overall business. At lower levels, performance standards become more specific until one reaches the asset under review.

The author will refer the readers to ISO 14224 as a useful guideline for determining asset hierarchy and taxonomy. An operating context must be developed for the asset on which an RCM3 analysis will be conducted. The operating context describes the overall process, the part played by the asset in the process, the business impact of the asset, and the asset function. The steps in each one are detailed below:

- Overall process:
  - Describe the overall processes directly related to the asset or system.
  - Describe the environment in which the asset is operated. Is it inside a building or outside? Is the environment clean or dirty? Dry or humid? Neutral, corrosive, or acidic?
  - Also mention here who will be viewed as the *operating crew*.
- Part played by the asset or system in the overall process:
  - Describe the main purpose of the asset (why does it exist?) and how it fits into the overall process noted above as well as other processes in the business unit or division, etc.
  - Describe the desired performance required from this asset with quantitative performance standards.
- Business impact of the asset:
  - Describe the overall consequences if this asset either does not operate at all or does not operate at the required desired performance level.

- Describe the possible environmental impact related to the failure of the asset, taking into account environmental laws and regulations.
- Describe the possible safety impact related to the failure of the asset.
- Describe the possible economic impact related to the failure of the asset. Quantify the economic losses related to equipment downtime and resulting production losses as lost revenue per hour downtime, etc.
- Describe possible impact on product or service quality and resulting impact on customers and customer service.
- Asset function:
  - Develop a functional description of the asset. The description must be based on the logical process flow as far as possible.
  - List the major components of this asset (e.g., polysystem, dry feeder, mixing tank, mixer, feed pumps, etc.).
  - Describe the capability as well as the expected levels of performance of the component—nameplate data, data from manuals or pump curves, etc.
  - Include the presence of standby and/or redundancy capabilities.
  - Describe the protective device circuits associated with this asset (alarm/shutdown circuits) and include tag numbers.
  - Describe the control circuits associated with this asset (flow, pressure, etc.) and include tag numbers where possible.
  - Describe other sensors associated with this equipment (gauges, level indicators, etc.) and include tag numbers where possible.
  - Determine if the operation of the asset is manual or automatic, local or remote.
  - Indicate any special materials of construction.
  - Indicate any special operating conditions.
  - Include what safety procedures, policies, and guidelines are applicable.

- Documentation to be used as input to the operating context description:
  - Process flow diagrams (PFDs)
  - Mechanical flow diagrams (MFDs)
  - Piping and instrumentation drawings (P&IDs)
  - Operation and maintenance manuals

CHAPTER 4

Functions

Most people become engineers because they feel at least some affinity for things, be they mechanical, electrical, or structural. This affinity leads them to derive pleasure from assets in good condition and to feel offended by assets in poor condition.

These reflexes have always been at the heart of the concept of preventive maintenance. They have given rise to concepts such as "asset care," which, as the name implies, seeks to take care of assets. They have also led some maintenance strategists to believe that maintenance is all about preserving the inherent reliability or built-in capability of any asset.

In fact, this is not so.

As we gain a deeper understanding of the role of assets in business, we begin to appreciate the significance of the fact that any physical asset is put into service because someone wants it to do something. So it follows that when we maintain an asset, the state that we wish to preserve must be one in which it continues to do whatever its users want it to do. Later in this chapter, we will see that this state—what the users want—is fundamentally different from the built-in capability of the asset.

This emphasis on what the asset does rather than what it is provides a whole new way of defining the objectives of maintenance for any asset—one that focuses on what the user wants the asset to do. This is the most important single feature of the RCM process, and it is why many people regard RCM as "TQM applied to physical assets."

When we maintain an asset, the state which we wish to preserve must be one in which it continues to do whatever its users want it to do. This emphasis on what the asset does rather than what it is provides a whole new way of defining the objectives of maintenance for any asset—one which focuses on what the user wants. Clearly, in order to define the objectives of maintenance in terms of user requirements, we must gain a crystal-clear understanding of the functions of each asset together with the associated performance standards. This is why the RCM3 process asks the following question:

What are the functions and associated performance standards of the asset in its present operating context?

This chapter considers this question in detail. It describes how functions should be defined, explores the two main types of performance standards, reviews different categories of functions, and shows how functions should be listed.

4.1 Describing Functions

It is a well-established principle of value engineering that a function statement should consist of a verb and an object. It is also helpful to start such statements with the word *to* ("*to* pump water," "*to* transport people," etc.). However, as explained at length in the next part of this chapter, users not only expect an asset to fulfill a function. They also expect it to do so to an acceptable level of performance. So a function definition—and by implication the definition of the objectives of maintenance for the asset—is not complete unless it specifies as precisely as possible the level of performance desired by the user (as opposed to the built-in capability).

A function statement should consist of a verb, an object, and a desired standard of performance.

If we consider our earlier example of the pump in Chapter 3, the primary function of the pump in Figure 3.1 would be listed as:

To pump water from tank X to tank Y at not less than 300 gallons per minute

This example shows that a complete function statement consists of a verb ("to pump"), an object ("water"), and the standard of performance ("not less than 300 gallons per minute") desired by the user.

4.2 Performance Standards

The objective of maintenance is to ensure that assets continue to do what their users want them to do. The extent to which any user wants any asset to do anything can be defined by a minimum standard of performance. If we could build an asset that could deliver that minimum performance without deteriorating in any way, then that would be the end of the matter. The machine would run continuously with no need for maintenance.

However, in the real world, things are not that simple. Any organized system that is exposed to the real world will deteriorate. The end result of this deterioration is total disorganization (also known as "chaos" or" entropy"), unless steps are taken to arrest whatever process is causing the system to deteriorate.

For example, the pump in Figure 3.1 is pumping water into a tank from which the water is drawn at a rate of 300 gallons per minute. One process that causes the pump to deteriorate (failure mechanism) is impeller wear. This happens regardless of whether the pump is pumping acid or lubricating oil, and regardless of whether the impeller is made of titanium or mild steel. The only question is how fast it will wear to the point that it can no longer deliver 300 gallons per minute.

So if deterioration is inevitable, it must be allowed for. This means that when any asset is put into service, it must be able to deliver more than the minimum standard of performance desired by the user. What the asset is able to deliver is known as its initial capability (or inherent reliability). Figure 3.1 illustrates the right relationship between this capability and desired performance.

For instance, in order to ensure that the pump shown in Figure 3.1 does what its users want and to allow for deterioration, the system designers must specify a pump that has an initial built-in capability of something greater than 300 gallons per minute. In the example shown in the figure, this initial capability is 400 gallons per minute.

This means that performance can be defined in two ways:

- Desired performance (what the user wants the asset to do)
- Built-in capability (what it can do)

Later chapters look at how maintenance helps ensure that assets continue to fulfill their intended functions, either by ensuring that their capability remains above the minimum standard desired by the user or by restoring something approaching the initial capability if it drops below this point. When considering the question of restoration, bear in mind that:

- The initial capability of any asset is established by its design and by how it is made (inherent reliability).
- Maintenance can only restore the asset to this initial level of capability—it cannot go beyond it.

In practice, most assets are adequately designed and built, so it is usually possible to develop maintenance programs that ensure that such assets continue to do what their users want.

In short, such assets are maintainable, as shown in Figure 4.1.

On the other hand, if the desired performance exceeds the initial capability, no amount of maintenance can deliver the desired performance.

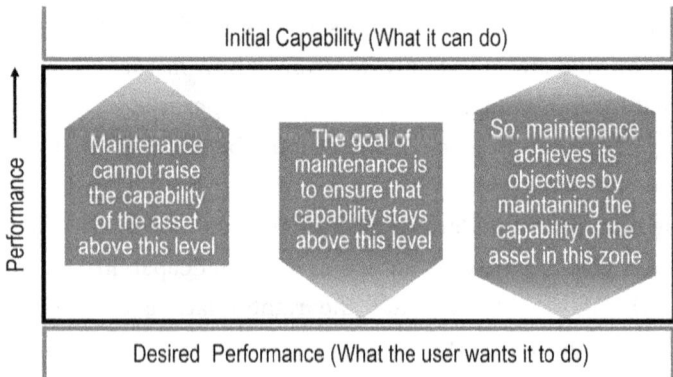

FIGURE 4.1 A maintainable asset

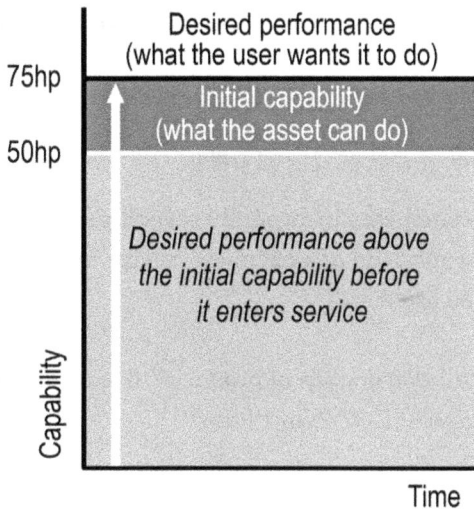

FIGURE 4.2 Not maintainable asset

In other words, such assets are not maintainable, as shown in Figure 4.2.

For example: If we try to draw 75 hp (desired performance) from a 50-hp electric motor (initial capability), the motor will keep tripping out and will eventually burn out prematurely. No amount of maintenance will make this motor big enough. It may be perfectly adequately designed and built in its own right; it just cannot deliver the desired performance in the context in which it is being used.

Two conclusions can be drawn from the above examples:

• For any asset to be maintainable, the desired performance of the asset must fall within the envelope of its initial capability.
• In order to determine whether this is so, not only do we need to know the initial capability of the asset, but we also need to know exactly what minimum performance the user is prepared to accept in the context in which the asset is being used.

This underlines the importance of identifying precisely *what the users want* when starting to develop a maintenance program. The following paragraphs explore key aspects of performance standards in more detail.

Multiple Performance Standards

Many function statements incorporate several performance standards. For example, one function of a chemical reactor in a batch-type chemical plant might be listed as:

> To heat up to 1,000 pounds of product X from ambient temperature to boiling point (257°F) in 1 hour

In this case, the *weight* of the product, the *temperature* ranges, and the *time* all present different performance expectations.

Quantitative Performance Standards

Performance standards should be quantified where possible, because quantitative standards are inherently much more precise than qualitative standards. Special care should be taken to avoid qualitative statements like "To produce as many widgets as required by production," or "To go as fast as possible." Function statements of this type are meaningless, if only because they make it impossible to define exactly when the item is failed.

In reality, it can be extraordinarily difficult to define precisely what is required, but just because it is difficult does not mean that it cannot or should not be done. One major user of RCM summed up this point by saying, "If the users of an asset cannot specify precisely what performance they want from an asset, they cannot hold the maintainers accountable for sustaining that performance."

Qualitative Standards

In spite of the need to be precise, it is sometimes impossible to specify quantitative performance standards, so we have to live with qualitative statements.

For instance, the primary function of housekeeping and cleaning is usually for the plant "to look acceptable." What is meant by "acceptable" is impossible to quantify and varies widely from person to person. As a result, user and maintainer need to take care to ensure that they share a common understanding of what is meant by words like "acceptable" before setting up a system intended to preserve that acceptability. Similarly, defining the standard of free movement may also be different from person to person.

Absolute Performance Standards

A function statement that contains no performance standard at all usually implies an absolute.

For instance, the concept of containment is associated with nearly all enclosed systems. Function statements covering containment are often written as follows:

To contain liquid X

The absence of a performance standard suggests that the system must contain *all* the liquid and that any leakage at all amounts to a failed state. In cases where an enclosed system can tolerate some leakage, the amount that can be tolerated should be incorporated as a performance standard in the function statement.

Variable Performance Standards

Performance expectations (or applied stress) sometimes vary infinitely between two extremes.

Consider, for example, a truck used to deliver loads of assorted goods to urban retailers. Assume that the actual loads vary between 0 (empty) and 5 tons, as shown in Figure 4.3. To allow for deterioration, the initial capability of the truck must be more than the "worst-case" load, which in this case is 5 tons. The maintenance program, in turn, must ensure that the capability does not drop below this level, in which case it would automatically satisfy the full range of performance expectations.

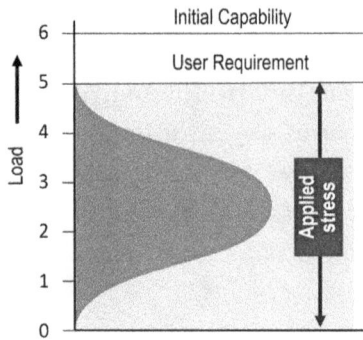

FIGURE 4.3 Variable performance standards

Another example of variable performance standards can be found in the water and wastewater industry. Daily and seasonal demands vary constantly, and water supply and wastewater treatment requirements vary accordingly. The daily demand changes between peak flows in the morning and evening (when people shower and use washers) to almost nothing when people are at work. A wastewater pump station must be capable of handling peak flows as well as being able to operate efficiently at low flows. A variable performance standard best defines the desired performance for these assets with the peak flow condition being the upper limit. The station may further be subject to huge inflows during storm events, and depending on how often these occur, the water flow during the storm event may become the upper limit for the variable performance standard.

Upper and Lower Limits

In contrast to variable performance expectations, some systems exhibit variable capability. These are systems that simply cannot be set up to function to exactly the same standard every time they operate.

For example, a grinding machine used to finish-grind a crankshaft will not produce exactly the same finished diameter on every journal. The diameters will vary, if only by a few microns. Similarly, a filling machine in a food factory will not fill two successive containers with exactly the same weight of food. The weights will vary, if only by a few ounces.

Figure 4.4 indicates that capability variations of this nature usually vary about a mean. In order to accommodate this variability, the associated desired standards of performance incorporate upper and lower limits.

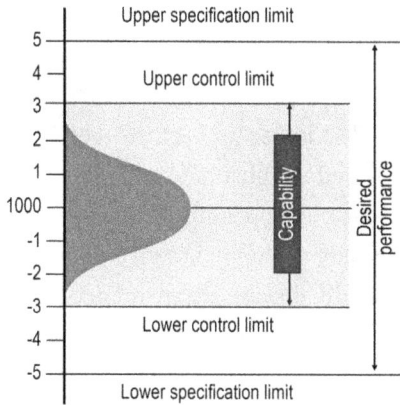

FIGURE 4.4   Upper and lower limits

For instance, the primary function of a sweets packing machine might be:

- To pack 12 ounces ± 0.05 ounce of sweets into bags at a minimum rate of 75 bags per minute

The primary function of a grinding machine might be:
- To finish-grind main bearing journals in a cycle time of 3.00 minutes ± 0.03 minute to a diameter of 1.75 inch ± 1/16 inch with a surface finish of Ra 0.2

The desired performance limits are known as the upper and lower specification limits. The limits of capability are known as the upper and lower control limits.

In practice, this kind of variability is usually unwanted for a number of reasons. Ideally, processes should be so stable that there is no variation at all and hence no need for two limits. In pursuit of this ideal, many industries are spending a great deal of time and energy on designing processes that vary as little as possible. However, this aspect of design and development is beyond the scope of this book. Right now, we are concerned purely with variability from the viewpoint of maintenance.

How much variability can be tolerated in the specification of any product is usually governed by external factors.

For instance, the lower limit that can be tolerated on the crankshaft journal diameter is governed by factors such as noise, vibration, and harshness, and the upper limit by the clearances needed to provide adequate lubrication. The lower limit of the weight of the bag of sweets (relative to the advertised weight) is usually governed by trading standards legislation, while the upper limit is governed by the amount of product that the company can afford to give away.

In cases like these, the desired performance limits are known as the upper and lower specification limits. The limits of capability (usually defined as being three standard deviations either side of the mean) are known as the upper and lower control limits. Quality management theory suggests that in a well-managed process, the difference between the control limits should ideally be half the difference between the specification limits. This multiple should allow a more than adequate margin for deterioration from a maintenance viewpoint.

Upper and lower limits not only apply to product quality. They also apply to other functional specifications such as the accuracy of gauges and the settings of control systems such as temperature controllers in HVAC systems and protective devices. Protective devices are discussed in much more detail in Chapter 7.

4.3 Different Types of Functions

Every physical asset has at least two—and often several—functions. If the objective of maintenance is to ensure that the asset can continue to fulfill these functions, then they must *all* be identified together with their current desired standards of performance. At first glance, this may seem to be a fairly straightforward exercise. However, in practice it nearly always turns out to be the single most challenging and time-consuming aspect of the maintenance strategy formulation process.

This is especially true of older facilities. Products change, plant configurations change, people change, technology changes, and performance expectations change—but still we find assets in service that have been there since the plant was built. Defining precisely what they are supposed to be doing now requires very close cooperation between maintainers and users. It is also usually a profound learning experience for everyone involved.

Functions are divided into two main categories: primary and secondary functions, and then further divided into various subcategories. These are reviewed below, starting with primary functions.

Primary Functions

Organizations acquire physical assets for one, possibly two, seldom more than three main reasons. These reasons are defined by suitably worded function statements. Because they are the main reasons why the asset is acquired, they are known as *primary functions*. They are the reasons why the asset exists at all, so care should be taken to define them as precisely as possible.

Primary functions are usually fairly easy to recognize. In fact, the names of most industrial assets are based on their primary functions. For instance, the primary function of a packing machine is to pack things, of a crusher to crush something, and so on.

As mentioned earlier, the real challenge lies in defining the current performance expectations associated with these functions. For most types of equipment, the performance standards associated with primary functions concern speeds, volumes, and storage capacities. Product quality also usually needs to be considered at this stage.

Chapter 3 mentioned that our ability to achieve and sustain satisfactory quality standards depends increasingly on the capability and condition of the assets that produce the goods. These standards are usually associated with primary functions. As a result, take care to incorporate product quality criteria into primary function statements where relevant. These include dimensions for machining, forming, or assembly operations; purity standards for food, chemicals, and phar-maceuticals; hardness in the case of heat treatment; filling levels or weights for packaging; and so on.

Functional Block Diagrams. If an asset is very complex or if the interaction between different systems is poorly understood, it is sometimes helpful to clarify the operating context by drawing up functional block diagrams. These are simply diagrams showing all the primary functions of an enterprise at any given level. Functional block diagrams are the subject of many other publications and will not be discussed in more detail in this book, however Appendix II provides more detail.

Multiple Independent Primary Functions. An asset can have more than one primary function. For instance, the very name of a military fighter/bomber suggests that it has two primary functions. In such cases, both should be listed in the functional specification.

A similar situation is often found in manufacturing, where the same asset may be used to perform different functions at different times. For instance, a single reactor vessel in a chemical plant might be used at different times to reflux (boil continuously) three different products under three different sets of conditions, as follows:

Product

	1	2	3
Pressure (psi)	30	145	80
Temperature (°F)	356	250	285
Batch size (gallons)	125	160	200

(It could be said that this vessel is not performing three different functions, but that it is performing the same function to different standards of performance. In fact, the distinction does not matter because we arrive at the same conclusion either way.)

In cases like this, one could list a separate function statement for each product. This would logically lead to three separate maintenance programs for the same asset. Three programs may be feasible—perhaps even desirable—if each product runs continuously for very long periods. However, if the interval between long-term maintenance tasks is longer than the changeover intervals, then it is impractical to change the tasks every time the machine is changed over to a different product.

Another example is when a wastewater treatment plant performs two distinct functions—the first being to treat wastewater only during the winter months (when little or no stormwater enters the system) and the second being to treat stormwater during springtime and summer when the snow melts and summer storms bring lots of stormwater into the plant. Inflow volumes may triple during the summer, and treatment policies for grit settlement and disinfection may vary dramatically. In this case, some equipment in the treatment plant will have different primary functions.

One way around this problem is to combine the worst-case standards associated with each product into one function statement. In the above example, a combined function statement could be "To reflux up to 200 gallons of product at temperatures up to 356°F and pressures up to 145 psi." This will lead to a maintenance program that might embody some overmaintenance some of the time, but that will ensure that the asset can handle the worst stresses to which it will be exposed.

Serial or Dependent Primary Functions. One often encounters assets that must perform two or more primary functions in series. These are known as serial functions.

For instance, the primary functions of a machine in a food factory may be "To fill 300 cans with food per minute" and then "To seal 300 cans per minute."

The distinction between multiple primary functions and serial primary functions is that in the former case, each function can be performed independently of the other, while in the latter, one function must be performed before the other. In other words, for the canning machine to work properly, it must fill the cans before it seals them.

Secondary Functions

Most assets are expected to fulfill one or more functions in addition to their primary functions. These are known as secondary functions.

For example, the primary function of an automobile might be described as follows:

> To transport up to five people at speeds of up to 70 mph along made roads

If this was the only function of the vehicle, then the only objective of the maintenance program for this car would be to preserve its ability to carry up to five people at speeds of up 70 mph along made roads. However, this is only part of the story, because most car owners expect far more from their vehicles, ranging from the ability to carry luggage, to the ability to indicate how much fuel is in the fuel tank.

To help ensure that none of these functions are overlooked, they are divided into eight categories:

- Environmental integrity
- Safety, structural integrity
- Control, containment, comfort, cleanliness
- Appearance
- Protection
- Economy/efficiency
- Recycling, repurposing, reuse, regulations, and regulatory requirements
- Superfluous functions

Note that the first letters of each line in this list form the word "ESCAPERS."

Although secondary functions are usually less obvious than primary functions, the loss of a secondary function can still have serious consequences—sometimes more serious than the loss of a primary function. As a result, secondary functions often need at least as much maintenance as primary functions, so they too must be clearly identified. The following pages explore the main categories of these functions in more detail.

Environmental Integrity. The previous chapter explained how society's environmental expectations have become a critical feature of the operating context of many assets. RCM begins the process of compliance with the associated standards by incorporating them in appropriately worded function statements.

For instance, one function of a car exhaust or a factory smokestack might be "To contain no more than X micrograms of a specified chemical per cubic foot." The car exhaust system might also be the subject of environmental restrictions dealing with noise, and the associated functional specification might be "To emit no more than X dB measured at a distance of Y feet behind the exhaust outlet."

Safety. Most users want to be reasonably sure that their assets will not hurt or kill them. In practice, most safety hazards emerge later in the RCM process as failure modes. However, in some cases it is necessary to write function statements that deal with specific threats to safety. For instance, two safety-related functions of a toaster are "To prevent users from touching electrically live components" and "To not burn the users."

Many processes and components are unable to fulfill the safety expectations of users on their own. This has given rise to additional functions in the form of protective devices. These devices pose some of the most difficult and complex challenges facing the maintainers of the modern industrial plant. As a result, they are dealt with separately below.

A further subset of safety-related functions consists of those that deal with product contamination and hygiene. These are most often found in the food and pharmaceutical industries. The associated performance standards are usually tightly specified and lead to rigorous and comprehensive maintenance routines (cleaning and testing/validation).

Structural Integrity. Many assets have a structural secondary function. This usually involves supporting some other asset, subsystem, or component.

For example, the primary function of the wall of a building might be to protect people and equipment from the weather, but the wall might also be expected to support the roof (and bear the weight of shelves and pictures).

Large, complex structures with multiple load-bearing paths and high levels of redundancy need to be analyzed using a specialized version of RCM. Typical examples of such structures are airframes, the hulls of ships, and the structural elements of offshore oil platforms.

Structures of this type are rare in industry in general, so the relevant analytical techniques are not covered in this book. However, straightforward, single-celled structural elements can be analyzed in the same way as any other function described in this chapter.

Control. In many cases, not only do users want assets to fulfill functions to a given standard of performance, but they also want to be able to regulate the performance. This expectation is summarized in separate function statements.

For instance, the primary function of a car, as suggested earlier, was "To transport up to five people at speeds of up to 70 mph along paved roads." One control function associated with this function could be "To enable the driver to regulate speed at will between −10 mph (reverse) and +70 mph."

Indication or feedback forms an important subset of the control category of functions. This includes functions that provide operators with real-time information about the process (gauges, indicators, telltales, VDUs, and control panels) or that record such information for later analysis (digital or analog recording devices, cockpit voice recorders in aircraft, etc.). Performance standards associated with these functions not only relate to the ease with which it should be possible to read and assimilate or to play back the information, but also cover the accuracy of the information.

For instance, the function of the speedometer of a car might be described as "To indicate the road speed to the driver to within +5%–0% of the actual speed."

Containment. In the case of assets used to store things, a primary function is to contain whatever is being stored. However, containment should also be acknowledged as a secondary function of all devices used to transfer material of any sort—especially fluids. This includes tanks, vessels, pipes, pumps, conveyors, chutes, hoppers, and pneumatic and hydraulic systems.

Risk-based inspection (RBI) has one primary objective, which is containment. The integration of the RCM and RBI processes shares this objective.

Containment is also an important secondary function of items like gearboxes and transformers. (In this context, note again the remarks on absolute standards and containment earlier in this chapter.)

Comfort. Most people expect their assets not to cause them anxiety, grief, or pain. These expectations are listed under the heading of "comfort" because the major English dictionaries define "comfort" as being free from anxiety, pain, grief, and so on. (These expectations can also be classified under the heading of "ergonomics.")

Too much discomfort affects morale, so it is undesirable from a human viewpoint. It is also bad business, because people who are anxious or in pain are more likely to make incorrect decisions. Anxiety is caused by poorly explained, unreliable, or unintelligible control systems, be they for domestic appliances or for oil refineries. Pain is caused by assets—especially clothing and furniture—that are incompatible with the people using them.

The best time to deal with these problems is, of course, at the design stage. However, deterioration and/or changing expectations can cause this category of functions to fail like any other. The best way to set about ensuring that this doesn't happen is to define appropriate functional specifications.

For example, one function of a control panel might be "To indicate clearly to a color-blind operator up to 5 feet away whether pump A is operating or shut down." A control-room chair might be expected "To allow operators to sit comfortably for up to 1 hour at a time without inducing drowsiness."

Cleanliness and Contamination. An important expectation of maintenance personnel is to reduce or eliminate the number of induced failures, such as the introduction of dirt and contaminants into lubrication systems. Although these requirements are best dealt with during design, it is possible to list this requirement as a function of the system.

For example, one function of the gearbox or oil reservoir may be "To allow free movement of air in response to temperature changes of the oil without allowing dirt into the system."

Appearance. The appearance of many items embodies a specific secondary function. For instance, the primary function of the paintwork on most industrial equipment is to protect it from corrosion, but a bright color might be used to enhance its visibility for safety reasons. Similarly, a public utility must be seen as being conservative with spending ratepayers' money, while also projecting an image of value and success.

Protective Devices. As physical assets become more complex, the number of ways they can fail is growing almost exponentially. This has led to corresponding growth in the variety and severity of the risks associated with the failure. In an attempt to eliminate (or at least to reduce) these risks, increasing use is being made of automatic protective devices. These work in one of five ways:

- To draw the attention of the operators to abnormal conditions (Warning lights and audible alarms that respond to failure effects. The effects are monitored by a variety of sensors including level switches, load cells, overload or overspeed devices, vibration or proximity sensors, temperature or pressure switches, etc.)

- To shut down the equipment in the event of a failure (these devices also respond to failure effects, using the same types of sensors and often the same circuits as alarms, but with different settings)
- To eliminate or relieve abnormal conditions that follow a failure and that might otherwise cause much more serious damage (firefighting equipment, safety valves, rupture discs or bursting discs, emergency medical equipment)
- To take over from a function that has failed (standby plant of any sort, redundant structural components)
- To prevent dangerous situations from arising in the first place (guards and safety signs)

The purpose of these devices is to protect people from failures or to protect machines or to protect products—and in some cases all three.

Protective devices ensure that the failure of the function being protected is much less serious than it would be if there were no protection. The presence of protection also means that the maintenance requirements of a protected function are often less stringent than they would be otherwise.

Consider a milling machine whose milling cutter is driven by a toothed belt. If the belt were to break in the absence of any protection, the feed mechanism would drive the stationary cutter into the workpiece (or vice versa) and cause serious secondary damage. This can be avoided in two ways:

- By implementing a comprehensive proactive maintenance routine designed to prevent the failure of the belt.
- By providing protection such as a broken belt detector to shut down the machine as soon as the belt breaks. In this case, the only consequence of a broken belt is a brief stoppage while it is replaced, so the most cost-effective maintenance policy might simply be to let the belt fail. But this policy is only valid if the broken belt detector is working, and steps must be taken to ensure that this is so.

The maintenance of protective devices—especially devices that are not fail-safe—is discussed in much more detail in later chapters. However, this example demonstrates two fundamental points:

- That protective devices often need more routine maintenance attention than the devices they are protecting
- That we cannot develop a sensible maintenance program for a protected function without also considering the maintenance requirements of the protective device

It is only possible to consider the maintenance requirements of protective devices if we understand their functions. So when listing the functions of any asset, we must list the functions of all protective devices.

A final point about protective devices concerns the way their functions should be described. These devices act by exception (in other words, when something else goes wrong), so it is important to describe them correctly. In particular, protective function statements should include the words "if" or "in the event of," followed by a very brief summary of the circumstances or the event that would activate the protection.

For instance, if we were to describe the function of a tripwire as being "To stop the machine," anyone reading this description could be forgiven for thinking that the tripwire is the normal stop-start device. To remove any ambiguity, the function of a tripwire should be described as follows:

To be capable of stopping the machine in the event of an emergency at any point along its length

The function of a safety valve may be described as follows:

To be capable of relieving the pressure in the boiler if the pressure exceeds 250 psi

Economy/Efficiency. Those who use assets of any sort only have finite financial resources. This leads them to put a limit on what they are prepared to spend on operating and maintaining the asset. How much they are prepared to spend is governed by a combination of three factors:

- The actual extent of their financial resources
- How much they want whatever the asset will do for them
- The availability and cost of competitive ways of achieving the same end

At the operating context level, functional expectations concerning costs are usually spelled out in the form of expenditure budgets. At the asset level, economic issues can be addressed directly by function statements that define what users expect in areas such as fuel economy and loss of process materials.

For instance, a car might be expected "to travel at least 35 miles per gallon of fuel at a constant 65 mph, and at least 50 miles per gallon of fuel at 35 mph." A fossil fuel power station might be expected "to export at least 45% of the latent energy in the fuel as electrical power." A plant using an expensive solvent might want "to lose no more than 0.5% of solvent X per month."

Recycling, Repurposing, Reuse   Modern plants are designed and built to be sustainable and use as much sustainable technology and material as possible. This has become an important consideration that has specific maintenance requirements and special treatment expectations. In the world of rising costs, organizations are not just encouraged to make use of secondhand or previously used equipment, but sometimes forced to do so. This leads to improved planning and maintenance to make sure equipment is preserved as best as possible. There is also a growing market for used equipment. Condition assessments on equipment are also becoming increasingly popular for assets that are considered for reuse. It is no longer enough to put a fresh coat of paint on the equipment—more detailed assessments including *qualitative* and *quantitative assessments* reveal true asset health.

Regulations and Regulatory Requirements.   A requirement may be compliance and adherence to regulations and statutory requirements.

Like reliability, these regulations and requirements are not functions themselves, but they constitute certain mandatory actions to be taken. Safety signs, barriers, and information about hazardous material are a few examples that fall into this category.

Regulations and Regulatory Requirements. A requirement may be compliance and adherence to regulations and regulatory requirements which constitute certain mandatory actions to be taken. Regulations associated with capacity assurance and mandatory testing of protective circuits fall into this category and must be included in the function statement. Failure to comply may result in violations and penalties. This may not necessarily fall into any other category covered by the other secondary functions.

Superfluous Functions. Items or components are sometimes encountered that are completely superfluous. This usually happens when equipment has been modified frequently over a period of years or when new equipment has been overspecified. (These comments do not apply to redundant components built in for safety reasons, but to items that serve no purpose at all in the context under consideration.)

For example, a pressure-reducing valve was built into the supply line between a gas manifold and a gas turbine. The original function of the valve was to reduce the gas pressure from 120 psi to 80 psi. The system was later modified to reduce the manifold pressure to 80 psi, after which the valve served no useful purpose.

It is sometimes argued that items like these do no harm and that it costs money to remove them, so the simplest solution may be to leave them alone until the whole plant is decommissioned. Unfortunately, this is seldom true in practice. Although these items have no positive function, they can still fail and so reduce overall system reliability. To avoid this, they need maintenance, which means that they still consume resources.

It is not unusual to find that between 5% and 20% of the components of complex systems are superfluous in the sense described above.

If they are eliminated, it stands to reason that the same percentage of maintenance problems and costs will also be eliminated. However, before this can be done with confidence, the functions of these components first need to be identified and clearly understood.

A Note on Reliability. There is often a temptation to write "reliability" function statements such as "To operate 7 days a week, 24 hours per day." In fact, reliability is not a function in its own right. It is a performance expectation that pervades all the other functions. It is properly dealt with by dealing appropriately with each of the failure modes that could cause each loss of function. This issue is discussed further in Chapter 12.

Using the ESCAPERS Categories. There will often be doubt about which of the ESCAPERS categories some functions belong to. For instance, should the function of a seat reclining mechanism be classified under the heading of "control" or "comfort"?

In practice the precise classification does not matter. What does matter is that we identify and define all the functions that are likely to be expected by the user. The list of categories merely serves as an aid to help ensure that none of these expectations are overlooked.

4.4 How Functions Should Be Listed

A properly written functional specification—especially one that is fully quantified—precisely defines the objectives of the enterprise. This ensures that everyone involved knows exactly what is wanted, which, in turn, ensures that maintenance activities remain focused on the real needs of the users (or "customers"). It also makes it easier to absorb changes triggered by changing expectations without derailing the whole enterprise.

Figure 4.5 shows an RCM Information Worksheet. As you can see, functions are listed in the left-hand column. The functions apply to the exhaust system of a 5-megawatt gas turbine. A complete Information Worksheet is presented at the end of Chapter 5.

Primary functions are listed first, and the functions are numbered numerically, as Figure 4.5 shows. There is no relative ranking between different functions. It is not unusual that secondary functions may have more severe consequences and therefore higher relative importance, e.g., containment of hazardous fluids.

RCM3™ Information Worksheet	Location ID
	Location Description

© 2018 ALADON

Functions	
1	To channel all the hot turbine exhaust gas without restriction to a fixed point 30 feet above the roof of the turbine hall
2	To reduce exhaust noise levels to ISO Noise Rating 30 at 150 feet
3	To ensure that the surface temperature of the ducting inside the turbine hall does not exceed 140°F
4	To transmit a warning signal to the turbine control system if the exhaust gas temperature exceeds 890°F and a shutdown signal if it exceeds 930°F at a point 12 feet from the turbine
5	To allow free movement of the ducting in response to temperature changes

FIGURE 4.5   Describing functions

CHAPTER 5

Failed States

Chapter 2, Section 2.3 explained that the RCM3 process entails asking eight questions about selected assets, as follows:

1. What are the operating conditions (how the equipment or system is being used)?
2. What are the functions and associated performance standards of the asset in its present operating context?
3. In what ways does it fail to fulfill its functions (failed states)?
4. What causes each failed state (failure modes)?
5. What happens when each failure occurs (failure effects and consequence severity)?
6. What are the risks associated with each failure (inherent risk quantified)?
7. What *must* be done to reduce intolerable risks to a tolerable level (using proactive risk management strategies)?
8. What *can* be done to reduce or manage tolerable risks in a cost-effective way?

Previous chapters dealt at length with the first two questions. After a brief introduction to the general concept of failure, this chapter considers the third question, which deals with functional failures.

5.1 Failures

In the previous chapter, we saw that people or organizations acquire assets because they want the assets to do something, and they also expect the assets to fulfill the intended functions to an acceptable standard of performance.

Chapter 4 went on to explain that for any asset, both to do what its users want and to allow for deterioration, the initial capability of the asset must exceed the desired standard of performance. Thereafter, as long as the capability of the asset continues to exceed the desired standard of performance, the user will be satisfied.

On the other hand, if for any reason the asset is unable to do what the user wants, the user will consider it to have failed.

This leads to a basic definition of failure:

Failure is the inability of any asset to do what its users want it to do.

This is illustrated in Figure 5.1.

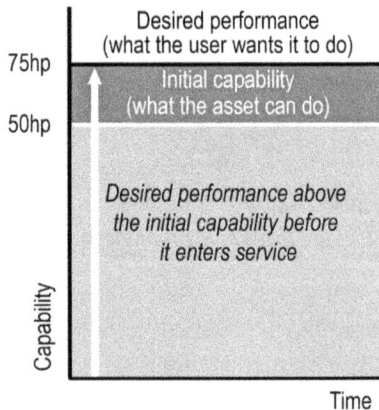

FIGURE 5.1 The general failed state

For instance, if the pump shown in Figure 3.1 in Chapter 3 is unable to pump 300 gallons per minute, it will not be able to keep the tank full, and so its users will regard it as "failed" or similarly, a 50hp motor subject to a 75hp load as illustrated in Figure 3.1. The motor will continue to trip and eventually burn out or fail.

5.2 Failed States (Functional Failures)

The above definition of failure treats the concept of failure as if it applies to an asset as a whole. In practice, this definition is vague, because it does not distinguish clearly between the failed state (functional failure) and the events that cause the failed state (failure modes). It is also simplistic, because it does not take into account the fact that each asset has more than one function, and each function often has more than one desired standard of performance. The implications are explored in the following paragraphs.

For instance, the pump in Figure 3.1 has at least two functions. One is to pump water at not less than 300 gallons per minute, and the other is to contain the water. It is perfectly feasible for such a pump to be capable of pumping the required amount (not failed in terms of its primary function) while leaking excessively (failed in terms of the secondary function).

Conversely, it is equally possible for the pump to deteriorate to the point where it cannot pump the required amount (failed in terms of its primary function), while it still contains the liquid (not failed in terms of the secondary function).

This shows why it is more accurate to define failure in terms of the loss of specific functions rather than the failure of an asset as a whole. It also shows why the RCM process uses the term "failed state" to describe functional failures, rather than the word "failure" on its own. However, to complete the definition of failure, we also need to look more closely at the question of performance standards.

Performance Standards and Failure

The boundary between satisfactory performance and failure is specified by a performance standard. Given that performance standards apply to individual functions, "failure" can be defined precisely by defining a functional failure as follows:

> A functional failure is the inability of any asset to fulfill a function to a standard of performance that is acceptable to the user.

Risk is defined as any deviation from what is expected. Failed states are deviations from what is expected (what the user wants). The failed state defines the risk associated with owning and operating assets.

The following paragraphs discuss different aspects of functional failure under the following headings:

- Partial and total failures
- Upper and lower limits
- Variable performance standards
- Gauges and indicators
- The operating context

Partial and Total Failures.　The above definition of functional failure covers the complete loss of function. It also covers situations where the asset still functions but performs outside acceptable limits.

For example, the primary function of the pump discussed earlier is "To pump water from tank X to tank Y at not less than 300 gallons per minute." This function could suffer from two functional failures:

- Fails to pump any water at all
- Pumps water at less than 300 gallons per minute

Partial failure is nearly always caused by different failure modes from total failure, and the consequences are different. This is why all the functional failures that could affect each function should be recorded.

Record all the functional failures associated with each function.

Note that partial failure should not be confused with the situation where the asset deteriorates slightly but its capability remains above the level of performance required by the user.

For example, the initial capability of the pump in Figure 3.1 is 400 gallons per minute. Impeller wear is inevitable, so this capability will decline. As long as it does not decline to the point where the pump is unable to pump 300 gallons per minute, it will still be able to fill the tank and so keep the users satisfied in the context described (see Figure 5.2).

However, if the capability of the asset deteriorates so much that it falls below the desired performance, its users will consider it to have failed.

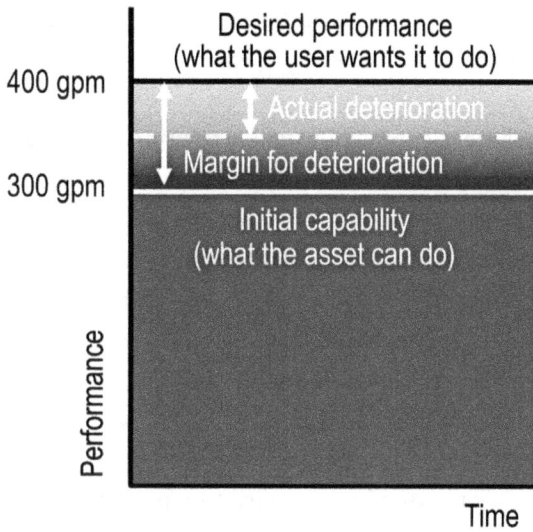

FIGURE 5.2 Asset operating satisfactorily despite some deterioration

Upper and Lower Limits. The previous chapter explained that the performance standards associated with some functions incorporate upper and lower limits. Such limits mean that the asset has failed if it produces products that are over the upper limit or below the lower limit. In these cases, the breach of the upper limit usually needs to be identified *separately* from the breach of the lower limit. This is because the failure modes and/ or the consequences associated with going over the upper limit are usually different from those associated with going below the lower limit.

For example, the primary function of a sweets packing machine is listed as "To pack 0.5 pound (8 ounces) ± 0.05 ounce of sweets into bags at a minimum rate of 75 bags per minute." This machine has failed:

- If it is completely unable to pack at all (stopped completely)
- If it packs more than 8.05 ounces of sweets into any bags
- If it packs less than 7.95 ounces into any bags
- If it packs at a rate of less than 75 bags per minute

The function of a crankshaft grinding machine is listed as "To finish mill a work piece in a cycle time of 2.25 ± 0.03 minutes to a depth of

0.46 ± 0.01 inch with a flatness tolerance of 0.1 and a surface finish of Ra 1.6 μm." Failure occurs when:

- The machine is completely unable to grind the workpiece.
- The machine grinds the workpiece in a cycle time longer than 2.28 minutes.
- The machine grinds the workpiece in a cycle time less than 2.22 minutes.
- The diameter exceeds 0.47 inch.
- The diameter is below 0.45 inch.
- The surface finish is too rough.

Of course, if only one limit applies to a particular parameter, then only one failed state is possible. For instance, the absence of a lower limit on the roughness specification in the above example suggests that it is not possible to make the item too smooth. In some circumstances, this may not actually be true, so care needs to be taken to verify this point when analyzing functions of this type.

Figure 5.3 shows a process that is in control and meeting performance standards. In practice, the failed states associated with upper and lower limits can manifest themselves in two ways. First, the spread

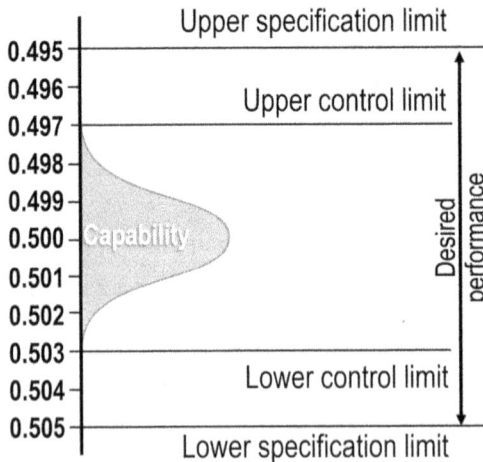

FIGURE 5.3   Upper and lower limits

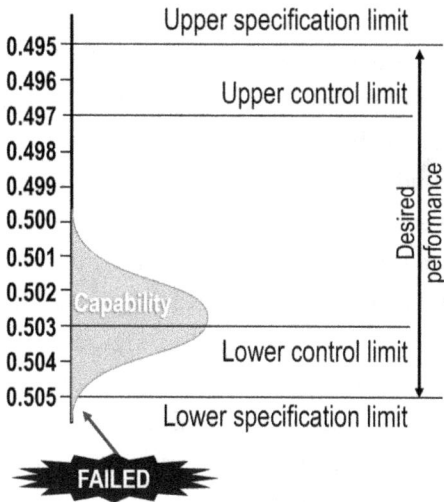

FIGURE 5.4   Capability breach, lower limit only

of capability could breach the specification limits in one direction only (either direction), as illustrated in Figure 5.4.

The second failed state occurs when the spread of capability is so broad that it breaches both the upper and the lower specification limits, as shown in Figure 5.5.

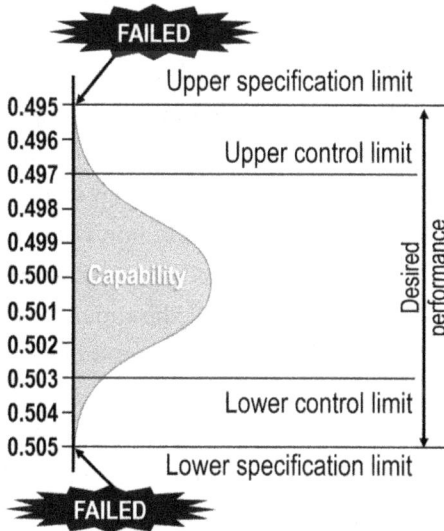

FIGURE 5.5   Capability breaches, upper and lower limits

Note that in both of the above cases, not all of the products produced by the processes in question will be failed. If the breach is minor, only a small percentage of out-of-spec products will be produced. However, the further off-center the grouping in the first case, or the broader the spread in the second case, the higher will be the percentage of failures.

Limits of this sort also apply to the performance standards associated with gauges, indicators, protection, and control systems.

In Chapter 4, Figure 4.4 ("upper and lower limits") illustrated a process that is in control and in specification. Figures 5.4 and 5.5 show that processes that are out of control and out of spec are in a failed state. The failure modes that can cause these failed states are discussed in the next chapter. (Chapter 6 deals with the implications of a process that is out of control but within specification.)

Variable Performance Standards. When systems are subject to variable performance standards (i.e., in the case of a wastewater pumping station), the station may require multiple pumps to effectively meet the variable performance standards. Pump(s) are required to transfer the water during variable flow conditions and although some of the variance can be dealt with by fitting variable speed drives, not all the conditions will be met by doing this. Two pumps may be required to meet the varying demand (daily and seasonal variations) and will have to operate as lead and lag (primary and secondary), which is different than duty/standby, which was discussed earlier. Let's assume the pumps are equal in size and capable of delivering 80 gpm each (in this example). The nominal flow is 65 gpm and maximum inflow has been recorded at 135 gpm at times that lasted less than one hour. Based on the operating context, it is rare for both pumps to operate simultaneously. One pump will be able to cope with the inflow for 90–95% of the time, and the second pump will be required during peak flows (mornings and evenings when water use increases) and during the wet season (because of some storm water infiltrating the system).

Typically, the wetwell level will be maintained via level switches (level sensor) and operation of the pumps will be as follows (for a station with two pumps installed):

Pump A – Lead pump will automatically start when the water level inside the wetwell rises to (say) 6'

Pump B – Lag pump will automatically start when the water level inside the wetwell rises to (say) 8'

A high-level alarm will be generated when the level in the wetwell rises to (say) 10'

Both pumps will shut down when the water level in the wetwell drops below (say) 4'

Low-level alarm will be generated in the control room when water level drops to (say) 2.5'

The control logic (PLC) will automatically alternate the pumps after each drawdown cycle.

The primary function will be recorded as follows:

To transfer up to 135 gallons per minute wastewater from the collection system to the forced main line (nominal 65 gallons per minute)

Failed states will be recorded as follows:

- Unable to transfer any wastewater at all (total failure)
  - Failures to components, equipment, or subsystems common to both pumps will be considered, i.e., power failure, discharge header failure, etc.
- Unable to transfer 135 gallons per minute (partial failure with impact during peak flows)
  - Failure of one pump (i.e., lag pump) cause a loss of function to transfer peak flows; failures of one pump will be considered, i.e., lag pump bearing seized, lag pump discharge check valve plugged, etc.

In this example, the lead pump will operate and transfer up to 80 gallons per minute and only during peak flow periods will there be an intolerable risk of breaching environmental standards due to sewer overflows as illustrated in Figure 5.6.

- Unable to transfer 65 gallons per minute (partial failure with impact during all times)

– Partial failure: failures that would affect both pump systems will be considered, i.e., solids build up in the bottom of the wet well causing restricted suction lines, ragging of the pumps, etc.

In this example, even during low flows, failures may pose an intolerable risk of a sewer overflow.

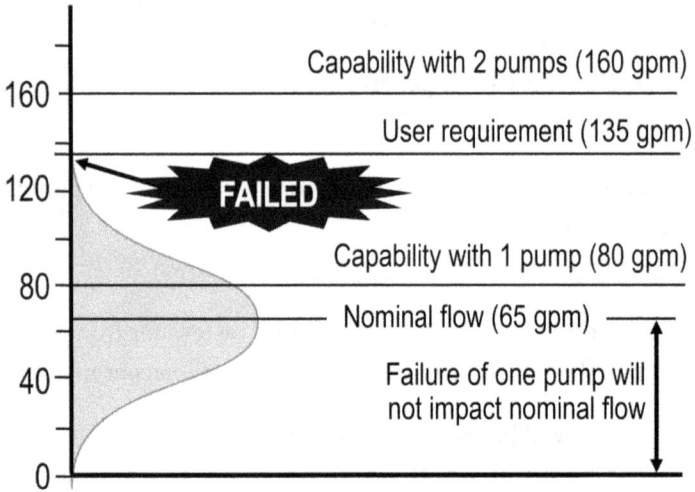

FIGURE 5.6 Defining failure of variable performance standards

Gauges and Indicators. The above discussion has tended to focus on product quality. Chapter 4 mentioned that upper and lower limits also apply to the performance standards associated with gauges, indicators, protection, and control systems. Depending on failure modes and consequences, it may also be necessary to treat the breach of these limits separately when listing functional failures.

For instance, the function of a temperature gauge could be listed as "To display the temperature of process X to within 2% of the actual process temperature." This gauge can suffer from three functional failures:

- Fails altogether to display the process temperature
- Displays a temperature more than 2% higher than the actual temperature
- Displays a temperature more than 2% lower than the actual temperature

Functional Failures and the Operating Context. The exact definition of failure for any asset depends very much on its operating context. This means that in the same way that we should not generalize about the functions of identical assets, so we should take care not to generalize about functional failures.

For example, we saw how the pump shown in Figure 3.1 fails if it is completely unable to pump water and also fails if it is unable to pump at least 300 gallons per minute. If the same pump is used to fill a tank from which water is drawn at 350 gallons per minute, the second failed state occurs if the throughput drops below 350 gallons per minute.

Who Should Set the Standard?

An issue that needs careful consideration when defining functional failures is the "user." To this day, most maintenance programs in use around the world are compiled by maintenance people working on their own. These people usually decide for themselves what is meant by "failed." In practice, their view of failure often turns out to be quite different from that of the users, with sometimes disastrous consequences for the effectiveness of their programs.

For example, one function of a hydraulic system is to contain oil. How well it should fulfill this function can be subject to widely differing points of view. There are production managers who believe that a hydraulic leak only amounts to a functional failure if it is so bad that the equipment stops working altogether. On the other hand, a maintenance manager might suggest that a functional failure has occurred if the leak causes excessive consumption of hydraulic oil over a long period of time. Then again, a safety officer might say that a functional failure has occurred if the leak creates a pool of oil on the floor in which people could slip and fall or which might create a fire hazard. This is illustrated in Figure 5.7.

The maintenance manager (who controls the hydraulic oil budget) may ask the operators for access to the hydraulic system to repair leaks "because oil consumption is excessive." However, access may be denied because the operators think the machine "is still working

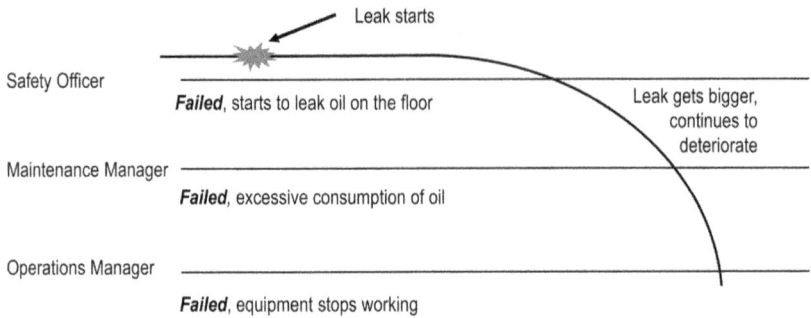

FIGURE 5.7   Different views about failure

OK." When this happens, the maintenance people (1) record that the machine "was not released for preventive maintenance" and (2) form the opinion that their production colleagues "don't believe in PM." For similar reasons, the maintenance manager might not release a maintenance person to repair a small leak when requested to do so by the safety officer.

In fact, all three parties almost certainly do believe in prevention. The real problem is that they have not taken the trouble to agree on exactly what is meant by "failed," and so they do not share a common understanding of what they are seeking to prevent.

This example illustrates three key points:

- The performance standard used to define functional failure—in other words, the point where we say "so far and no further"—defines the level of proactive maintenance needed to avoid that failure (in other words, to sustain the required level of performance).
- Much time and energy can be saved if these performance standards are clearly established before the failures occur.
- The performance standards used to define failure must be set by operations and maintenance people working together with any-one else who has something legitimate to say about how the asset should behave.

It is important, however, to realize that the failed state does not imply that the equipment must be taken out of service in the example above. It simply means that the consequences and associated risk need to be managed to tolerable levels. For example, when the safety officer is concerned about the hazard of slipping or a fire, the risk can be mitigated by placing secondary containment underneath the gearbox or by screening the area. Dealing with the consequences allows continued safe operation of the asset.

How Failed States Should Be Recorded

Failed states are listed in the second column of the RCM3 Information Worksheet. They are coded alphabetically, as shown in Figure 5.8.

RCM3™ Information Worksheet	Location ID
	Location Description
© 2015 ALADON	

	Functions		Failed States
1	To channel *all* the hot turbine exhaust gas without restriction to *a fixed point 30 feet above the roof* of the turbine hall	A	Unable to channel gas *at all*
		B	Gas flow restricted
		C	Fails to channel *all the gas*
		D	Fails to convey gas to a fixed point 30 feet above the roof
2	To reduce exhaust noise levels to *ISO Noise Rating 30 at 150 feet*	A	Noise level exceeds ISO Noise Rating 30 at 150 feet
3	To ensure that the surface temperature of the ducting inside the turbine hall does not rise above *140°F*	A	Duct surface temperature exceeds *140°F*

FIGURE 5.8   Describing failed states

4	To transmit a warning signal to the turbine control system if the exhaust gas temperature exceeds *890°F* and a shutdown signal if it exceeds *930°F at a point 12 feet from the turbine*	A	Incapable of sending a warning signal if exhaust temperature *exceeds 890°F*
		B	Incapable of sending a shutdown signal if exhaust temperature *exceeds 930°F*
5	To allow *free* movement of the ducting in response to temperature changes	A	Does not allow *free* movement of ducting

FIGURE 5.8 Describing failed states (*continued*)

CHAPTER 6

Failure Modes and Effects Analysis

By defining the functions and desired standards of performance of any asset, we define the objectives of maintenance with respect to that asset. Defining failed states (functional failures) enables us to spell out exactly what we mean by "failed." These two issues were addressed by the second and third steps of the RCM3 process.

The next two questions seek to identify the failure modes that are reasonably likely to cause each failed state in order to ascertain the failure effects (consequence severity) associated with each failure mode and to assess the inherent risk each failure mode poses. This is done by performing a failure modes and effects analysis (FMEA) for each failed state.

This chapter describes the main elements of an FMEA, starting with a definition of the term "failure mode."

6.1 What Is a Failure Mode?

A failure mode could be defined as any event that is likely to cause an asset (or system or process) to fail. However, Chapter 5 explained that it is both vague and simplistic to apply the term "failure" to an asset as a whole. It is much more precise to distinguish between a "functional failure" (a failed state) and a "failure mode" (an event and the circumstances leading up to an event that could cause a failed state). This distinction leads to the following, more precise definition of a failure mode:

A failure mode is any event that causes a failed state and the mechanisms leading up to the failed state.

The best way to show the connection and the distinction between failed states and the events that could cause them is to list the failed states first, then to record the failure modes that could cause each failed state, as shown in Figure 6.1.

RCM3™ Information Worksheet © 2017 Aladon V0.0	Location ID **P-03-001-00H2O** Location Description **Water Storage and Supply System**		
Function	**Failed State**	**Failure Mode Cause**	**Failure Mode Mechanism**
1 To transfer water at a minimum rate of 300 gpm	A Unable to transfer water at all	1 Pump bearing seized	Normal wear
			Lack of lubrication
			Axial thrust too high
			Incorrect installation

FIGURE 6.1   Examples of a failure mode cause and failure mode mechanisms of a pump

Figure 6.1 also indicates that a failure mode is the combination of the *failure mode cause* and *failure mode mechanism*. The description of a failure mode cause should consist of a noun and a verb. The failure mode mechanism is the event leading up to the failure mode cause. Many different failure mode mechanisms may actually lead up to a failure mode cause (failed state). The description should contain enough detail for it to be possible to select an appropriate risk management strategy, but not so much detail that excessive amounts of time are wasted on the analysis process itself.

In particular, the verbs used to describe failure modes should be chosen with care, because they strongly influence the subsequent risk management strategy selection process. For instance, verbs such as "fails" or "breaks" or "malfunctions" should be used sparingly, because they give little or no indication about what an appropriate way might be for managing the failure. The use of more specific verbs makes it possible to select from the full range of failure management options.

For example, a term like "coupling tails" provides no clue about what might be done to anticipate or prevent the failure. However, if we say, "Coupling bolts come loose due to incorrect torque setting," or "Coupling hub fails due to fatigue," then it becomes much easier to identify a possible proactive risk management strategy.

In the case of valves or switches, one should also indicate whether the loss of function is caused by the item failing in the open or closed position—"Valve jams closed" says more than "Valve fails." In the interests of complete clarity, it may sometimes be necessary to take this one step further.

For instance, "Valve jams closed due to rust on lead screw" or "Valve jams closed due to buildup of debris" is clearer than "Valve jams closed." Similarly, one might need to distinguish between "Bearing seizes due to normal wear and tear" and "Bearing seizes due to lack of lubrication." The RCM3 process therefore lists the failure cause as the direct reason for the failed state, and the mechanisms are the events leading up to the failure.

Figure 6.2 describes the failure modes (causes and mechanisms) best.

These issues are discussed at length later in this chapter, but first we look at why we need to analyze failure modes at all.

FAILED STATE	FAILURE MODES	
	FAILURE CAUSE	FAILURE MECHANISM
Pressure drops to 45 psi	Filter blocked	Dirt build up
Unable to maintain control over automobile	Tire burst	Sharp object in the road
Unable to supply any water	Pump bearings seized	Lack of lubrication
Unable to shut down the turbine in the event of low oil pressure	Low oil pressure switch circuit failed	
Unable to contain the coal on the conveyor belt	Conveyor belt torn	Normal wear and tear
Unable to prevent leaks	Tank leaks	Tank shell corroded
Unable to remove particles greater than 8μm from the lubricating oil	Oil filter holed	Improper installation

FIGURE 6.2   Examples of a failure mode cause and failure mode mechanisms

6.2 Why Analyze Failure Modes

A single machine can fail for dozens of reasons. A group of machines or system such as a production line can fail for hundreds of reasons. For an entire plant, the number can rise into the thousands or even tens of thousands.

Most managers shudder at the thought of the time and effort likely to be involved in identifying all these failure modes. Many decide that this type of analysis is just too much work, and they abandon the whole idea entirely. In doing so, these managers overlook the fact that on a day-to-day basis, *maintenance is really managed at the failure mode level.* For instance:

- Work orders or job requests are raised to cover specific failure modes.
- Day-to-day maintenance planning is all about making plans to deal with specific failure modes.
- In most industrial undertakings, maintenance and operations people hold meetings every day. The meetings usually con-sist almost entirely of discussions about what has failed, what caused the failure (and who is to blame), what is being done to

repair it, and—sometimes—what can be done to stop it from happening again. In short, the entire meeting is spent discussing failure modes.

- To a large extent, technical history recording systems record individual failure modes (or at least what was done to rectify them).

In too many cases, these failure modes are discussed, recorded, or otherwise dealt with *after* they have occurred. Dealing with failures after they have happened is, of course, the essence of *reactive maintenance*.

Proactive management, on the other hand, means dealing with events *before* they occur—or at least deciding how they should be dealt with if they were to occur. In order to do this, we need to know beforehand what events are likely to occur. The events leading up to failures in this context are failure mechanisms. So if we wish to apply truly proactive maintenance to any physical asset, we must try to identify all the failure modes (causes and mechanisms) that are reasonably likely to affect that asset. Ideally, they should be identified before they occur at all, or if this is not possible, before they occur again.

Once each failure mode has been identified, it becomes possible to consider what happens when it occurs, then to assess its consequences and associated risk, and then to decide what (if anything) should be done to anticipate, prevent, detect, or correct it or perhaps even to design it out.

So the maintenance task selection process—and much of the subsequent management of these tasks—is carried out at the failure mode level. This is briefly illustrated in the following example.

Consider again the RCM Information Worksheet shown in Figure 6.1. This applies to the primary function of the pump first shown in Figure 3.1. Figure 6.3 shows that the pump is a direct coupled single-stage back-pull-out end-suction volute pump sealed by a mechanical seal. In this example, we look more closely at the three failure mechanisms that are thought to be likely to affect the impeller only. These are discussed in some detail below and summarized in Figure 6.3.

Here we take a closer look at each of the three failure mechanisms in the figure:

Impeller worn out

Manage this failure through performance monitoring or changing impeller before reaching the "end of useful life."

Impeller jammed by foreign object

Manage this failure by installing an inlet screen or preventing foreign objects being left in the system after maintenance.

Impeller adrift

Manage this failure through installation procedure and training of maintenance staff to fit impellers properly.

FIGURE 6.3   Typical failures of the impeller of a centrifugal pump

- **Impeller wear.** This is likely to be an age-related phenomenon. As shown in Figure 6.3, this means that it is likely to conform to the second of the six failure patterns introduced in the first chapter (failure pattern B). So, if we know roughly what the useful life of the impeller is, and if the consequences of the failure are serious enough, then we may decide to prevent this failure by changing the impeller just before the end of the useful life. We may also apply some form of condition-based maintenance strategy through monitoring of the pump performance (e.g., flow, pressure, etc.) and replace the impeller based on the condition.
- **Impeller jammed by foreign object.** The likelihood of a foreign object appearing in the suction line will almost certainly have nothing to do with how long the impeller has been in service. As

a result, it stands to reason that this failure mode will occur on a random basis (pattern E). There would also be no warning that the failure was about to occur. So if the consequences were serious enough, and the failure happened often enough, we would be likely to consider modifying the system, perhaps by installing some sort of filter or screen in the suction line.

- **Impeller adrift.** If the impeller fastening mechanism is adequately designed and it keeps coming adrift, this would almost certainly be because it wasn't put on properly in the first place. (If we knew that this was so, then perhaps the failure mode should actually be described as "Impeller fitted incorrectly.") This, in turn, means that the failure mode is most likely to occur soon after start-up, as shown in Figure 6.3 (pattern F), and we would probably deal with it by improving the relevant training or procedures.

This example reinforces the point that the level at which we manage the maintenance of any asset is not at the level of the asset as a whole (in this case, the pump), and not even at the level of any component (in this case, the impeller), but at the level of each failure mode mechanism. So before we can develop a systematic, proactive maintenance management strategy for any asset, *we must identify what these failure modes (causes and mechanisms) are (or could be)*.

The example also suggests that one of the failure modes could be eliminated by a design change and another by improving training or procedures. *So not every failure mode is dealt with by scheduled maintenance.* Chapters 8 and 9 describe an orderly approach to deciding what is likely to be the most suitable way of dealing with each failure.

Note also that the failure management solutions proposed in Figure 6.3 represent only one of several possibilities in each case.

For instance, we could monitor impeller wear by monitoring the pump performance (as stated earlier) and only change the impeller when it needs it. We also need to bear in mind that adding a screen to the suction line adds three more failure possibilities, which need to be analyzed in turn (it could block up, it could be holed and therefore cease to screen, and it could disintegrate and damage the impeller).

Chapter 8 examines these alternatives in more detail.

These points all indicate that the identification of failure modes is one of the most important steps in the development of any program intended to ensure that any asset continues to fulfill its intended functions. In practice, depending on the complexity of the item, its operating context, and the level at which it is being analyzed, between one and thirty failure modes are usually listed per functional failure.

The next two sections of this chapter consider:

- Categories of failure modes
- The level of detail required

Thereafter, the last three parts of the chapter consider failure effects, sources of information for an FMEA, and how failure modes and effects should be listed.

6.3 Categories of Failure Modes

Some people regard maintenance as being all about—and only about—dealing with deterioration. Some even go so far as to specify that FMEAs carried out on their assets should deal only with failure modes caused by deterioration and should ignore other categories of failure modes (such as human errors and design flaws). This is unfortunate, because it often transpires that deterioration causes a surprisingly small proportion of failures. In these cases, restricting the analysis just to deterioration can lead to a woefully incomplete maintenance strategy.

On the other hand, if one accepts that maintenance means ensuring that physical assets continue to do whatever their users want them to do, then a comprehensive maintenance program must address *all* the events that are reasonably likely to threaten that functionality. Failure modes can be classified into one of three groups, as follows:

- When capability falls below the desired performance
- When the desired performance rises above initial capability
- When the asset is not capable of doing what is wanted from the outset

Each of these categories is discussed in the following paragraphs.

Falling Capability

The first category of failure modes covers situations where capability is above desired performance to begin with, but then it drops below desired performance after the asset is put into service, as illustrated in Figure 6.4.

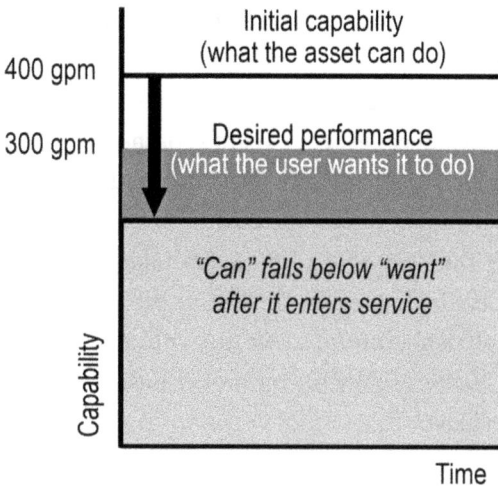

FIGURE 6.4 Failure mode Category 1: falling capability

The five principal failure mechanisms of reduced capability are:

- Deterioration
- Lubrication failures
- Dirt buildup or accumulation of dirt
- Disassembly
- Human errors that reduce capability

Each of these mechanisms is examined in turn.

Deterioration. Any physical asset that fulfills a function that brings it into contact with the real world is subject to a variety of stresses. These stresses cause the asset to deteriorate by lowering its capability, or more accurately, its resistance to stress. Eventually resistance drops so much that the asset can no longer deliver the desired performance—in other words, it fails. Deterioration covers all forms of wear and tear (fatigue, corrosion, abrasion, erosion, evaporation, degradation of insulation, etc.). These failure mechanisms should, of course, be included in a list of failure modes wherever they are thought to be reasonably likely.

Lubrication Failures. Lubrication is associated with two types of failure modes. The first concerns lack of lubricant, and the second the failure of the lubricant itself.

With regard to lack of lubrication, things have changed considerably in the last two decades. Fifty years ago, the majority of lubrication points were replenished manually. The cost of lubricating each of these points was tiny compared with the cost of not doing so. It was also tiny compared with the cost of analyzing the lubrication requirements of each point in detail. This meant that it was just not worth carrying out an in-depth analytical exercise to set up a lubrication program. Instead, these programs were usually set up on the basis of a quick survey by a lubrication specialist.

Nowadays, however, "sealed-for-life" components and centralized lubrication systems have become the norm in most industries. This has led to a massive reduction in the number of points where a human has to apply oil or grease to a machine—and has led to a massive increase in the consequences of failure (especially the failure of centralized lubrication systems). From the analytical viewpoint, this means that it is now cost-effective to:

- Use RCM to analyze centralized lubrication systems in their own right
- Consider the loss of lubricant in the few remaining manually lubricated points as individual failure modes

The second category of failures associated with lubrication concerns deterioration of the lubricant itself. It is caused by phenomena such as shearing of the oil molecules, oxidation of the base oil, and additive depletion. In some cases, deterioration of the oil may be aggravated by the buildup of sludge or the presence of water or other contaminants.

A lubricant may also fail to do its job simply because the wrong lubricant has been used. If any or all of these failure modes are considered to be likely in the context under consideration, then they should be recorded and subjected to further analysis. (This also applies to transformer oil and hydraulic oil.)

Dirt and Contamination. Dirt or dust is a very common and usually preventable cause of failure, so failures caused by dirt should be listed in the FMEA whenever they are thought to be likely to cause any functional failure. Dirt interferes directly with machines by causing them to block, stick, or jam. It is also a principal cause of the failure of functions that deal with the appearance of assets (things that should look clean look dirty). Dirt can also cause product quality problems, either by getting into the clamping mechanisms of machine tools and causing misalignment or by getting directly into products such as food, pharmaceuticals, or the oilways of engines.

Disassembly. If components fall off machines, if assemblies fall apart, or if whole machines come adrift, the consequences are usually serious, so the relevant failure modes should be listed. These include the failure of welds, soldered joints, rivets, bolts, electrical connections, or pipefittings that can also fail due to fatigue or corrosion or that simply come undone.

Also take care to record the functions and associated failure modes of locking mechanisms such as split pins and lock nuts when considering the integrity of assemblies.

Human Errors That Reduce Capability. The final subset of the "falling capability" category of failure modes consists of those caused by human error. As the name implies, these refer to errors that reduce the capability of the process to the extent that it is unable to function as required by the user.

Examples include manually operated valves left shut, causing a process to be unable to start, parts incorrectly fitted by maintenance craftspeople, or sensors set in such a way that a machine trips out when nothing is wrong.

If failure modes of this type are known to occur, they should be recorded in the FMEA so that appropriate failure management decisions can be made later in the process. However, when listing failure modes caused by people, take care simply to record what went wrong and not who caused it. If too much emphasis is placed on "who" at this stage, the analysis could become unnecessarily adversarial, and people begin to lose sight of the fact that it is an exercise in avoiding or solving problems, not attaching blame. For instance, it is enough to say "Control valve set too high," not "Control valve incorrectly set by instrument technician."

Increase in Desired Performance (or Increase in Applied Stress)

The second category of failure modes occurs when desired performance is within the envelope of the capability of the asset when it is first put into service, but then the desired performance increases until it falls outside the capability envelope. This causes the asset to fail in one of two ways:

- The desired performance rises until the asset can no longer deliver it.
- The increase in stress causes deterioration to accelerate to the extent that the asset becomes so unreliable that it is effectively useless.

An example of the first case occurs if the users of the pump shown in Figure 3.1 were to increase the offtake from the tank to 425 gallons per minute. Under these circumstances, the pump is unable to keep the tank full. (Note that in this case the users are not forcing the pump to work any faster—they have simply opened a valve a bit wider somewhere else in the system.)

The second case occurs, for instance, if the owner of a motor car whose engine is "redlined" at 6,000 rpm persists in revving the motor to 7,000 rpm. This causes the engine to deteriorate more quickly than if the user keeps the revs within the prescribed limits, so the engine fails more often.

This phenomenon is illustrated in Figure 6.5. It occurs for four reasons, the first three of which embody some kind of human error:

- Sustained, deliberate overloading
- Sustained, unintentional overloading
- Sudden, unintentional overloading
- Incorrect process material

Each of these reasons is examined in turn.

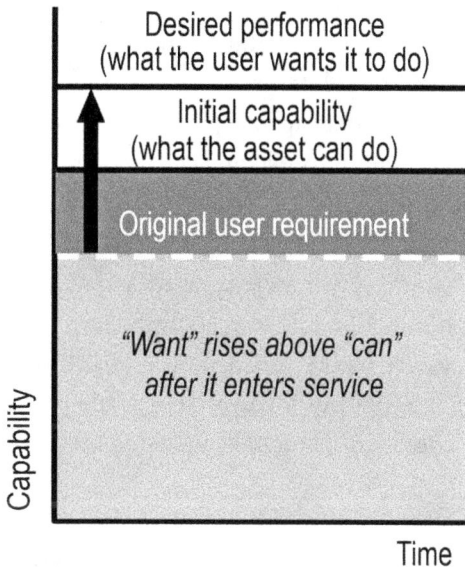

FIGURE 6.5   Failure mode Category 2: Increase in desired performance

Sustained, Deliberate Overloading. In many industries, users quickly give in to the temptation simply to speed up equipment in response to increased demand for existing products. In other cases, assets acquired for one product are used to process a product with different characteristics. This often reduces reliability and/or availability, especially when the increased stresses begin to approach or exceed the ability of the asset to withstand them.

(This phenomenon causes some of the most ferocious disputes between maintenance and operations people. When it occurs, operations people tend to claim that "there must be something wrong with our maintenance," while maintenance accuses operations of "flogging the machine to death." These disputes occur because operations people usually focus on what they want out of each asset, while maintenance people tend to think in terms of what it can do. Neither of them is "wrong"—they are simply considering the problem from two different points of view.)

In these cases, implementing "better" maintenance procedures will do little or nothing to solve the problem. In fact, maintaining a machine that cannot deliver the desired performance has been likened to rearranging the deck chairs on the Titanic. In such cases we need to look beyond maintenance for solutions. The two options are to modify the asset to improve its inherent capability, or to lower our expectations and operate the machine within its existing capabilities.

Sustained, unintentional overloading. Many industries respond to increased demand by undertaking formal "debottlenecking" programs. These programs entail increasing the capability of a production facility such as a production line to accommodate a new level of desired performance. However, it is surprising how often a few small subsystems or components get left out of the overall upgrade program, this means that part of the "upgraded" plant ends up being incapable of doing what its users want.

Demand for the products produced by the facility illustrated in the example has increased to the extent that its users wish to increase output from 400 to 500 tons per week. The dotted lines represent the capability of each operation. They show that most of the operations are already capable of meeting the new requirement. However, operations 3, 8, and 10 are capable of less than 500 tons, so they are the "bottlenecks." To achieve the new target, the users "debottleneck" these operations by installing new machines or components that are capable of producing well over 500 tons per week. They also upgrade the power supplies to match.

However, in this example the need to upgrade the instrument air supply was overlooked, so the plant begins to suffer intermittent instrument

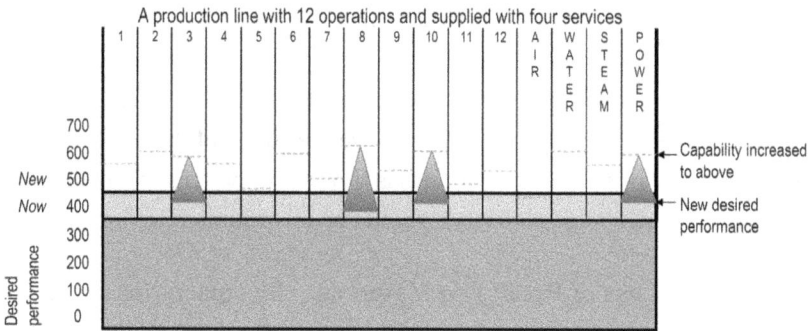

FIGURE 6.6 The destabilizing impact of debottlenecking

problems when demand for instrument air is at a maximum. (Note also that although the unchanged operations were already capable of more than 500 tons, their margin for deterioration is reduced by the upgrade program, so they also begin to fail more often.)

Clearly, if a plant is suffering from failure modes of this type, the failures should be recorded in the FMEA so that they can be dealt with appropriately.

(Some industrial organizations have found that despite the best efforts of their engineers, debottlenecking usually causes so much instability that it is forbidden in all but the most tightly controlled and heavily restricted circumstances. In these cases, growth is handled by allowing for it in the design of the original plant and/or by building new plants.)

Sudden, Unintentional Overloading. Many failures are caused by sudden and (usually) unintentional increases in applied stress, usually caused in turn by one of the following:

- Incorrect operation (for instance, if a machine is put into reverse while moving forward)
- Incorrect assembly (for instance, overtorqueing a bolt)
- External damage (for instance, if a forklift truck smashes into a pump or lightning strikes a poorly protected electrical installation)

These are not actually increases in desired performance, because no one wants the operator to put the machine into reverse at the wrong moment or the forklift to smash the pump. However, they belong in this category because applied stress rises above the ability of the asset to withstand it. If any of these failure modes is thought to be reasonably likely in the context under consideration, it should be incorporated in the FMEA.

Incorrect Process or Packaging Materials. Manufacturing processes often suffer functional failures caused by process materials that are out of specification (in terms of such variables as consistency, hardness, or pH). Similarly, packaging plants often suffer from inadequate or incompatible packaging materials.

In both cases, the machines fail or run badly because they cannot handle the out-of-spec material. This can be seen as an increase in applied stress.

In practice, these "failure modes" are seldom the result of a failure of the asset under review but are nearly always the effect of a failure elsewhere in the system. This means that remedial action has to be applied to a different asset. However, acknowledging these failures in the analysis of the affected asset helps to ensure that they will receive attention when the system that is really causing the problem is analyzed.

Example: In section 8.16 we will see how a water utility that uses lime slurry to treat drinking water for pH control and to prohibit rust in the pipes and steel tanks. Lime powder is mixed with water to form the slurry, and the slurry is pumped into the drinking water clear wells through a diffuser. From the clear wells, drinking water is distributed to reservoirs and homes.

The utility started to experience reliability issues in the lime feed system. The lime started to "bridge" inside the silos and not flow freely to the slurry mixing tanks. The company found that the vibrator pads mounted on the outside of the silo to assist with the transfer worsened the situation. The operators had to hammer the silo walls to loosen the lime and allow it to flow to the mixer. It became standard practice for the operators to bang on the silo when they walked passed the silos, and a hammer was left permanently on top of the gate valve below the silo. They started to damage the silo cone by the constant beating on the wall of the silo.

What actually caused the problem was not a design issue but "incorrect process material." The lime is sourced from different mines in the United States, and the utility changed suppliers. The "new" lime was delivered with a higher moisture content than before, which caused the powder to cake and form lumps, resulting in bridging in the outlet. Changing to another supplier and clearly specifying what the maximum allowable moisture content should be resolved the issue.

Failure modes of this sort should be incorporated in the FMEA where they are known to affect the asset under review, with a comment in the "Failure Effects" column that directs attention to the real source of the problem.

Initial Incapability

Chapter 3 explained that for any asset to be maintainable, its desired performance must fall within the envelope of its initial capability. It went on to mention that the majority of assets are, in fact, built this way. However, situations do arise where desired performance is outside the envelope of the initial capability right from the outset, as shown in Figure 6.7.

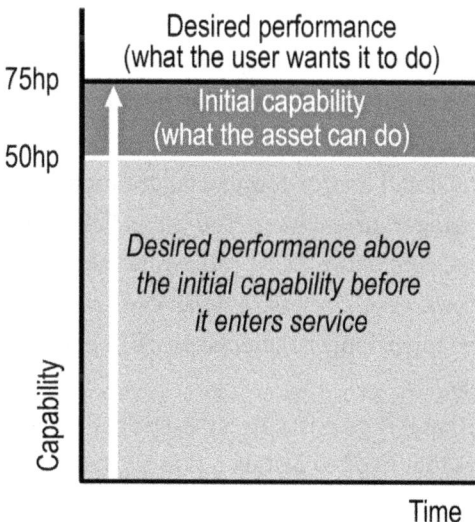

FIGURE 6.7   Failure mode Category 3: General failed state

This incapability problem seldom affects entire assets. It usually affects just one or two functions of one or two components, but these weak links upset the operation of the whole chain. The first step toward rectifying design problems of this nature is to list them as failure modes in an FMEA.

> From the above categories it is clear that everybody who is involved in the process is responsible for reliability (designers, engineers, builders, operators, and maintainers).

6.4 How Much Detail

Earlier in this chapter, it was mentioned that failure modes (cause and mechanism) should be described in enough detail for it to be possible to select an appropriate failure management strategy, but not in so much detail that excessive amounts of time are wasted on the analysis process itself.

> Failure modes (cause and mechanism) should be defined in enough detail for it to be possible to select a suitable failure management policy.

In practice, it can be surprisingly difficult to find an appropriate level of detail. However, it is important to do so, because the level of detail profoundly affects the validity of the FMEA and the amount of time needed to do it. Too little detail and/or too few failure modes lead to superficial and sometimes dangerous analyses. Too many failure modes and/or too much detail causes the entire RCM process to take much longer than it needs to. In extreme cases, excessive detail can cause the process to take two or even three times longer than necessary (a phenomenon known as analysis paralysis).

This means that it is essential to try to strike the right balance. Some of the key factors that need to be taken into account are discussed in the following paragraphs.

Causation

The causes or mechanisms of any failed state can be defined to almost any level of detail, and different levels are appropriate in different situations. At one extreme, it is sometimes enough to summarize the causes of a functional failure in one statement, such as "Machine fails." At the other, we may need to consider what goes wrong at the molecular level and/or explore the remoter corners of the psyche of the operators and maintainers in a bid to define so-called root causes of failure.

The extent to which failure modes can be described at different levels of detail is illustrated in Figure 6.8.

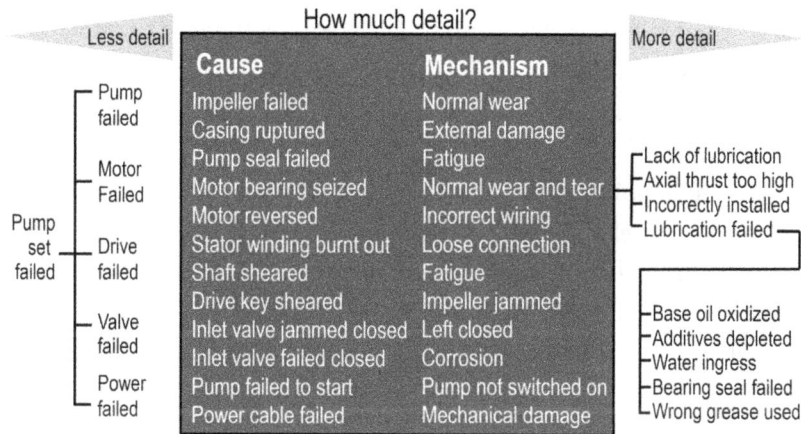

FIGURE 6.8   Showing how to list causes and mechanisms at different levels of detail

Figure 6.9 is based on the pump set shown in Figure 6.3. Figure 6.9 lists ways in which the pump set might suffer from the functional failure "Unable to transfer any water at all." These failure modes are considered at seven different levels of detail, as indicated on the following pages.

The top level (level 1) is the failure of the pump set as a whole. Level 2 recognizes the failure of the five major components of the pump set—the pump, the drive shaft, the motor, the switchgear, and the inlet/

outlet. Thereafter failures are considered in progressively more detail. When considering this example, please note that:

- Levels have been defined and failure modes allocated to each level for the purpose of this example only. They are not any kind of universal classification.
- Figure 6.9 does not show all failure possibilities at each level, so don't use this example as a definitive model.
- It is possible to analyze some of the failure modes at even lower levels than level 5, but it would very seldom be necessary to do so in practice.
- The failure modes listed only apply to the functional failure "Unable to transfer water at all." Figure 6.9 does not show failure modes that would cause partial failure or other functional failures, such as loss of containment or loss of protection.

The first point to emerge from this example is the connection between the level of detail and the number of failure modes listed. The example shows that the further one drills down in an FMEA, the larger the number of failure modes that can be listed.

For instance, there are five failure modes listed at level 2 for the pump set in Figure 6.9 but many more at level 3 and even more at levels 5 and 6.

Two more key issues that arise from Figure 6.9 concern root causes and human error. They are discussed in more detail in a later part of this chapter.

Failure Causes—Root Cause of Failure

The term "root cause" is often used in connection with the analysis of failures. It implies that if one drills down far enough, it is possible to arrive at a final and absolute level of causation. In fact, this is seldom the case.

For instance, the failure mode "Impeller nut overtightened" is caused by an assembly error. If we were to go down one level further,

LEVEL 1	LEVEL 2	LEVEL 3	LEVEL 4	LEVEL 5	LEVEL 6	LEVEL 7
Pump set fails	Pump fails	Impeller fails	Impeller comes adrift	Mounting nut undone	Nut not tightened correctly	Assembly error
				Mounting nut worn away	Nut eroded/corroded away	Wrong material specified
					Nut made of wrong material	Wrong material specified
				Impeller nut cracked	Impeller nut overtightened	Assembly error
					Nut made of wrong material	Wrong material specified
						Wrong material specified
				Impeller key sheared	Wrong key steel specified	Design error
						Procurement error
					Wrong key steel specified	Storekeeping error
						Requisitioning error
			Object smashes impeller	Part in system after maintenance	Assembly error	See Appendix 2
				Foreign object enters system	Suction strainer not installed	Assembly error
					Strainer holed by corrosion	
		Casing ruptured	Casing bolts come loose	Casing bolts undertightened	Assembly error	See Appendix 2
				Bolt loosened by vibration		
				Casing bolts corroded away		
				Bolts fail due to fatigue		
			Casing joint fails	Joint incorrectly fitted	Assembly error	See Appendix 2
				Joint fails due to fretting		
			Casing smashed	Casing smashed by vehicle	Operating error	See Appendix 2
					Pump in vulnerable position	Design error
				Smashed by object from sky	Casing hit by meteorite	
					Casing hit by part of aircraft	
		Pump seal fails	Normal wear and tear	Seal abraded		
			Pump runs dry	See "water supply fails" below		
			Seal misaligned	Assembly error	See Appendix 2	
			Seal faces dirty	Assembly error	See Appendix 2	
			Wrong seal fitted	Wrong seal supplied	Procurement error	See Appendix 2
					Storekeeping error	See Appendix 2
				Wrong seal supplied	Design error	See Appendix 2
			Damaged seal installed	Pump seal dropped in stores	Storekeeping error	See Appendix 2
				Pump seal damaged in transit	Procurement error	See Appendix 2

FIGURE 6.9 Failure modes at different levels of detail

LEVEL 1	LEVEL 2	LEVEL 3	LEVEL 4	LEVEL 5	LEVEL 6	LEVEL 7
Pump set fails	Motor fails	Bearings seize	Normal wear and tear	Subsurface fatigue on outer race		
				Balls worn away		
			Axial thrust too great	Motor undersized		
			Lubrication failure	Bearing seals fail	Seals damaged on installation	Assembly error
					Seals poorly fitted to bearing	Manufacturing error
				Grease fails	Base oil oxidised	
					Grease liquified	
					Additives depleted	
				Wrong lubricant installed	Manufacturing error	
			Bearing wrongly installed	Damaged prior to installation	Bearing dropped in store	Storekeeping error
					Bearing damaged in transit	Procurement error
				Damaged during installation	Bearing hit with hammer	Assembly error
				Bearing misaligned	Assembly error	See Appendix 2
				Defective bearing installed	Defective bearing supplied	Manufacturing error
					Bearing corroded in store	Storekeeping error
				Wrong bearing installed	Wrong bearing specified	Design error
					Wrong bearing supplied	Procurement error
		Motor reverses	Motor wired incorrectly	Assembly error	See Appendix 2	
		Stator winding burns out	Motor insulation fails	Insulation deteriorates	Normal wear and tear	
				Motor operated at high load	Operating error	See Appendix 2
				Insulation damp	Motor casing gasket fails	Normal deterioration
						Gasket fitted incorrectly
					Motor casing damaged	Motor dropped in store
						Motor hit by foreign object
					Motor stored in damp area	Storekeeping error
					Casing gasket not fitted	Assembly error
					Water sprayed on motor	Operating error
					Motor casing bolts loose	Assembly error
			Motor overheated	Fan grille blocked by dirt		
			Motor fan fails	Fan fitted the wrong way round	Assembly error	
				Fan not fitted	Assembly error	
		Not switched on	Operating error	See Appendix 2		

FIGURE 6.9 Failure modes at different levels of detail (continued)

LEVEL 1	LEVEL 2	LEVEL 3	LEVEL 4	LEVEL 5	LEVEL 6	LEVEL 7
Pump set fails	Driveline fails	Shaft shears	Shears due to fatigue	Stress raisers in steps in shaft	Sharp radii specified	Design error
					Wrong radii cut	Manufacturing error
				Defective steel supplied	Steel manufacturing error	
		Drive key shears	Wrong key steel specified	Design error	See Appendix 2	
			Wrong key steel supplied	Procurement error	See Appendix 2	
			Key cut too short	Assembly error	See Appendix 2	
	Valve closed	V/v jammed shut	Valve handle missing	Handle cannibalized	Assembly error	See Appendix 2
			Valve spindle seized	Spindle seized due to corrosion	Grease worn off valve spindle	
					Wrong grease used	Assembly error
		Valve left shut	Operating error	See Appendix 3		
	Power fails	Switchgear fails	Contactor fails open	Contactor points worn		
				Contactor coil burns out		
				Contactor spring fails due to fatigue		
				Contactor points dirty	Points dirty on installation	Assembly error
					Switchbox cover lets in dirt	Cover badly fitted
						Cover not closed properly
		Spurious trip	Overload c/b set too low	Assembly error	See Appendix 2	
				Overload setting drifts		
			Spurious failure of fuse	Wrong fuse fitted	Procurement error	See Appendix 2
				Defective fuse fitted	Defective fuse supplied	Manufacturing error
					Fuse damaged on installation	Assembly error
			Switched off accidentally	Operating error	See Appendix 2	
	Power cable fails	Power cable fails	Cable insulation fails	Insulation deteriorates		
				Insulation manufacturing defect		
			Cable damaged	Cable damaged by impact	Operating error	See Appendix 2
				Cable abraded	Cable too long	Assembly error
					Cable poorly restrained	Design error
			Connection fails	Connection loose	Loosens in service	
					Installed too loose	Assembly error
					Terminal box fails	Box damaged by impact
				Connection corroded		Box cover loose
						Box cover seal not fitted
Incoming power fails						

FIGURE 6.9   Failure modes at different levels of detail (*continued*)

| 137

the assembly error might have occurred because the "fitter was distracted." He might have been distracted because his "child was ill." This failure might have occurred because the "child ate bad food in restaurant."

Clearly, this process of drilling down could go on almost forever—way beyond the point at which the organization doing the FMEA has any control over the failure modes. This is why this chapter stresses repeatedly that the level at which any failure mode should be identified is the level at which it is possible to identify an appropriate failure management policy. (This is equally true whether one is carrying out an FMEA before failures occur or a root cause analysis after a failure has occurred.)

The fact that the level that is appropriate varies for different failure modes means that we do not have to list all failure modes at the same level on the Information Worksheet. Some failure modes might be identified at level 2, others at level 5, and the rest somewhere in between.

For instance, in one particular context, it may be appropriate to list all likely failure modes and mechanisms. In another context, it may be appropriate for an entire FMEA for an identical pump set to consist of the single failure mode "Pump set fails." Another context may call for yet another selection.

Obviously, in order to be able to stop at an appropriate level, the people doing such analyses need to be aware of the full range of risk management policy options. These are discussed at length in Chapters 8 and 9.

Other factors that influence the level of detail are considered in the rest of this part of this chapter.

Failure Mechanisms

We know that the failed state defines the loss of function. "Failure mode" (cause and mechanism) is a term used to describe any event that causes a failed state, listed at an appropriate level of detail. "Failure mechanism" describes the processes leading up to the cause of the failed state, e.g., dirt buildup, "normal wear," or corrosion.

Normal Deterioration Mechanisms

Normal deterioration methods include:

- **Corrosion.** The general corrosion process occurring under atmospheric conditions where carbon steel (Fe) is converted to iron oxide (Fe_2O_3).
- **Wear/abrasion.** The process of material removal because of contact between two surfaces or a surface and fluids. The more abrasive the fluids, the quicker the rate of wear.
- **Metal fatigue.** Failure of a component by cracking after the continued application of cyclic stress, which exceeds the material's endurance limit.
- **Gradual buildup of dirt, scale, etc.**

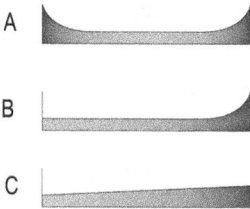

A

B

C

Other Mechanisms

Other mechanisms can typically be human-induced processes such as:

- Incorrect installation
- Incorrect operation
- Inadequate maintenance

F

Then there are also "random" processes such as:

- "Mechanical" damage caused by a falling object or foreign object
- Tire punctured by a foreign object
- Impeller jammed by a foreign object
- Equipment overload

E

Identifying the Root Cause

For each failure mechanism, one needs to apply a different risk management strategy. This means that when we list the failure mode, we need to identify the actual underlying mechanism of the failure (the root cause).

This is achieved by listing the failure mode cause as well as a mechanism. Listing a failure mode cause and a mechanism will assist people in identifying the actual underlying cause of the failure. The mechanism of a failure mode can actually be broken up in two more levels of detail, namely the trigger and the process. Fatigue is a common example. Fatigue can be described as follows:

"Fatigue occurs when a material is subjected to repeat loading and unloading (trigger). If the loads are above a certain threshold, microscopic cracks will begin to form at the stress concentrators such as the surface, persistent slip bands (PSBs), and grain interfaces. Eventually a crack will reach a critical size, the crack will propagate suddenly, and the structure will fracture. The shape of the structure will significantly affect the fatigue life; square holes or sharp corners will lead to elevated local stresses where fatigue cracks can initiate. Round holes and smooth transitions or fillets will therefore increase the fatigue strength of the structure."

The *trigger* for fatigue can be described as cyclic stress. The *process* can be described as metallic bonds loosen.

The *mechanism* associated with this trigger and process can be described as metal fatigue. The end result of this mechanism can mostly be described as a catastrophic fracture.

It can be seen based on the above discussion that it is very important for us to understand the technical characteristics of the failure mode, because these technical characteristics will determine the most effective way (risk management strategy) through which we can eliminate completely, or manage to a tolerable level, the risk of the failure mode.

The reason for identifying failure modes is that our every effort with a proactive maintenance program is to prevent failures from occurring or manage the risk of the failure mode to a level that will be tolerable to the organization. We can only do this if we can identify the failure mode causes and failure mode mechanisms at the appropriate level and direct our attention to these.

In fact, each and every task to be performed in the name of PM should be focused on preventing a failure mode, or at the least, managing the effect and consequence of a failure mode.

Clearly, we see from this that maintenance policies (or other risk management strategies) can only be determined and managed on a cause and mechanism (failure mode) level. Identifying, preventing, or managing the risks of failure modes is the heart of any maintenance program.

When listing failure modes, we need to take into consideration that there are different categories of failure modes. The three main failure mode categories are physical failures, human error, and latent causes.

Physical Failures. These are usually associated with normal deterioration such as corrosion, erosion, fatigue, etc. These phenomena can be accelerated through incorrect operation. These failures can often successfully be managed through some form of maintenance. Earlier on in the chapter, we looked at examples of physical failures that included corrosion, wear/abrasion, metal fatigue, and a gradual buildup of dirt, scale, etc.

Human Error. Human error implies that failures are caused by specific human actions such as incorrect operation or incorrect maintenance. These failures are managed through actions that are directed at the human errors, such as improved training. Appendix III provides a brief summary of key issues involved in the classification and management of such errors.

Failure pattern F, as discussed before, is mostly the result of some form of human error. Figure 6.10 lists failures typically caused through human error.

Detailed human behavior analysis is beyond the scope of this book, but we mentioned a number of general ways in which human error could cause machines to fail. The RCM3 process goes on to suggest that if the associated failure modes are thought to be reasonably likely, they should be incorporated in the FMEA so that an appropriate risk management strategy can be identified and implemented. Where failure modes end with the word "error," some form of human error caused the failure.

- incorrect functional specification
- poor design specification
- poor quality manufacturing
- incorrect installation
- incorrect commissioning
- incorrect operation (gross overload of equipment)
- unnecessary maintenance
- excessive invasive maintenance
- bad workmanship and quality control

FIGURE 6.10 Start-up failures or human error

Human errors can be grouped into four categories:

- Anthropometric factors
- Human sensory factors
- Physiological factors
- Psychological factors

Human error is discussed in detail in Appendix IV.

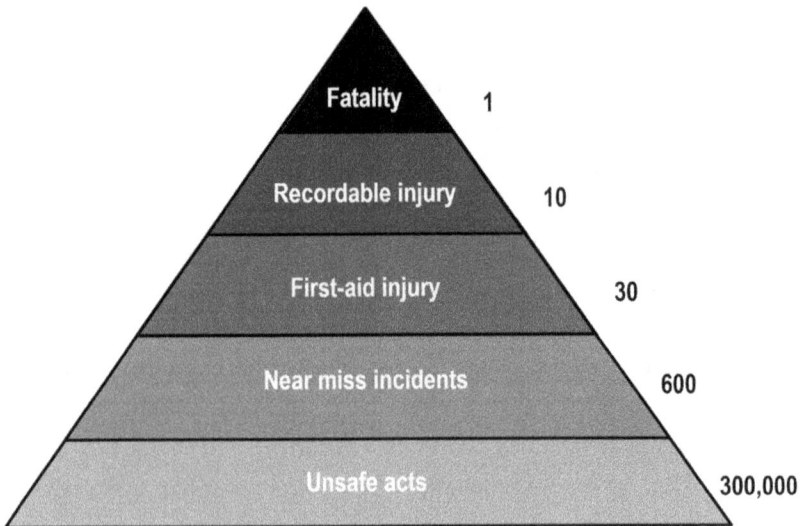

FIGURE 6.11 Heinrich's Law

Latent Causes

Latent causes are normally associated with organizational procedures such as operating procedures, maintenance procedures, spares purchasing and issuing procedures, etc. These failures are normally dealt with through the development of procedures where they do not exist or improvement where they do but are inadequate.

As discussed earlier, RCM3 deals with all the categories where they are defined as reasonably likely.

Behavior-Based Safety

According to Herbert W. Heinrich, the basis for the theory of behavior-based safety holds that as many as 95% of all workplace accidents are caused by unsafe acts. Heinrich came to this conclusion after reviewing thousands of accident reports completed by supervisors, who generally blamed workers for causing accidents without conducting detailed investigations into the root causes. While Heinrich's figure that 88% of all workplace accidents and injuries/illnesses are caused by "man-failure" is perhaps his most oft-cited conclusion, his book actually encouraged employers to control hazards, not merely focus on worker behaviors. "No matter how strongly the statistical records emphasize personal faults or how imperatively the need for educational activity is shown, no safety procedure is complete or satisfactory that does not provide for the correction or elimination of- physical hazards," Heinrich wrote in his book. Emphasizing this aspect of workplace safety, Heinrich devoted much of his work to the subject of machine guarding. His research is criticized by some for being outdated and unscientific. One critic thinks that Heinrich's Law should be replaced by a model that emphasizes safety in design, compared to the former's emphasis on behavior.

It is the author's opinion that much can be done through reliability in design (i.e., using Poka-Yoke principles), but human error remains of concern and proper risk management through procedures, training, and certification should not be underestimated. Poka-Yoke is discussed in more detail in Section 8.20 (page 337).

Heinrich's Law is discussed in more detail in Chapter 7, Section 7.5.

Probability

Different failure modes occur at different frequencies. Some may occur regularly, at average intervals measured in months, weeks, or even days. Others may be extremely improbable, with mean times between occurrences measured in millions of years. When preparing an FMEA, decisions must be made continuously about what failure modes are so unlikely that they can safely be ignored. This means that we do not try to list every single failure possibility regardless of its likelihood.

> When listing failure modes, do not try to list every single failure possibility regardless of its likelihood.

In other words, only failure modes that might reasonably be expected to occur in the context in question should be recorded. A list of "reasonably likely" failure modes should include the following:

- **Failures that have occurred before on the same or similar assets.** These are the most obvious candidates for inclusion in an FMEA unless the asset has been modified so that the failure cannot occur again. As will be discussed later, sources of information about these failures include people who know the asset well (your own employees, vendors, or other users of the same equipment), technical history records, and data banks. In this context, note the comments later in this chapter about the shortcomings of most technical history records and in Chapter 12 about the danger of too much reliance on historical data.
- **Failure modes that are already the subject of proactive maintenance routines, and so would occur if no proactive maintenance was being done.** However, a review of existing schedules should only be carried out as a final check after the rest of the

RCM analysis has been completed in order to reduce the possibility of perpetuating the status quo. (Some users of RCM are tempted to assume that all reasonably likely failure modes are covered by their existing PM systems, and hence that these are the only failure modes that need to be considered in the FMEA. This assumption leads these users to develop the entire FMEA by working backward from their existing maintenance schedules and then working forward through the last three steps of the RCM process. This approach is usually adopted in the belief that it will speed up or streamline the process. In fact, this approach is not recommended, because among other shortcomings, it leads to dangerously incomplete RCM analyses.)

- **Any other failure modes that have not yet occurred but that are considered to be real possibilities.** Identifying and deciding how to deal with failures that have not happened yet calls for a high degree of judgment. On the one hand, we need to list all reasonably likely failure modes, while on the other, we don't want to waste time on failures that have never occurred before and that are extremely unlikely (incredible) in the context in question.

For example, sealed-for-life bearings are installed on the motor driving the pump shown in Figure 6.3. This means that the likelihood of lubrication failure is low—so low that it would not be included in most FMEAs. On the other hand, failure due to lack of lubricant probably would be included in FMEAs prepared for manually lubricated components, centralized lube systems, and gearboxes. However, the decision not to list a failure mode should be tempered by careful consideration of the failure consequences.

Consequences

If the consequences are likely to be very severe indeed, then less likely failure possibilities *should* be listed and subjected to further analysis.

For example: A limestone crusher in a cement factory (see Figure 6.12) is driven by an electric motor through eight belts and two sheaves. A brass keyway locates the sheave onto the drive shaft and protects the drive mechanism (belts, shaft, and motor) against damage in the event of the crusher jamming. When discussing likely failure modes that could cause a loss of the crusher's primary function, the review group felt a sheared brass keyway was an unlikely failure mode (since it had not happened before, and it was also not something that was being prevented through an existing PM program). The members of the group did not consider the consequences before they discarded the failure mode! The facilitator and the review group decided to ignore the failure mode "Brass keyway sheared." Unfortunately, the review group also did not capture the very important secondary function of the brass keyway, which was "To act as a protective device."

The brass keyway actually sheared while the RCM analysis was being conducted. A metal part entered the crusher, causing the crusher to jam and shear the keyway. The drive motor continued to turn, and the feed conveyor that was interlocked with the crusher motor continued to feed the crusher. The limestone filled the crusher and started to spill onto the floor next to the crusher, quickly building up in a mound around the crusher. The sheave and drive belts that were still turning heated up until the belts actually caught fire. The shaft became red hot and was badly damaged in the process. By the time the operator observed the failure and stopped the process by activating the emergency stop circuit, the crusher building was filled with limestone.

The limestone had to be removed manually, and the downtime (lost production) and repair costs (including fabrication of a new shaft) were in excess of $1.5 million. If the members of the review group considered these consequences of the failure mode, they would have included it in the analysis and would have recommended the appropriate risk management strategy—which was to fit a motion sensor onto the drive shaft that would shut the feed conveyor drive and crusher drive motor down in the event the shaft stopped turning.

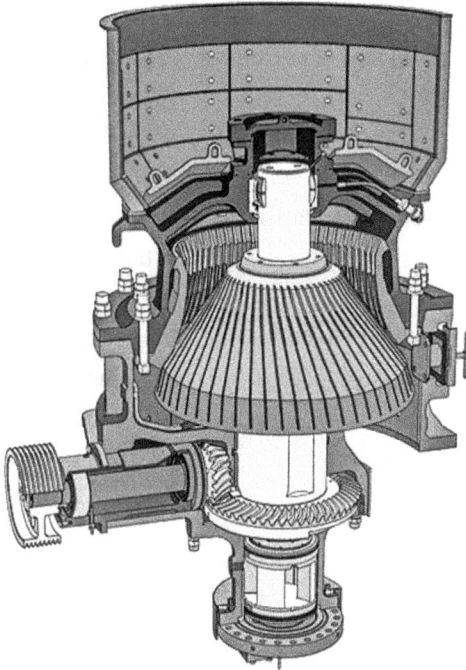

FIGURE 6.12 Rotary cone crusher

From the above, it can be seen that it is important to list all failure modes reasonably likely to cause each failed state.

What Is Meant by Reasonably Likely?

"Reasonable likelihood" means just that—a likelihood that meets the test of reasonableness when applied by trained and knowledgeable people. (A term often used instead of "reasonable" in this context is "credible.") If people who are trained to use RCM3, and who are knowledgeable about the asset in its operating context, agree that the probability that a specific failure mode could occur is sufficiently high to warrant further analysis, then that failure mode should be listed.

This step of the RCM3 process needs to be managed carefully and brings up the need for an RCM3 facilitator. In order to manage the recording of reasonably likely failure modes, the process must be facilitated by a well-trained and experienced RCM facilitator.

Cause Versus Effect

Care should be taken not to confuse causes and effects when listing failure modes. This is a subtle mistake most often made by people who are new to the RCM process.

For example, one plant had some 200 gearboxes, all of the same design and all performing more or less the same function on the same type of equipment. Initially, the following failure modes were recorded for one of these gearboxes:

- Gearbox bearings seize.
- Gear teeth stripped.

These failure modes were listed to begin with, because the people carrying out the review recalled that each failure had happened in the past to their knowledge (some of the gearboxes were 20 years old). The failures did not affect safety, but they did affect production. So the implication was that it might be worth doing preventive tasks like "Check gear teeth for wear," or "Check gearbox for backlash," or "Check gearbox bearings for vibration." However, further discussion revealed that both failures had occurred because the oil level had not been checked when it should have been, so the gearboxes had actually failed due to lack of oil. What is more, no one could recall that any of the gearboxes had failed if they had been properly lubricated. As a result, the failure mode was eventually recorded as:

- Gearbox fails due to lack of oil.

This underlined the importance of the obvious proactive task, which was to check the oil level periodically. (This is not to suggest that all gearboxes should be analyzed in this way. Some are much more complex or much more heavily loaded, and so they are subject to a wider variety of failure modes. In other cases, the failure consequences may be much more severe, which would call for a more defensive view of failure possibilities.)

Failure Modes and the Operating Context

We have seen how the functions and functional failures of any item are influenced by its operating context. This is also true of failure modes in terms of causation, probability, and consequences.

For example, consider the three pumps shown in Figure 3.2 (in Chapter 3). The failure modes that are likely to affect the standby pump (such as brinelling of the bearings, stagnation of water in the pump casing, and even the "borrowing" of key components to use elsewhere in an emergency) are different from those that might affect the duty pump, as set out in Figure 6.9.

Similarly, electric mining shovels operating in mines in Canada would be subject to different failure modes from the same model shovel operating in mines in Chile. Similarly, a gas turbine powering a jet aircraft would have different failure modes from the same type of turbine acting as a prime mover on an oil platform.

These differences mean that great care should be taken to ensure that the operating context is identical before applying an FMEA developed in one set of circumstances to an asset that is used in another. (Note also the comments regarding the use of generic FMEAs in Section 6.6 of this chapter.)

The operating context affects levels of analysis as well as the causes and consequences of failure. As discussed earlier, it might be appropriate to identify failure modes for two identical assets at one level in one operating context and at another level in another.

6.5 Failure Effects and Consequence Severity

The fifth question of the RCM3 review process entails listing what happens when each failure mode occurs. These are known as failure effects.

> Failure effects describe what happens when a failure
> mode occurs.

Note that failure effects are not the same as failure consequences. A failure effect answers the question "What happens?" whereas a failure consequence answers the question "(How) does it matter?"

A description of failure effects should include all the information needed to support the evaluation of the consequences of the failure. Specifically, when describing the effects of a failure, the answers to the following should be recorded:

- When is the failure mode most likely to occur?
- How often would the failure mode occur if nothing is done to prevent it?
- What evidence (if any) is there that the failure has occurred?
- In what ways (if any) does it pose a threat to safety or the environment?
- In what ways (if any) does it affect production or operations?
- What physical damage (if any) is caused by the failure?
- Does it cause any secondary damage?
- What is the revenue loss (if any)?
- What must be done to repair the failure?

These issues are reviewed in the following paragraphs. Note that one of the objectives of this exercise is to establish whether proactive maintenance is necessary. If we are to do this correctly, we cannot assume that some sort of proactive maintenance is being done already, so the effects of a failure should be described as if nothing were being done to prevent it.

When Is the Failure Mode Most Likely to Occur?   In order to identify the effects of the failure with the appropriate amount of detail so that the consequences and subsequent risk can be assessed, it is important to describe when the failure is most likely to occur.

Some failures are most likely to occur during or after start-up, or they may occur during normal operation, at landing (in the case of aircraft), during peak flow (storm event), or during high seasonal demand. Failures that occur immediately after start up may be an indication of human error (incorrect installation or incorrect parts) or incorrect start-up procedures.

Failures that occur during maximum demand periods, such as stormwater pump stations dealing with storm events, may be an indication of design flaws (equipment inherent reliability below what the user expects), and similarly failures that occur during high seasonal demand (bottling facility in a beer brewing company) may be a result of inadequate design capability.

The consequences of a plane's landing gear failing when it taxis on the runway will be vastly different when it fails while it is landing. In order to assess the worst case, the failure effect description should consider when the failure is most likely to happen. This does not mean the absolute worst case. It is also important to consider whether the failure mechanism takes time to develop or whether the failure is a sudden event.

How Often Would the Failure Mode Occur If Nothing Is Done to Prevent It?

The RCM3 process is based on the assumption that nothing is done to prevent the failure from happening and that no one interferes with the failure while it is in the process of happening. A true *zero-based* analysis can only be performed when assuming no maintenance is present and nobody is attending to the failure while it happens—even if we know that this is not a realistic assumption. RCM3 also considers the potential worst case, that is, when any protection associated with protected functions is not available or fails to operate on demand. The assumption that existing maintenance programs or operators in control rooms would prevent consequences from happening is dangerous and should not be considered.

The inherent risk *calculation* is based on these assumptions. If the inherent risk is intolerable, we must do something to mitigate the risk. This will be discussed in much more detail in later chapters.

What Evidence (If Any) Is There That the Failure Has Occurred?

Failure effects should be described in a way that enables the team doing the RCM analysis to decide whether the failure will become evident to the operating crew under normal circumstances.

For instance, the description should state whether the failure causes warning lights to come on or alarms to sound (or both), and it should specify whether the warning is given on a local panel or in a central control room (or both).

Similarly, the description should state whether the failure is accompanied (or preceded) by obvious physical effects such as loud noises, fire, smoke, escaping steam, unusual smells, or pools of liquid on the floor. It should also state whether the machine shuts down as a result of the failure.

For example, if we are considering the seizure of the bearings (due to normal wear) of the pump shown in Figure 3.1, the failure effects might be described as follows (the italics describe what would make it evident to the operators that a failure has occurred):

- **Local effect.** The bearing wears over time, and the bearing temperature, noise, and vibration will rise. Eventually the bearing will seize or collapse.
- **Next-higher-level effect.** *The motor trips on overload, and the trip alarm sounds in the control room. The tank Y low-level alarm sounds after 20 minutes, and the tank runs dry after 30 minutes.*
- **End effect.** The downtime to replace the bearings takes 4 hours.
- **Potential worst-case effect.** In the event that the motor trip circuit failed, the motor could be damaged, which would result in longer downtime and additional repair costs.

In the case of a stationary gas turbine, a failure mode that occurred in practice was the gradual buildup of combustion deposits on the compressor blades. These deposits could be partially removed by the periodic injection of special materials into the airstream, a process known as jet blasting. The failure effects were described accordingly as follows (we will later see how to record failure effects correctly):

- **Local effect.** Combustion deposits build up on the blades over time, causing imbalance and vibration to increase. The compressor efficiency declines, and the governor compensates to sustain power output, causing the exhaust temperature to rise

- **Next-higher-level effect.** *The exhaust temperature is displayed on the local control panel and in the central control room.* If no action is taken, the exhaust gas temperature rises above 885°F under full power. *A high exhaust gas temperature alarm sounds on the local control panel, and a warning light comes on in the central control room. Above 935°F, the control system shuts down the turbine.* (Running at temperatures above 885°F shortens the creep life of the turbine blades.)
- **End effect.** The blades can be partially cleaned by jet blasting, and jet blasting takes about 30 minutes.
- **Potential worst-case effect.** In the event that the high exhaust gas temperature alarm fails to go off in the control room and the control system fails to shut down the turbine when the temperature rises above 935°F, the turbine will be damaged and the downtime to replace it will be up to 12 weeks.

This is an unusually complex failure mode, so the description of the failure effects is somewhat longer than usual. The average description of a failure effect usually amounts to between 20 and 60 words.

When describing failure effects, do not prejudge the evaluation of the failure consequences by using the word "hidden" or "evident." The words are part of the consequence evaluation process, and using them prematurely could bias this evaluation incorrectly.

When dealing with protective devices, failure effect descriptions should state briefly what would happen if the protected device were to fail while the protective device was unavailable (failed also). This is recorded as potential worst-case failure effects, which will be discussed in more detail later. An important aspect to raise at this point is:

> The RCM3 process asks the question, "What makes the failure evident?" This refers to the failed state and not the failure mode. It further does not place a time limit on when the failure becomes evident. A failure that becomes evident over a reasonable period of time will be an evident failure.

In What Ways (If Any) Does It Pose a Threat to Safety or the Environment? Modern industrial plant design has evolved to the point that only a small proportion of failure modes present a direct threat to safety or the environment. However, if there is a possibility that someone could get hurt or killed as a direct result of the failure, or that an environmental standard or regulation could be breached, the failure effect should describe how this could happen. Examples include:

- Increased risk of fire or explosions
- The escape of hazardous chemicals (gases, liquids, or solids)
- Electrocution
- Falling objects
- Pressure bursts (especially pressure vessels and hydraulic systems)
- Exposure to very hot or molten materials
- The disintegration of large rotating components
- Vehicle accidents or derailments
- Exposure to sharp edges or moving machinery
- Increased noise levels
- The collapse of structures
- The growth of bacteria
- Ingress of dirt into food or pharmaceutical products
- Flooding

When listing these effects, do not make qualitative statements like "This failure has safety consequences" or "This failure affects the environment." Simply state what happens and leave the evaluation of the consequences to the next stage of the RCM process.

Note also that we are concerned not only about possible threats to our own staff (operators and maintainers), but also about threats to the safety of customers and the community as a whole. This may call for some research by the team doing the analysis into the environmental and safety standards that govern the process under review.

In What Ways (If Any) Does It Affect Production or Operations? Failure effect descriptions should also help with decisions about operational and nonoperational failure consequences. To do so, they should indicate how production is affected (if at all) and for how long. This is usually given by the amount of downtime associated with each failure.

In this context, downtime means the total amount of time the asset would normally be out of service as a result of this failure, from the moment it fails until the moment it is fully operational again. As indicated in Figure 6.13, this is usually much longer than the repair time.

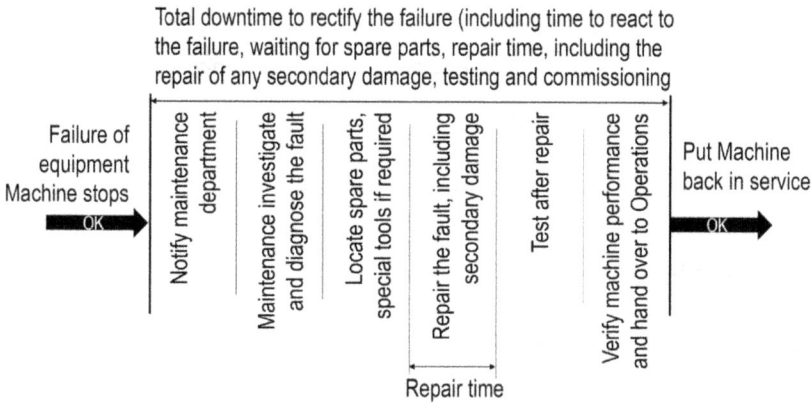

FIGURE 6.13 Downtime versus repair time

Downtime as defined above can vary greatly for different occurrences of the same failure, and the most serious consequences are usually caused by the longer outages. Since it is the consequences that are of most interest to us, the downtime recorded on the Information Worksheet should be based on the typical worst case.

For instance, if the downtime caused by a failure that occurs late on a weekend night shift is usually much longer than it is when the failure occurs on a normal day shift, and if such night shifts are a regular occurrence, we list the downtime that occurs on a weekend night shift.

It is, of course, possible to reduce the operational consequences of a failure by taking steps to shorten the downtime, most often by reducing the time it takes to get hold of a spare part. However, as discussed in Chapter 4, we are still in the process of defining the problem at this stage, so the analysis should be based—at least initially—on current spares-holding policies.

Note that if the failure affects operations, it is more important to record the downtime than the mean time to repair the failure (MTTR, for two reasons:

- In the minds of many people, the term "repair time" has the meaning shown in Figure 6.13. If this term is used instead of "downtime," it could upset the subsequent assessment of the operational consequences of failure.
- We should base the assessment of consequences on the typical worst case and not the mean, as discussed above.

If the failure does not cause a process stoppage, then the average amount of time it takes to repair the failure should be recorded, because this can be used to help establish workforce requirements.

What physical damage (if any) is caused by the failure? Physical damage (if any) should be described with appropriate level of detail. These may be symptoms or observable evidence of failure causes and mechanisms. The physical damage will also capture the information that may be necessary to determine the spare parts and work required to repair the asset or system. Physical damage will be visible in many different forms such as:

- Holed tank
- Broken shaft
- Smashed impeller
- Burnt motor windings
- Etc

Does it cause any secondary damage? In order to determine the full impact of the failure (safety, environmental, and operational capability), any secondary damage directly related to the failure should be captured. Secondary damage will extend downtime and increase the risk associated with the failure. It may actually be the difference between a tolerable and intolerable risk. The potential-worst case description should also capture information about secondary damage that would occur in the event the protective device directly related to the failure fails to operate (or is not available when needed). Secondary damage is an important consideration to determine whether or not maintenance is worth doing. Examples of secondary damage are:

- Shaft shear (lack of lubrication causing bearing failure)
- Check valve damaged (due to water hammer)
- Turbine bearing seized (lubrication oil filter collapsed)
- Etc

What Must Be Done to Repair the Failure?   Failure effects should also state what must be done to repair the failure. The corrective action may require process interruptions that will further delay restoring the system or assets. This can be included in the statement about downtime, as shown in italics in the following examples:

- Downtime to *replace bearings*—about 4 hours
- Downtime to *clear the blockage* and *reset the trip switch*—about 30 minutes
- Downtime to *strip the turbine and replace the disc*—about 2 weeks

What Is the Revenue Loss (If Any)?   In most cases, regardless of the consequences of failures (safety, environmental, or operational), failures have some negative economic impact. In private-sector companies, these must be avoided in order to stay competitive. When it is easy to calculate, the revenue loss should be included in the failure effect description. When managers with overall responsibility for the assets have to make risk mitigation decisions, e.g., whether or not to keep long lead time spare parts, having access to the details could assist with making those decisions almost intuitively. Additionally, this may also drive one-time change decisions in order to reduce the overall loss in income or possible penalties.

When considering public utilities or service companies, failures may not directly impact revenue, but failures and the inability to provide service have a negative impact on the public's perception, which may result in tough negotiations and justification for future rate increases. Failures may also lead to permit violations and penalties. Additionally, failures could cause insurance premiums to increase. All these economic consequences should be considered to make the right decision at the decision-making stage.

6.6 Sources of Information about Failure Modes and Effects

When considering where to get the information needed to draw up a reasonably comprehensive FMEA, remember the need to be proactive. This means that as much emphasis should be placed on what could happen as on what has happened. The most common sources of information are discussed in the following paragraphs, together with a brief review of their main advantages and disadvantages.

The Manufacturer or Vendor of the Equipment

When carrying out an FMEA, the source of information that usually springs to mind first is the manufacturer. This is especially so in the case of new equipment. In some industries, this has reached the point where manufacturers or vendors are routinely asked to provide a comprehensive FMEA as part of the equipment supply contract. Apart from anything else, this request implies that manufacturers know everything that needs to be known about how the equipment can fail and what happens when this occurs.

This is seldom the case in reality.

In practice, few manufacturers are involved in the day-to-day operation of the equipment. After the end of the warranty period, almost no manufacturers get regular feedback from the equipment users about what fails and why. The best that many of them can do is try to draw conclusions about how their equipment is performing from a combination of anecdotal evidence and an analysis of spares sales. (This is true except when a really spectacular failure occurs, in which case lawyers tend to take over from engineers. At this point, rational technical discussion about root causes often ceases.)

Manufacturers also have little access to information about the operating context of the equipment, desired standards of performance, failure consequences, and the skills of the user's operators and maintainers. More often the manufacturers know nothing about these issues. As a result, FMEAs compiled by these manufacturers are usually generic and often highly speculative, which greatly limits their value.

The small minority of equipment manufacturers that are able to produce a satisfactory FMEA on their own usually fall into one of two categories:

- They are involved in maintaining the equipment throughout its useful life, either directly or through closely associated vendors. For instance, most privately owned motor vehicles are maintained by the dealers that sold the vehicles. This enables the dealers to provide the manufacturers with copious failure data.
- They are paid to carry out formal reliability studies on prototypes as part of the initial procurement process. This is a common feature of military procurement but much less common in industry.

In most cases, the best way to access whatever knowledge manufacturers possess about the behavior of the equipment in the field is to ask them to supply experienced field technicians to work alongside the people who will eventually operate and maintain the asset, to develop FMEAs that are satisfactory to both parties. If this suggestion is adopted, the field technicians should, of course, have unrestricted access to specialist support to help them answer difficult questions.

When adopting this approach, issues such as warranties, copyrights, languages that the participants should be able to speak fluently, technical support, confidentiality, and so on should be handled at the contracting stage, so that everyone knows clearly what to expect of each other.

Note the suggestion to use field technicians rather than designers. Designers are often surprisingly reluctant to admit that their designs can fail, which reduces their ability to help develop a sensible FMEA.

Generic Lists of Failure Modes

Generic lists of failure modes are lists of failure modes—or sometimes entire FMEAs—prepared by third parties. They may cover entire systems, but more often they cover individual assets or even single components. These generic lists are touted as another method of speeding up or streamlining this part of the maintenance program development process. In fact, they should be approached with great caution, for the following reasons:

- **The level of analysis may be inappropriate.** A generic list may identify failure modes at a lower-level equivalent to, say, level 5 in Figure 6.9 when all that may be needed is level 2. This means that far from streamlining the process, the generic list would condemn the user to analyzing many more failure modes than necessary. Conversely, the generic list may focus on level 3 or 4 in a situation where some of the failure modes really ought to be analyzed at level 5 or lower.
- **The operating context may be different or even ignored.** The operating context of your asset may have features that make it susceptible to failure modes that do not appear in the generic list. Conversely, some of the modes in the generic list might be extremely improbable (if not impossible) in your context.
- **Performance standards may differ.** Your asset may operate to different standards of performance, which means that your whole definition of failure may be completely different from that used to develop the generic FMEA.

These three points mean that if a generic list of failure modes is used at all, it should only be used to supplement a context-specific FMEA, and never be used on its own as a definitive list.

Other Users of the Same Equipment

Other users are an obvious and very valuable source of information about what can go wrong with commonly used assets, provided, of course, that competitive pressures permit the exchange of data. This is often done through industry associations (as in the offshore oil industry), through regulatory bodies (as in civil aviation), or between different branches of the same organization. However, note the above comments about the dangers of generic data when considering these sources of information.

Technical History Records. Technical history records can also be a valuable source of information. However, they should be treated with caution for the following reasons:

- They are often incomplete.
- More often than not, they describe what was done to repair the failure ("Replaced main bearing") rather than what caused it.
- They do not describe failures that have not yet occurred.
- They often describe failure modes that are really the effect of some other failure.

These drawbacks mean that technical history records should only be used as a supplementary source of information when preparing an FMEA, and never as the sole source.

The People Who Operate and Maintain the Equipment

In nearly all cases, by far the best sources of information for preparing an FMEA are the people who operate and maintain the equipment on a day-to-day basis. They tend to know the most about how the equipment works, what goes wrong with it, how much each failure matters, and what must be done to fix it, and if they don't know, they are the ones who have the most reason to find out.

The best way to capture and to build on their knowledge is to arrange for them to participate formally in the preparation of the FMEA as part of the overall RCM process. The most efficient way to do this is with an

RCM3 review group under the guidance of a suitably trained facilitator (see Figure 6.14) at a series of meetings. (The most valuable source of additional information at these meetings is a comprehensive set of process and instrumentation drawings, coupled with ready access to process and/or technical specialists on an ad hoc basis.) This approach to RCM was introduced in Chapter 1 and is discussed at much greater length in Chapter 12.

FIGURE 6.14 Typical RCM review group

Resnikoff Conundrum

Some assume that an analysis will always start with statistical data for every possible failure mode, so it isn't surprising that aspiring RCM adopters can be deterred by the time and complexity of the exercise. Is it really necessary to collect historical data and to subject them to complex analysis before starting an RCM project?

If this were true, then very few RCM analyses would ever be undertaken. Very often, detailed data simply do not exist. Leaving aside for a moment the quality of historical reporting in the average maintenance management system, there are at least two reasons why the search for data may be fruitless.

First, unless large fleets of equipment have been in use for years, there simply may not be enough data to be useful. For anything other than simple mean time between failure and availability statistics, a few data points are unlikely to be sufficient; tens or even hundreds may be needed to establish a trend.

Second, there is a fundamental problem with collecting maintenance failure data, which was pointed out by H. L. Resnikoff. Suppose
we would like to use historical failure data recorded in some form of
asset management system. Data on expensive, high-consequence failures would be particularly valuable to us, because we may be able to
prevent those failures from happening, or it may be possible to reduce
the (high) cost of maintenance if we can do it differently. The problem—
Resnikoff's conundrum—is this:

> We need failure data to analyze trends and to improve
> maintenance, but our existing maintenance schedules
> probably do a good job of preventing failures.

This is especially true of maintenance that is intended to prevent high-
consequence failures. The overall result is that while there may be plenty
of failure data for failures that don't matter, there is very little for the
failures that we need to prevent.

All of this leaves us in a difficult position. Somehow, we need to develop
maintenance schedules for real equipment with only limited failure history. The alternative is to continue to carry out existing maintenance tasks
and hope that nothing significant has been missed.

Fortunately, RCM recognizes that it simply isn't practical to carry out
long statistical exercises to determine failure rates and failure development
characteristics. The equipment that we have—the equipment that we are
operating today—needs a maintenance schedule now. Few organizations
have the luxury of shutting down operations while statisticians gather reliability data. The RCM3 decision diagram focuses attention on the data that
are necessary. Where there is uncertainty (and there often is a great deal of
uncertainty), it is recognized during the analysis; and where necessary, the
decisions are based on worst-case estimates, resulting in robust and defensible maintenance. Data requirements are driven by what is appropriate for
the analysis; there is no need to carry out long statistical exercises before
the process can start.

In many cases, we will have to make decisions with inadequate and
sometimes obsolete data.

6.7 Levels of the Analysis and the Information Worksheet

Earlier in this chapter we showed how failure modes can be described at almost any level of detail. The level of detail that is ultimately selected should enable a suitable failure management policy to be identified. In general, higher levels (less detail) should be selected if the component or subsystem is likely to be allowed to run to failure or subject to failure-finding, while lower levels (more detail) need to be selected if the failure mode is likely to be subjected to some sort of proactive maintenance.

The detail used to describe failure modes on Information Worksheets is also influenced by the level at which the FMEA as a whole is carried out. This, in turn, is governed by the level at which the entire RCM analysis is performed. For this reason, we review the principal factors that influence the overall level of analysis (which is also known as "level of indenture") before considering how this affects the detail with which failure modes should be described.

Level of Analysis

RCM is defined as a process used to determine what must be done to ensure that any physical asset continues to do whatever its users want it to do in its present operating context. In the light of this definition, we have seen that it is necessary to define the context in detail before we can apply the process. However, we also need to define exactly what the physical asset is to which the process will be applied.

For example, if we apply RCM to a truck, is the entire truck the asset? Or do we subdivide the truck and analyze, say, the drive train separately from the braking system, the steering, the chassis, and so on? Or should we further subdivide the drive train and analyze the engine separately from the gearbox, propshaft, differentials, axles, and wheels? Or should the engine be divided into engine block, engine management system, cooling system, fuel system, and so on before starting the analysis? What about subdividing the fuel system into tank, pump, pipes, and filters?

This issue needs careful thought, because an analysis carried out at too high a level becomes too superficial, while one done at too low a level can become unmanageable and unintelligible. The following paragraphs explore the implications of carrying out the analysis at different levels.

Starting at a Low Level. One of the most common mistakes in the RCM process is carrying out the analysis at too low a level in the equipment hierarchy.

For example, when thinking about the failure modes that could affect a motor vehicle, a possibility that comes to mind is a blocked fuel line. The fuel line is part of the fuel system, so it seems sensible to address this failure mode by raising a worksheet for the fuel system. Figure 6.15 indicates that if the analysis is carried out at this level, the blocked fuel line might be the seventh failure mode to be identified out of a total of perhaps a dozen that could cause the functional failure "Unable to transfer any fuel at all."

RCM3™ Information Worksheet © 2017 Aladon V0.0	Location ID **Engine**		
	Location Description **Fuel System**		

Function	Failed State	Failure Mode Cause	Failure Mode Mechanism
1 To transfer fuel from the fuel tank to the engine at a rate of 0.25 gallons per minute	A Unable to transfer fuel at all	1 No fuel in tank	Tank holed by sharp object
		2 Fuel filter plugged	Contamination
		5 Fuel line plugged	Foreign object
		7 Fuel pump failed ...... etc.	Normal wear

FIGURE 6.15   Failure modes of a fuel system

When the decision worksheet has been completed for this subsystem, the RCM review group proceeds to the next system, and so on, until the maintenance requirements of the entire vehicle have been reviewed. This seems to be straightforward enough until we consider that the vehicle can actually be subdivided into literally dozens—if not hundreds—of subsystems at this level. If a separate analysis is carried out for each subsystem, the following problems begin to arise:

- The further down the hierarchy one progresses, the more difficult it becomes to conceptualize and define performance standards. (One could also ask who actually cares about the precise amount of fuel passing through the fuel system as long as the fuel economy of the vehicle is within reasonable limits and the vehicle has enough power.)
- At a low level, it becomes equally difficult to visualize and hence to analyze failure consequences.
- The lower the level of the analysis, the more difficult it becomes to decide which components belong to which system. For instance, is the accelerator part of the fuel system or the engine control system?
- Some failure modes can cause many subsystems to cease to function simultaneously (such as a failure in the supply of power to an industrial plant). If each subsystem is analyzed on its own, failure modes of this type are repeated again and again.
- Control and protective loops can become very difficult to deal with in a low-level analysis, especially when a sensor in one subsystem drives an actuator in another through a processor in a third. For instance, a rev limiter that reads a signal off the flywheel in the engine block subsystem might send a signal through a processor in the engine control subsystem to a fuel shutoff valve in the fuel subsystem. If special attention is not paid to this issue, the same function ends up being analyzed three times in slightly different ways, and the same failure-finding task is prescribed more than once for the same loop.

- A new worksheet has to be raised for each new subsystem. This leads to the generation of vast quantities of paperwork for the analysis of the entire vehicle or the consumption of equally large amounts of computer memory space. The associated manual or electronic filing systems have to be very carefully structured if the information is to remain manageable. In short, the whole exercise starts to become much more extensive and much more intimidating than it needs to be.

FMEAs are often carried out at too low a level in the equipment hierarchy because of a belief that there is a correlation between the level at which we identify failure modes and the level at which the FMEA (or the RCM analysis as a whole) should be performed. In other words, it is often said that if we want to identify failure modes in detail, then we ought to carry out a separate FMEA for each replaceable component or subsystem.

In fact, this is not so. The level at which failure modes can be identified is independent of the level at which the analysis is performed, as shown in the next section of this chapter.

Starting at the Top. Instead of starting the analysis toward the bottom of the equipment hierarchy, we could start at the top.

For example, the primary function of a truck was listed as follows: "To transport up to 40 tons of material at speeds of up to 75 mph (average 60 mph) from X to Y on one tank of fuel." The first functional failure associated with this function is "Unable to move at all." The four failure modes shown in Figure 6.15 could all cause this functional failure, so instead of being listed on an Information Worksheet for the fuel system, they could have been listed on a worksheet covering the entire truck, as shown in Figure 6.16.

The main advantages of starting the analysis at this level are as follows:

- Functions and performance expectations are much easier to define.
- Failure consequences are much easier to assess.

RCM3™ Information Worksheet	Location ID
	40 Ton Truck
© 2017 Aladon V0.0	Location Description

Function	Failed State	Failure Mode Cause	Failure Mode Mechanism
1 To transfer up to 40 tons of material from Sartsville to Endburg with speeds up to 75 mph (average 60 mph) on one tank of fuel	A Unable to transfer material at all	18 No fuel in tank	Tank holed by sharp object
		22 Fuel filter plugged	Contamination
		55 Fuel line plugged	Foreign object
		73 Fuel pump failed ...... etc.	Normal wear

FIGURE 6.16 Failure modes of a truck

- It is easier to identify and analyze control loops and circuits as a whole.
- There is less repetition of functions and failure modes.
- It is not necessary to raise a new Information Worksheet for each new subsystem, so analyses carried out at this level consume far less paper.

However, the main disadvantage of performing the analysis at this level is that there are hundreds of failure modes that could render the truck effectively unable to move. These range from a flat front tire to a sheared crankshaft. So if we were to try to list all the failure modes at this level, it is highly likely that several would be overlooked altogether.

For instance, we have seen how the blocked fuel system might have been the seventh failure mode out of twelve to be identified in the analysis carried out at the fuel system level. However, at the truck level,

Figure 6.16 shows that it might have been seventy-third out of several hundred failure modes.

Intermediate Levels. The problems associated with high- and low-level analyses suggest that it may be sensible to carry out the analysis at an intermediate level. In fact, we are almost spoiled for choice, because most assets can be subdivided into many levels and the RCM process applied at any one of these levels.

For example, Figure 6.17 shows how the 40-ton truck could be divided into at least five levels. It traces the hierarchy from the level of the truck as a whole down to the level of the fuel lines. It goes on to show how the primary function of the asset might be defined at each level on an RCM Information Worksheet and how the blocked fuel line could appear at each level.

Given the choice of five (sometimes more) possibilities, how do we select the level at which to perform the analysis?

We have seen that the top level usually embodies too many failure modes per function to permit sensible analysis. In spite of this, however, we still need to identify the main functions of the asset or system at the highest levels in order to provide a framework for the rest of the analysis.

For example, an operator acquires a truck to carry goods from A to B, not to pump fuel along a fuel line. Although the latter function contributes to the former, the overall performance of the asset—and hence of its maintenance—tends to be judged at the highest levels. For instance, the chief executive of a truck fleet is much more likely to ask, "How is truck X performing?" than "How is the fuel system on truck X performing?" (unless the fuel system is known to be causing a problem).

Chapter 3 explained that in practice a statement of the operating context provides a record of the main functions and associated performance standards of any asset or system at levels above the level at which the RCM analysis is to be carried out.

On the other hand, we have seen that the initial inclination is nearly always to start too low in the asset hierarchy. For this reason, a good general rule (especially for people new to RCM) is to carry out the analysis

40 Ton Truck

	Function		Failed State		Failure Mode Cause	Failure Mode Mechanism
1	To transfer up to 40 tons of material from Sartsville to Endburg with speeds up to 75 mph (average 60 mph) on one tank of fuel	A	Unable to transfer material at all	18	No fuel in tank	Tank holed by sharp object
				22	Fuel filter plugged	Contamination
				55	Fuel line plugged	Foreign object
				73	Fuel pump failed	Normal wear
					...... etc.	

Drive Train   Brake System   Steering System

	Function		Failed State		Failure Mode Cause	Failure Mode Mechanism
1	To propel the vehicle carrying a load of up to 40 tons on made roads with speeds up to 75 mph	A	Unable to propel the vehicle at all	1	Tank holed	Sharp object
				3	Fuel pump failed	Sharp object
				5	Fuel filter plugged	Contamination
				7	Fuel line plugged	Foreign object
				9	Fuel line severed	Foreign object
					...... etc.	

Engine   Gearbox   Propshaft

	Function		Failed State		Failure Mode Cause	Failure Mode Mechanism
1	To provide up to 545 HP of power at 2500 rpm to the input shaft of the gearbox	A	Unable to provide any power at all	4	No fuel in tank	Tank holed by sharp object
				12	Fuel filter plugged	Contamination
				25	Fuel line plugged	Foreign object
				63	Fuel pump failed	Normal wear
					...... etc.	

Fuel System   Engine Block   Cooling System

	Function		Failed State		Failure Mode Cause	Failure Mode Mechanism
1	To transfer fuel from the fuel tank to the engine at a rate of 0.25 gallons per minute	A	Unable to transfer fuel at all	1	No fuel in tank	Tank holed by sharp object
				2	Fuel filter plugged	Contamination
				5	Fuel line plugged	Foreign object
				7	Fuel pump failed	Normal wear
					...... etc.	

Fuel Lines   Fuel Tank   Fuel Filter

	Function		Failed State		Failure Mode Cause	Failure Mode Mechanism
1	To carry fuel from the fuel tank to the engine without leaks	A	Unable to carry the fuel from the fuel tank at all	1	Fuel filter plugged	Contamination
				7	Fuel line plugged	Foreign object
				9	Fuel line severed	Foreign object
					...... etc.	

Etc.....

FIGURE 6.17 The 40-ton truck can be divided into five levels or more

one level—or even two levels—higher than at first seems sensible. This is because it is always easier to break complex subsystems out of a high-level analysis than it is to go up a level when one has started too low. This is discussed in more detail in the next section of this chapter.

With a bit of practice (especially concerning what is meant by "a level at which it is possible to identify a suitable failure management policy"), the most suitable level at which to carry out any analysis eventually becomes intuitively obvious. In this context, note that it is not necessary to analyze every system at the same level throughout the asset hierarchy.

For instance, the entire braking system could be analyzed at level 2, as shown in Figure 6.17, but it may be necessary to analyze the engine at level 3 or even level 4.

How Failure Modes and Effects Should Be Recorded

Once the level of the entire RCM analysis has been established, we then have to decide what degree of detail is necessary to define each failure mode within the framework of that analysis. There is no technical reason why all the failure modes cannot be listed (together with their effects) at a level that enables a suitable failure management policy to be selected.

However, even intermediate-level analyses sometimes generate too many failure modes per functional failure, especially for primary functions. This usually happens when the asset incorporates complex subassemblies that could themselves suffer from a large number of failure modes. Examples of such subassemblies include small electric motors, small hydraulic systems, small gearboxes, control loops, protective circuits, and complex couplings.

Depending as usual on context and consequences, these subassemblies can be handled in one of four different ways.

Option 1. List all the reasonably likely failure modes of the subassembly individually as part of the main analysis—in other words, at levels equivalent to level 3, 4, 5, or 6 in Figure 6.9.

For example, consider an asset that could stop completely as a result of the failure of a small gearbox. On the Information Worksheet for this asset, this gearbox failure could be listed as shown in Figure 6.18.

		FAILURE MODES		FAILURE EFFECTS
Record all subsystem failure modes as part of the main analysis	1	Gearbox seal failed	Normal wear and tear	Gearbox loses its lubricant and seizes. Motor trips, and alarm sounds in control room. 3 hours downtime to replace the gearbox. New bearings fitted in the workshop
	2	Gearbox teeth stripped	Normal wear and tear	Gearbox teeth wear and strip. Motor may not trip, but machine stops. 3 hours downtime to replace the gearbox. New gear set fitted in the workshop

FIGURE 6.18   Record all subsystem failure modes as part of the main analysis

In general, the failure modes that could affect a subassembly should be incorporated in a higher-level analysis if the subassembly is likely to suffer from no more than about six failure modes that are considered to be worth identifying and that will cause any one functional failure of the higher-level system.

Option 2. List the failure of the subassembly as a single failure mode on the Information Worksheet to begin with; then raise a new worksheet to analyze the functions, functional failures, failure modes, and effects of the subassembly as a separate exercise.

For example, the failure of the gearbox discussed above could have been listed as shown in Figure 6.19.

		FAILURE MODES		FAILURE EFFECTS
Analyze the subsystem separately	1	Gearbox failed		Motor trips, and alarm sounds in control room (Gearbox analyzed separately)

FIGURE 6.19   Analyze subsystems separately

A subassembly is usually worth treating this way if more than 10 failure modes of the subassembly could cause the loss of any one function of the main assembly. (If there are between seven and nine failure modes per functional failure, use option 1 or option 2, bearing in mind that separate analyses mean more analyses, but fewer failure modes per analysis.)

A subassembly is usually worth treating this way if more than 10 failure modes of the subassembly could cause the loss of any one function of the main assembly.

Option 3. List the failure of the subassembly on the Information Worksheet as a single failure mode—in other words, at a level equivalent to level 1 or 2 in Figure 6.9—and then record its effects and leave it at that.

For example, if it was considered appropriate to treat the failure of the gearbox discussed above in this fashion, it would be listed as shown in Figure 6.20.

Treat the subsystem as a single failure mode		FAILURE MODES		FAILURE EFFECTS
	1	Gearbox failed		Motor trips, and alarm sounds in control room. 3 hours downtime to replace the gearbox

FIGURE 6.20   Treating subsystems as single failure modes

This approach should only be adopted for a component or subassembly that has the following characteristics:

- It is not subject to detailed diagnostic and repair routines when it fails, but is simply replaced and either discarded or subjected to later repair.
- It is quite small but quite complex.
- It does not have any dominant failure modes.
- It is not likely to be susceptible to any form of proactive maintenance.

Option 4.   In some cases, a complex subassembly might suffer from one or two dominant failure modes that are readily preventable, and a number of less common failures that may not be worth preventing because the frequency and/or the consequences of the failures do not warrant it.

For example, the gearbox would most certainly fail if it runs dry (gearbox operate without oil). The gearbox may also fail for other reasons that may not be known to the user and these are few and far between. In this case, the failure modes for this gearbox might be listed as follows (see Figure 6.21):

Analyze dominant failure mode(s) only, and treat the rest of the subsystem as a "black box"		FAILURE MODES		FAILURE EFFECTS
	1	Gearbox failed	Lack of lubrication	Gearbox runs dry and seizes. Motor trips, and alarm sounds in control room. 3 hours downtime to replace the gearbox. New bearings fitted in the workshop
	2	Gearbox failed		Motor trips, and alarm sounds in the control room. 3 hours downtime to replace the gearbox

FIGURE 6.21   Listing dominant failure modes only and treat the rest of the system as a black box

This option is really a combination of options 1 and 3.

Services.   The failure of services (power, water, steam, air, gases, vacuum, etc.) is treated as a single failure mode from the point of view of the asset that is supplied by that service, because detailed analysis of these failures is usually beyond the scope of the asset in question. Such failures are noted for information purposes ("Power supply fails"), their effects are recorded, and they are then analyzed in detail when the service is analyzed as a whole.

How to Record Failure Effects

The failure effects should allow the review group to assess the consequences and subsequent risk of each failure mode. The review group is made up of people who know the equipment and system best, and the group's members come from various disciplines, e.g., operations, maintenance, engineering, supplier companies, health and safety, etc. The people reviewing the analysis and signing off on the recommendations are the people with overall responsibility for the asset or system under review and may be the manager responsible for operations and the production manager or the manager responsible for maintenance, or even both. In order to make it possible for these different disciplines to participate effectively in the analysis and implementation process, the failure effects are split into four categories—local effect, next-higher-level effect, end effect, and potential worst-case effect. The following section explains the reason for doing so.

Local Effect. This effect describes what happens to the equipment directly or the impact the failure mode has on the equipment it is part of. Consider the failure of a benzene system transfer pump bearing due to normal wear and tear. The local effect could be something as follows:

> "Over a period of time the bearing wears, and bearing vibration and temperature will increase. If unattended, the bearing will eventually seize or collapse."

In the Information Worksheet it may look something like Figure 6.22.

Failure Effect	
Local Effect	Through normal use the bearing starts wear and vibration, noise and temperature will increase. Eventually the bearing will seize or collapse.
Next Higher-Level Effect	
End Effect	
Potential Worst-Case Effect	

FIGURE 6.22　Description of local effects

Next-Higher-Level Effect. The next-higher-level effect describes what would happen to the system if the pump is partly off. We will also consider the most likely time the failure would occur, whether or not the function is protected (through a protective device), and the sequence of events following the failure. Consider that nothing is done to prevent the failure from happening.

The next-higher-level effect for the pump bearing failure could be something like this:

> "The pump will stop, and the motor will overload, causing the motor breaker to trip. The trip will generate a 'motor-failed alarm' in the control room, and the 'Motor Overload' light will light up in the control room. With the pump down, no benzene will be transferred to the process, and the benzene supply tank high-level alarm will be generated when the benzene level reaches the high level in the tank. The benzene recovery system upstream of the tank will also shut down."

In the Information Worksheet it may look something like Figure 6.23.

Failure Effect	
Local Effect	Through normal use the bearing starts wear and vibration, noise and temperature will increase. Eventually the bearing will seize or collapse.
Next Higher-Level Effect	The pump **motor trips on overload** and the "Motor Overload" indicator lights up in the control room. With the pump down, the system is unable to provide benzene to the process. The benzene supply tank "high level alarm" will also sound when no benzene transfer is taking place.
End Effect	
Potential Worst-Case Effect	

FIGURE 6.23   Description of next-higher-level effects

When we deal with protective devices, the next-higher-level effect must describe what happens when the protective device fails to operate. We can use the failure of a low-level alarm in a tank as an example. The failure of the low-level alarm only matters in the event of another failure that will cause a low level in the tank, e.g., a hole in the tank, a failing supply pump, or a faulty level control circuit. The failure effect could be recorded as follows:

"The failure of the low-level alarm circuit only matters in the event of another failure that may cause the tank level to drop below normal operating levels. The alarm will not be generated, and operations will not be aware of low tank levels that may cause service interruptions."

And in the Information Worksheet it may look something like Figure 6.24.

End Effect. This describes the ultimate effect that the failure has on safety and/or the environment (if any) and any impact on production or operational capability.

Failure Effect	
Local Effect	Low-level alarm will be unavailable on demand.
Next Higher-Level Effect	**This only matters in the event of another failure** causing the level to drop below normal levels. The alarm will not be generated and operations may not be aware of the low level in the tank.
End Effect	
Potential Worst-Case Effect	

FIGURE 6.24   Description of next-higher-level effects—protective devices

The following need to be considered:

- Impact on health and safety (illness, injury, or death)
- Impact on environment (violations, pollution, spills, contamination, etc.)
- Production downtime or lost sales
- Increased cost of production and penalties
- Service-level disruption and eroding customer confidence
- Product and service quality
- Public perception
- Cost of insurance

The end effect further includes any consequential damage and costs such as:

- Overtime
- Additional resources (material and people)
- Increased energy costs and additional start-up costs
- Emergency procedures (e.g., repairs and spare parts)
- Scrap and disposal costs

For the failure of the benzene system pump bearing, the end effect description could be recorded as:

> "The bearing will need to be replaced, which will take 5 hours. One delivery cycle will be lost, and the company will lose $53,000 plus the cost of the repairs. Depending on the time of year (peak demand), a batch may have to be ordered from the competitors to fulfill contract requirements."

Figure 6.25 shows how it might look in the Information Worksheet.

Potential Worst-Case Effect. Failure effects describe what would happen if no specific task is done to anticipate, prevent, or detect the failure. Protective devices such as alarms, shutdowns, and relief systems,

Failure Effect	
Local Effect	Through normal use the bearing starts wear and vibration, noise and temperature will increase. Eventually the bearing will seize or collapse.
Next Higher-Level Effect	The pump **motor trips on overload** and the "Motor Overload" indicator lights up in the control room. With the pump down, the system is unable to provide benzene to the process. The benzene supply tank "high level alarm" will also sound when no benzene transfer is taking place.
End Effect	The time to repair / replace the pump may take up to 5 hours. One delivery cycle would be lost at a cost of $53,000. A batch may have to be ordered from the competitors in order to fulfil contractual requirements.
Potential Worst-Case Effect	

FIGURE 6.25   Description of end effects

which are designed to deal with the failed or failing state, are nothing more than built-in failure management systems. In order to perform a true zero-based analysis, the effects of the failure of the protected function should be assessed as if a protective device like these were not present (or not working).

Therefore, the inherent risk of the failure of the protected function is established as if the protective device(s) did not exist or failed to operate on demand.

> The potential worst-case effect (PWCE) is not necessarily the absolute worst case conceivable.

The potential worst-case effect is the plausible worst-case impact to the organization and its operations arising from a risk where all existing protective devices directly related to the failure mode under consideration are assumed to be unavailable. The protective devices directly related to the protected function or failure mode under consideration are those described in the next-higher-level effect.

For the benzene system the pump motor was equipped with an electric overload circuit, and PWCE could be described as follows:

"If the existing protection fails, the bearing will overheat and fail cat-astrophically. The risk of a fire is greatly increased. Although it is unlikely that personnel will be injured, the damage and downtime will be excessive and may result in a partial shutdown of the plant."

Figure 6.26 shows how the description might be entered into the Information Worksheet.

Failure Effect	
Local Effect	Through normal use the bearing starts wear and vibration, noise and temperature will increase. Eventually the bearing will seize or collapse.
Next Higher-Level Effect	The pump motor trips on overload and the "Motor Overload" indicator lights up in the control room. With the pump down, the system is unable to provide benzene to the process. The benzene supply tank "high level alarm" will also sound when no benzene transfer is taking place.
End Effect	The time to repair/replace the pump may take up to 5 hours. One delivery cycle would be lost at a cost of $53,000. A batch may have to be ordered from the competitors in order to fulfil contractual requirements.
Potential Worst-Case Effect	If the existing protection fails, the bearing will overheat and fail catastrophically. The risk of a fire is greatly increased. Although it is unlikely that personnel will be injured, the damage and downtime will be excessive and may result in a partial shutdown of the plant.

FIGURE 6.26   Description of PWCE

In the event of multiple layers of protection, the layers are com-bined and treated as a protective system or safety instrumented system (SIS) for determining the typical worst-case effects, as the example in Figure 6.27 illustrates.

Defining the PWCE and risk (at this level) ensures that the reli-ability of the protected function is addressed as a priority. The preferred risk management strategy (failure management policy) is to reduce the dependence on the protection (protection is always only a "backup" function).

It could be possible to increase the reliability of the protected func-tion although the resulting "revised risk" for the failure on its own may still be "intolerable" according to the risk assessment; however, with

Failure Effect	
Local Effect	**Level control switch** circuit fails to operate.
Next Higher-Level Effect	The level in the tank will continue to rise and **the high level switch, the high-high level switch or the ultimate high level switch** will be activated which will shut down the pump and raise an alarm in the control room.
End Effect	Time to investigate and repair or replace any one of the level switches can take up to 12 hours.
Potential Worst-Case Effect	If the *high level switch, the high-high level switch as well as the ultimate high-level switch all fail,* the level will continue to rise, the tank will overflow and caustic acid will flow into the nearby river causing environmental and health issues.

FIGURE 6.27   Description of PWCE for a protective system

existing protective devices, the consequences and ultimate risk could be reduced or eliminated to a tolerable level.

This "intolerability" is addressed when the failure management policies for the existing protective device(s) are optimized. This will be discussed in Chapter 8 (hidden functions).

6.8 A Completed Information Worksheet

Failure effects are listed in the last column of the Information Worksheet alongside the relevant failure mode, as shown in Figure 6.28.

RCM3™ Information Worksheet
© 2018 Aladon V2.0

Location ID		No	Compiled by	Date	Sheet
Location Description **Water Storage & Supply System**		Ref.	Reviewed by	Date	of

Aladon — The Risk Reliability GLOBAL NETWORK

Function		Failed State		Failure Mode		Failure Effect		Inherent Risk			
				Cause	Mechanism			Pted	P	C	R
1 To transfer water at a minimum rate of 300 gpm from tank X to tank Y in the presence of a standby pump	B	Transfer water in the absences of a standby pump	1	Duty pump bearing seized	Normal wear	**Local Failure Effect**	Over time, the bearing starts to wear. The bearing temperature and vibration increase and the bearing will eventually seize or collapse. This failure could occur on average once every 5 years (if no maintenance is being done).	4		3	17(S)
						Next Higher-Level Effect	The motor trips on electric overload and pumping stops. No water will be transferred but the water continues to drain from tank Y. The low-level alarm will sound in the Control Room when the level drops below 12,000 gallons. Operator will start the standby pump.				
						End Effect	It takes 4 hours to repair the pump (replace the pump bearing) or switch out the pump.				
						Potential Worst-Case Effect	In the event the standby pump fails to start, the process will come to a standstill. The tank has no more than 2.5 hours of water left in the tank and the process will be down for 1.5 hours. The downtime cost is $225,000 plus the cost of repairing the pump.				

FIGURE 6.28   Information Worksheet

CHAPTER 7

Failure Consequences and Risk

In previous chapters we explained that RCM3 asks the following eight questions from each asset or system:

1. What are the operating conditions (how the equipment or system is being used)?
2. What are the functions and associated performance standards of the asset in its present operating context?
3. In what ways does it fail to fulfill its functions (failed states)?
4. What causes each failed state (failure modes)?
5. What happens when each failure occurs (failure effects and consequence severity)?
6. What are the risks associated with each failure (inherent risk quantified)?
7. What *must* be done to reduce intolerable risks to a tolerable level (using proactive risk management strategies)?
8. What *can* be done to reduce or manage tolerable risks in a cost-effective way?

The answers to the first five questions were discussed at length in Chapters 3 to 6. These showed how RCM Information Worksheets are used to record the functions of the asset under review and to list the associated functional failures, failure modes, and failure effects. The next step is to determine how each failure mode matters by defining the inherent risk posed by each failure.

The next questions are asked about each individual failure mode. This chapter considers the sixth question in the RCM3 process:

> What are the risks associated with each failure (inherent risk quantified)?

7.1 Risk Management for Physical Risks

How Do We Manage Risk?

Most modern managers have adopted an attitude of risk averseness. Zero risk tolerance is simply not attainable, and because of this, risk should be managed at tolerable levels. We will discuss how to manage risk to tolerable levels in the following sections. The risk management process used in RCM3 is based on the criteria as set out in the different international risk standards, such as AS/NZS 4360 2009 Risk Management, ISO 31000 Risk Management, and IEC/FDIS 31010 Risk Management.

According to ISO 31000:

> "Risk is the 'effect of uncertainty on objectives,' and an effect is a positive or negative deviation from what is expected. Risk is the combination of the effect (consequence) of the event and the likelihood of the event happening (probability)." See Figure 7.1.

Risk = Consequence X Probability

FIGURE 7.1   Quantifying risk

Risk and reliability management starts by identifying and understanding the risks or hazards that are present (*consequences of failure*), then determining the likelihood of these risks actually being real (*probability of happening*), and lastly ascertaining what needs to be done to

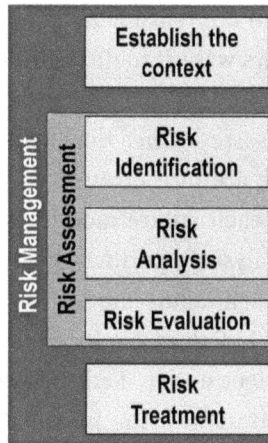

FIGURE 7.2 Risk management framework

mitigate (*reduce or eliminate*) the risks. Figure 7.2 illustrates how risk management should be approached according to the ISO 31000 and IEC/FDIS 31010 risk management standards. The steps are:

- Establish the context (operating context).
- Identify the risk.
- Determine (understand) the consequence of the failure and categorize the risk.
- Determine the probability of the failure mode.
- Quantify the risk.
- Identify risk management strategies to reduce or eliminate the risk (manage to tolerable level).

Establish the Context. The context within which risk is to be managed is established by defining the operating context, functions, and associated performance standards of the physical asset with regard to safety and environmental requirements as well as operational and financial expectations. The operating context was discussed at length in Chapter 3.

Identify and Quantify the Risk. In order to identify the risk, it is necessary to determine the failed states and what failure modes (causes and mechanisms) are reasonably likely to prevent the system from doing what the users want it to do. Once the likely failure modes have been identified, the effect and the associated consequences of each failure mode are assessed. Once the likelihood and consequences have been identified, we are in a position to determine the physical or operational risk that each failure mode poses to the organization. If a truly zero-based analysis is performed, this would represent the inherent risk posed by each failure mode. The RCM3 process records the failed states and likely failure modes (with their probabilities) in the Information Worksheet under "Failed State" and "Failure Mode." Figures 7.3 and 7.4 illustrate how the Information Worksheet is used to record the elements of risk.

FIGURE 7.3   Identify the risk—failed states and failure modes

FIGURE 7.4   Record the consequence severity through failure effects

In order to establish the severity of a failure mode, we need to understand the physical effect it will have.

Risks to the business are categorized in order to determine the effort (time and money) that will be required to manage the risk associated with the specific failure mode. Failures that pose serious risks will require a comprehensive approach in order to ensure that we will be able to manage the risks to a tolerable level.

Failures that pose less serious risks will require a less comprehensive approach. In fact, in some cases we may even decide to ignore the risk completely.

Managing the risks associated with physical assets and processes in industry has become a very high priority in industry as part of the overall business processes. When we consider the application of RCM3, the main objective is to identify the issues that influence the reliability of systems, processes, and equipment.

There is a growing formal recognition by the world in general, and regulatory authorities in particular, that although "zero" risk is a worthy objective, it is actually an unattainable ideal. Because of a much better understanding of the concept of risk management and the realization that there will always be inherent risk in everything we do, words like "tolerable," "practicable," and "defensible" are more generally used when making decisions on levels of risk. These terms acknowledge that despite our best efforts, safety and environmental incidents will still occur, but they also force us to keep the likelihood and severity of such incidents to what all the stakeholders in our organizations would consider to be a tolerable minimum.

Where does this lead us? It simply implies that there will always be inherent risk in everything we do and it does not matter what we decide; and while the decisions and assumptions will always be approximates, they need to be sensible and defensible.

The Question of Risk. Much as most people would like to live in an environment where there is no possibility of death or injury, it is generally accepted that there is an element of risk in everything we do. In other words, absolute zero is unattainable, even though it is a worthy target to keep striving for. This immediately leads us to ask what is attainable.

To answer this question, we first need to consider the question of risk in more detail.

Risk assessment consists of three elements. The first asks what could happen if the event under consideration did occur. The second asks how likely it is for the event to occur at all. The combination of these two elements provides a measure of the degree of risk. The third and often the most contentious element asks whether this risk is tolerable.

For example, consider a failure mode that could result in death or injury to 10 people (what could happen). The probability that this failure mode could occur is one in a thousand in any one year (how likely it is to occur). On the basis of these figures, the risk associated with this failure is:

$$10 \times (1 \text{ in } 1,000) = 1 \text{ casualty per } 100 \text{ years}$$

Now consider a second failure mode that could cause 1,000 casualties, but the probability that this failure could occur is 1 in 100,000 in any one year. The risk associated with this failure is:

$$1,000 \times (1 \text{ in } 100,000) = 1 \text{ casualty per } 100 \text{ years}$$

In these examples, the risk is the same although the figures upon which it is based are quite different. Note also that these examples do not indicate whether the risk is tolerable—they merely quantify it. Whether or not the risk is tolerable is a separate and much more difficult question that is dealt with later.

Note that throughout this book, the terms "probability" (1 in 10 chance of a failure in any one period) and "failure rate" (once in 10 periods on average, corresponding to a mean time between failures of 10 periods) are used as if they are interchangeable when applied to random failures. Strictly speaking, this is not true. However, if the MTBF is greater than about 4 periods, the difference is so small that it can usually be ignored.

The following paragraphs consider each of the two elements of risk in more detail.

What could happen if the failure occurred? Two issues need to be considered when determining what could happen if a failure were to occur. These are *what actually happens* and *whether anyone is likely to be hurt or killed* as a result.

What actually happens if any failure mode occurs should be recorded on the RCM Information Worksheet as its failure effects, as explained at length in Chapter 5. Chapter 5 also listed a number of typical effects that pose a threat to safety or the environment.

The fact that these effects *could* injure or kill someone does not necessarily mean that they *will* do so every time they occur. Some may even occur quite often without doing so. However, the issue is not whether such consequences are inevitable, but whether they are possible.

For example, if the hook were to fail on a traveling crane used to carry steel coils, the falling load would only hurt or kill anyone who happened to be standing under it or very close to it at the time. If no one was nearby, then no one would get hurt. However, the possibility that someone could be hurt means that this failure mode should be treated as a safety hazard and analyzed accordingly.

This example demonstrates the fact that the RCM process assesses safety consequences at the most conservative level. If it is reasonable to assume that any failure mode could affect safety or the environment, we assume that it can, in which case it must be subjected to further analysis. (We see later that the likelihood that someone will get hurt is taken into consideration when evaluating the tolerability of the risk.)

A more complex situation arises when dealing with safety hazards that are already covered by some form of built-in protection. We have seen that one of the main objectives of the RCM process is to establish the most effective way of managing each failure in the context of its consequences and associated risk. This can only be done if these consequences are evaluated to begin with as if nothing was being done to manage the failure (in other words, to predict or prevent it or to mitigate its consequences).

Protective devices that are designed to deal with the failed or the failing state (alarms, shutdowns, and relief systems) are nothing more than built-in failure management systems. As a result, to ensure that the rest of the analysis is carried out from an appropriate zero base, the consequences of the failure of protected functions should ideally be assessed as if protective devices of this type are not present.

For example, a failure that could cause a fire is always regarded as a safety hazard, because the presence of a fire extinguishing system does not necessarily guarantee that the fire will be controlled and extinguished.

The RCM process can then be used to validate (or revalidate) the suitability of the protective device itself from three points of view:

- Its ability to provide the required protection. This is done by defining the function of the protective device, as explained in Chapter 2.
- Whether the protective device responds fast enough to avoid the consequences, as discussed in Chapter 8.
- What must be done to ensure that the protective device continues to function in its turn, as discussed in Section 7.6 of this chapter and in Chapter 8.

How likely is the failure to occur? Chapter 6 mentioned that only failure modes that are reasonably likely to occur in the context in question should be listed on the RCM Information Worksheet. As a result, if the Information Worksheet has been prepared on a realistic basis, the mere fact that the failure mode has been listed suggests that there is some likelihood that it could occur and therefore that it should be subjected to further analysis.

(Sometimes it may be prudent to list a wildly unlikely but nonetheless dangerous failure mode in an FMEA, purely to place on record the fact that it was considered and then rejected. In these cases, a comment like "This failure mode is considered too unlikely to justify further analysis" should be recorded in the "Failure Effect" column.)

Identify Risk Management Strategies to Reduce or Eliminate the Risk (Manage to Tolerable Level). There are two main questions to consider: Is the risk tolerable, and who should evaluate risks?

Is the risk tolerable? One of the most difficult aspects of the management of safety is the extent to which beliefs about what is tolerable vary from individual to individual and from group to group. A wide variety of factors influence these beliefs, by far the most dominant of which is the degree of control that any individual thinks he or she has over the situation. People are nearly always prepared to tolerate a higher level of risk when they believe that they are personally in control of the situation than when they believe that the situation is out of their control.

For example, people tolerate much higher levels of risk when driving their own cars than they do as aircraft passengers. (The extent to which this issue governs perceptions of risk is given by the startling statistic that 1 person in 11,000,000 who travels by air between New York and Los Angeles in the United States is likely to be killed while doing so, while 1 person in 14,000 who makes the trip by road is likely to be killed. And yet some people insist on making this trip by road because they believe that they are "safer"!)

This example illustrates the relationship between the probability of being killed that any one person is prepared to tolerate and the extent to which that person believes he or she is in control. In more general terms, this might vary for a particular individual, as shown in Figure 7.5.

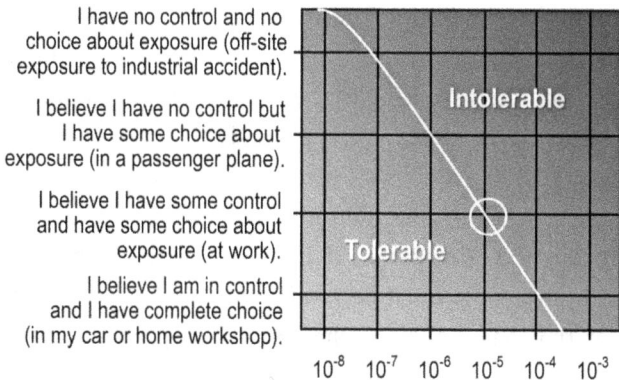

FIGURE 7.5 Tolerability of fatal risk

The figures given in this example are not meant to be prescriptive, and they do not necessarily reflect the views of the author—they merely illustrate what one individual might decide that he or she is prepared to tolerate. Note also that they are based on the perspective of one individual going about his or her daily business. This view then has to be translated into a degree of risk for the whole population (all the workers on a site, all the citizens of a town, or even entire population of a country).

In other words, if I tolerate a probability of 1 in 100,000 (10^{-5}) of being killed at work in any one year and I have 1,000 coworkers who all share the same view, then we all tolerate that on average 1 person per year on our site will be killed at work every 100 years—and that person may be me, and it may happen this year.

Bear in mind that any quantification of risk in this fashion can only ever be a rough approximation. In other words, if I say I tolerate a probability of 10^{-5}, it is never more than a ballpark figure. It indicates that I am prepared to tolerate a probability of being killed at work that is roughly 10 times lower than that which I tolerate when I use the roads (about 10^{-4}).

Always bearing in mind that we are dealing with approximations, the next step is to translate the probability that my coworkers and I are prepared to tolerate the outcome that any one of us might be killed by any event at work into a tolerable probability for each single event (failure mode or multiple failure) that could kill someone.

For example, continuing the logic of the previous example, the probability that any one of my 1,000 coworkers will be killed in any one year is 1 in 100 (assuming that everyone on the site faces roughly the same hazards). Furthermore, if the activities carried out on the site embody, say, 10,000 events that could kill someone, then the average probability that each event could kill one person must be reduced to 10^{-6} in any one year. This means that the probability of an event that is likely to kill ten people must be reduced to 10^{-7}, while the probability of an event that has a 1 in 10 chance of killing one person must be reduced to 10^{-5}.

The techniques by which one moves up and down hierarchies of probability in this fashion are known as probabilistic or quantitative risk assessments. This approach is explored further in Appendix IV. The key points to bear in mind at this stage are that:

- The decision about what is tolerable should start with the likely victim. How one might involve such "likely victims" in this decision in the industrial context is discussed later in this chapter.
- It is possible to link what one person tolerates directly and quantitatively to a tolerable probability of individual failure modes.

Although perceived degree of control usually dominates decisions about the tolerability of risk, it is by no means the only issue. Other factors that help us decide what is tolerable include the following:

- **Individual values.** To explore this issue in any depth is well beyond the scope of this book. Suffice it to contrast the views on tolerable risk likely to be held by a mountaineer with those of someone who suffers from vertigo, or those of an underground miner with those of someone who suffers from claustrophobia.
- **Industry values.** While every industry nowadays recognizes the need to operate as safely as possible, there is no escaping the fact that some are intrinsically more dangerous than others. Some even compensate for higher levels of risk with higher pay levels. The views of any individual who works in that industry ultimately boil down to his or her perception of whether the intrinsic risks are "worth it"—in other words, whether the benefit justifies the risk.
- **The effect on future generations.** The safety of children—especially unborn children—has an especially powerful effect on people's views about what is tolerable. Adults frequently display a surprising and even distressing disregard for their own safety. (Witness how much time has to be spent persuading some people to wear protective clothing.) However, threaten their offspring and their attitude changes completely.

    For example, a group that worked with hazardous chemicals treated words like "toxic" and "carcinogenic" with indifference, even though most of the members of this group were the people most at risk. However, as soon as it emerged that the chemical was also mutagenic and teratogenic, and the meaning of these words was explained to the group, the chemical was suddenly viewed with much greater respect.

- **Knowledge.** Perceptions of risk are greatly influenced by how much people know about the asset, the process it is part off, and the failure mechanisms associated with each failure mode. The more they know, the better their judgment. (Ignorance is often a two-edged sword. In some situations, people take the most appalling risks out of sheer ignorance, while in others, they wildly exaggerate the risks—also out of ignorance. On the other hand, we need to remind ourselves constantly of the extent to which familiarity can breed contempt.)

A great many other factors also influence perceptions of risk, such as the value placed on human life by different cultural groups, religious values, and even factors such as the age and marital status of the individual.

All these factors mean that it is impossible to specify a standard of tolerability that is absolute and objective. This suggests that the tolerability of any risk can only be assessed on a basis that is both relative and subjective—"relative" in the sense that the risk is compared with other risks about which there is a fairly clear consensus, and "subjective" in the sense that the whole question is ultimately a matter of judgment. But whose judgment?

Who should evaluate risks? The very diversity of the factors discussed above means that it is simply not possible for any one person—or even one organization—to assess risk in a way that will be universally acceptable. If the assessor is too conservative, people will ignore and may even ridicule the evaluation. If the assessor is too relaxed, he or she might end up being accused of playing with people's lives (if not actually killing them).

This suggests further that a satisfactory evaluation of risk can only be done by a group. As far as possible, this group should represent people who are likely to have a clear understanding of the failure mechanism, the failure effects (especially the nature of any hazards), the likelihood of the failure occurring, and the possible measures that can be taken to anticipate or prevent it. The group should also include people who have a legitimate view on the tolerability or otherwise of the risks. This means representatives of the likely victims (most often operators or maintainers in the case of direct safety hazards) and management people (who are usually held accountable if someone is injured or if an environmental standard is breached).

If it is applied in a properly focused and structured fashion, the collective wisdom of such a group will do much to ensure that the organization does its best to identify and manage all the failure modes that could affect safety or the environment. (The use of such groups is in keeping with the worldwide trend toward laws that say that safety is the responsibility of all employees, not just the responsibility of management.)

Groups of this nature can usually reach consensus quite quickly when dealing with direct safety hazards, because they include the people at risk. Environmental hazards are not quite so simple, because society at large is the likely victim and many of the issues involved are unfamiliar. So any group that is expected to consider whether a failure could breach an environmental standard or regulation must find out beforehand which of these standards and regulations cover the process under review.

7.2 Defining Risk and Developing a Risk Matrix

Before it is possible to apply any sort of risk management or mitigation strategy, it is necessary to develop a risk framework that is acceptable to *all* the people involved in the process. By "all" we mean most of the people dealing with the risk, i.e., stakeholders, managers, operators, and maintainers. We may even go beyond the organization and include the society as a whole. It is impossible to satisfy everyone, but it is in our best interest to obtain buy-in and the endorsement of the risk definitions prior to the start of the RCM3 analysis.

Similarly, to the people who must assess the risk, it is simply not possible for any one person—or any one organization—to create a risk framework or risk matrix that will be universally acceptable (or accepted by *all* in the organization). This suggests that the development of a risk matrix can only be done successfully by a group of people who have knowledge of the organization, the hazards associated with operations and maintenance, and the business as a whole. As far as possible, this group should represent people who are likely to have a clear understanding of the failure mechanism, the failure effects (especially the nature of any hazards), the likelihood of the failure occurring, and the possible measures that can be taken to anticipate or prevent it.

As stated before, the group should include people who have a legitimate view on the tolerability or otherwise of the risks.

If the risk matrix is developed in a properly focused and structured fashion, the collective wisdom of such a group will do much to ensure that the organization does its best to define the risk categories that will be sensible and defensible. (The use of such groups is again in keeping with the worldwide trend toward laws that say that safety is the responsibility of all employees, not just the responsibility of management.)

Groups of this nature can usually reach consensus quite quickly. All these factors mean that it is impossible to specify a standard of tolerability for any risk that is absolute and objective. This suggests that the tolerability of any risk can only be assessed on a basis that is both relative and subjective—"relative" in the sense that the risk is compared with other risks about which there is a fairly clear consensus, and "subjective" in the sense that the whole question is ultimately a matter of judgment. But whose judgment?

What Methodology Is Acceptable?

Several different methodologies for assessing risk are currently being used today. Most literature and standards on risk management divide risk management into three broad categories:

- Subjective risk assessment
- Quantitative risk assessment
- Qualitative (or semiquantitative) risk assessment

Subjective Risk Assessment. This approach requires the assessors to make a subjective decision based on their knowledge and experience. The more serious the risk, the less inclined we will be to use subjective risk assessment. Most RCM approaches have been based on this approach (formally or informally) when it comes to the question "Could the effect of this failure mode result in an intolerable risk to people or the environment?"

A group of people is simply asked if they think the risk associated with a specific failure is tolerable or not. The group should fully understand:

- The function(s) affected by the event
- The failure mode under consideration
- The effects and subsequent consequence of the failure mode

This group should include representatives of the people at risk (usually operators and maintainers):

- Representatives of the people who have to deal with the consequences (usually management)
- Any specialists who may be needed to help the group make a sensible and defensible decision

This is where RCM3 differs from the rest of the RCM approaches. RCM3 uses a combination of qualitative and quantitative risk assessment (as described below).

Quantitative Risk Assessment. This approach uses the likelihood or probability of the failure and the severity or consequence of the failure. In this case the likelihood and severity are formally determined through reliability analysis methods such as Monte Carlo analysis. Probabilistic risk analysis is used to determine probabilities of severity. The resulting risks are presented on "logarithmic" probability-severity plots. This type of risk assessment usually requires sophisticated validation software. Figure 7.6 shows an example.

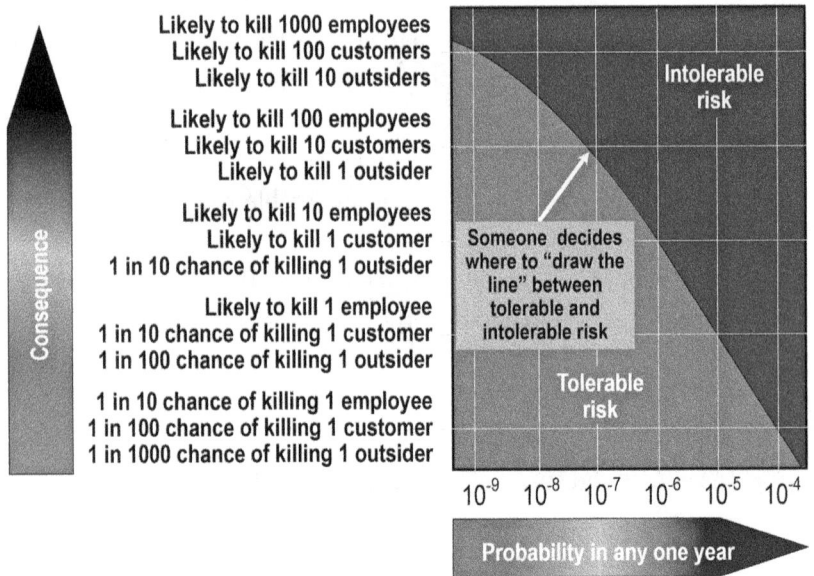

FIGURE 7.6 Tolerable and intolerable risk

Qualitative Risk Assessment. This approach uses the likelihood or probability of the failure mode and the severity or consequence of the failure mode plotted on a risk matrix to allocate the risk into different categories.

Note: In the context of RCM3, it is important to understand that each failure mode is plotted on the risk matrix to identify and categorize the risk of each individual failure mode and cause.

The likelihood and severity are divided into (typically) five different levels, which are expressed descriptively and in relative terms. The likelihood and consequence criteria need to be formalized throughout an organization for consistency. It is important to note that the examples used here are for illustration only and are not prescriptive in any way.

Figure 7.7 illustrates a typical risk matrix that can be used to assess the risk in a qualitative way.

Consequence or Loss Type
Additional *Loss Type* may exist for an event, identify and rate accordingly

(where an event has more than one *Loss Type* choose the *Consequence* with the highest ranking)
Consider the *Typical Worst Case* considering the relevant *Operating Context*

Consequence or Loss Type	1 Insignificant	2 Minor	3 Moderate	4 High	5 Major
SAFETY AND HEALTH Harm to People (Safety & Health)	First aid case / Exposure to minor health risk	Medical treatment case / Exposure to major health risk	Lost time injury / Reversible impact on health	Single fatality or loss of quality of life / Irreversible impact on health	Multiple fatalities / impact on health ultimately leading to a fatality
ENVIRONMENTAL Environmental Impact	Minimal environmental harm	Minor environmental harm - remedial short term	Serious environmental harm with irreversible impact during plant life	Major environmental harm with irreversible impact post plant life	Extreme environmental harm with irreversible impact
OPERATIONAL (ASSETS) Business Interruption / Material Damage & other Consequential Losses	No disruption to operations / Loss of production and assets totaling up to $25,000	Brief disruption to operations / Loss of production and assets totaling between $25,001 <> $150,000	Partial shutdown / Loss of production and assets totaling between $150,001 <> $500,000	Partial loss of operations / Loss of production and assets totaling between $500,001 <> $1,500,000	Substantial or total loss of operations / Loss of production and assets totaling more than $1,500,000
OPERATIONAL (L & R) Legal & Regulatory	Low level legal issue	Minor legal issue / noncompliance or breach of regulation	Serious breach of regulation / investigation & report to authority / prosecution and/or minor penalty possible	Serious breach of regulation / considerable prosecution & penalties	Very serious penalties and prosecutions / Multiple lawsuits & possible jail time
OPERATIONAL (R / S / C) Impact on Reputation / Social / Community	Slight impact / public awareness may exist but no public concern	Limited impact / local public concern	Considerable impact / regional public concern	Serious impact / national public concern	Extreme impact / international public concern

Risk Rating

Likelihood / Probability
Consider near hits as well as actual events

Likelihood / Probability		1 Insignificant	2 Minor	3 Moderate	4 High	5 Major
5 Almost Certain	The unwanted event occurred frequently; occurs one or more times a year and is likely to occur within 1 year	11 (M)	16 (S)	20 (S)	23 (H)	25 (H)
4 Likely	The unwanted event occured infrequently; occurs in the order of less than once a year and is likely to occur within 5 years	7 (M)	12 (M)	17 (S)	21 (H)	24 (H)
3 Possible	The unwanted event occurred sometimes and could occur within 10 years	4 (L)	8 (M)	13 (S)	18 (S)	22 (H)
2 Unlikely	The unwanted event occurred sometimes and could occur within 20 years	2 (L)	5 (L)	9 (M)	14 (S)	19 (S)
1 Rare	The unwanted event has never occurred in the business and is not likely to happen within 20 years	1 (L)	3 (L)	6 (M)	10 (M)	15 (S)

Legend:

1 – 5 (L) = Low Tolerable Risks – Can be tolerated without the need to do anything about the risk

6 – 12 (M) = Medium Tolerable Risks – Risks can be tolerated but should be reviewed periodically

13 – 20 (S) = Significant Risks – Risk Management Strategies must be developed to reduce the risk to medium (M) category

21 – 25 (H) = High Risks – Risk Management Strategies must be developed to reduce the risks to at least medium (M) category

FIGURE 7.7 Example of a risk matrix

Once the risk has been "plotted" on the risk matrix, or risk-ranking table, it is then possible to determine whether the risk is tolerable or not, as indicated in the table in Figure 7.8.

Risk Ranking	Risk Categories
1 - 5	Low (Tolerable) Risk
6 - 12	Medium (Tolerable) Risk
13 - 20	Significant (Intolerable) Risk
21 - 25	High (Intolerable) Risk

FIGURE 7.8   Qualitative risk ranking

Note: It is important to remember that the criteria shown in the figure are only a guideline—the way risk is perceived and quantified by different organizations will always be different.

As pointed out earlier, many other factors will influence perceptions of risk, such as the value placed on human life by different cultural groups, religious values, and even someone's age and/or marital status. Organizational norms are also influenced by laws and regulations, by the extent to which people at all levels are exposed to criminal proceedings if they break those laws, by the possibility of civil lawsuits, and by the cost of insurance.

Once the risk framework has been developed and agreement has been reached on tolerability, it is possible to assess the inherent risk posed by every failure mode. This is achieved by asking the review group to reach consensus on the probability of failure (later chapters deal with the criteria on how to determine likelihood or probability of failure) and the consequence severity as described in the failure effect description and defined by the risk definitions. The risk matrix is used to quantify the risk in relative terms and determine whether the risk is tolerable or not. Once the risk assessment has been performed, the RCM3 review group will be in a position to determine which failure modes require further consideration. The final two questions of the RCM3 process require the review group to consider the following:

What *must* be done to reduce intolerable risks to a tolerable level (using proactive risk management strategies)?

What *can* be done to reduce or manage tolerable risks in a cost-effective way?

The review group will determine the inherent risk through the steps listed in the previous section to judge how the final two questions will be dealt with. The facilitator may decide to focus on the intolerable risk only while the review group is present and deal with the tolerable risks at a later stage, thus saving time and valuable resources. The example in Figure 7.9 illustrates how the inherent risk is quantified in relative terms.

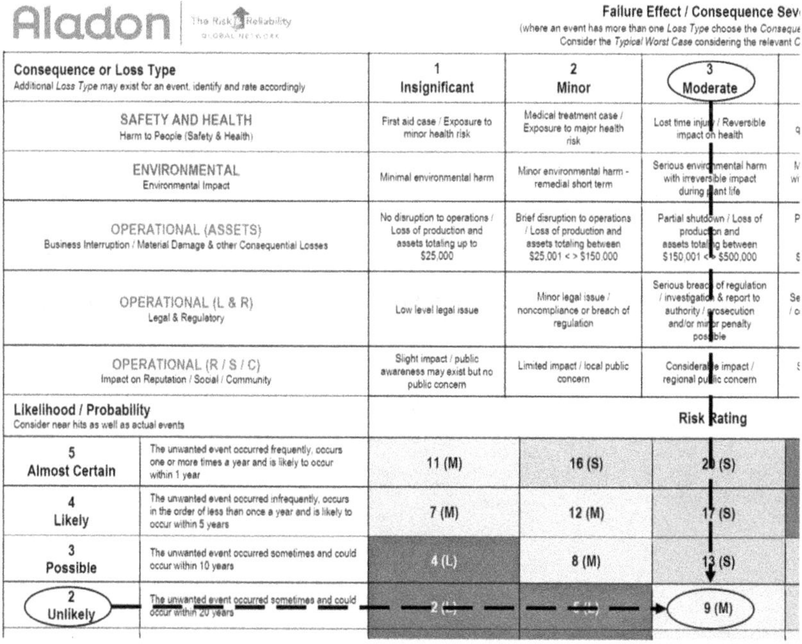

FIGURE 7.9 Determine inherent risk for all failure modes

From the failure effect description that was captured in the Information Worksheet that was shown in Figure 7.10, the review group determined the following:

Step 1. The bearing will seize on average every 10–20 years.

Step 2. The worst-case effect will occur when the protective device is not available or fails on demand, resulting in an operational consequence costing the organization in excess $150,000 (but less than $500,000).

Step 3. The risk associated with these criteria is a medium-tolerable risk 9 (M), as taken from the risk matrix.

Note: The failure effect description is used to assess the inherent risk.

The Information Worksheet will now be completed and will look like the one in Figure 7.10. The steps for completing the columns P_{ted}, P, C, and R are discussed in much more detail in Chapter 10, but for clarity they represent the following:

P_{ted}—probability of failure of protec*ted* functions

P—probability of failure of the functions that are not protected

C—consequence severity

R—inherent risk (combination of the probability of failure and consequence severity)

RCM3™ Information Worksheet © 2017 Aladon V0.0	Location ID			No	Compiled by	Date	Sheet	**Aladon** The Risk Reliability GLOBAL NETWORK
	Location Description			Ref	Reviewed by	Date	of	

Function	Failed State	Failure Mode Cause	Failure Mode Mechanism	Failure Effect		Inherent Risk			
						Pted	P	C	R
1 To transfer water at a minimum rate of 800 gpm from tank X to tank Y in the presence of a standby system	B Transfers water at a minimum rate of 800 gpm from tank X to tank Y in the absence of a standby system	1 Duty pump bearing seized	Normal wear	Local Failure Effect	Over time, the bearing starts to wear. The bearing temperature and vibration increase and the bearing will eventually seize or collapse. The motor trips on electric overload and pumping stops.	2		3	9(M)
			Lack of lubrication						
			Axial thrust too high	Next Higher Level Effect	The water continues to drain from tank Y and the standby pump will automatically start when the low-level in the tank is reached. The standby pump will restore the level in tank Y. The duty pump will be repaired.				
			Incorrect installation	End Effect	It will take 8 hours to repair the duty pump (replace the bearing). A replacement bearing can be obtained within one day and repairs will be done on-site at a cost of $3,500.				
			Dirt ingress	Potential Worst Case Effect	If the standby pump fails to start, transfer of the water will stop and tank Y will drain empty. The low-level alarm will be generated. There is no more than 2.5 hours of water left in tank Y. The process will be down for 1.5 hours. One batch will be lost with a cost of $150,000.				
			Etc.						

FIGURE 7.10 Completed RCM3 Information Worksheet

When assessing the risk and completing the Information Worksheet, the facilitator will only fill in one of the two columns that deal with the probability of failure, the one being the probability of failure of a protected function (where the function is protected by one or more protective devices) and the probability of failure of unprotected functions.

For example, the pump system described in Figure 7.10 transfers water from tank X to tank Y at a rate of up to 1,000 gallons per minute while water is supplied to the process at a constant rate of 800 gallons per minute. Tank Y is fitted with two limit switches, an upper-limit switch and a lower-limit switch that operate as follows:

- The lower-limit switch starts the pump and fills tank Y when the level drops to 120,000 gallons.
- The upper-limit switch shuts the pump down when the high level is reached at 240,000 gallons.

Tank Y is also fitted with a low-level alarm (just below the lower-level switch) that will generate an alarm in the central control room when the level drops below 120,000 gallons. In the event the pump fails to transfer the water, the low-level alarm will sound. The function "To transfer water at a minimum rate of 800 gpm from tank X to tank Y" is protected by a standby pump system, and the probability of failure (for any failure mode that will be considered for the duty pump system) that can cause the failed state: "Transfers water at a rate of 800 gallons per minute in the absence of a standby system" will be recorded in the "P_{ted}" column.

If a standby pump was not installed, the function will not be protected, and the probability of failure will be recorded in the "P" column.

The optimization of the protective device (standby pump system) will be treated together with the failure of the protected function.

It is important at this stage to mention that the protection is directly related to the function. It should not be confused with the protection of the motor, i.e., overload protection. The overload protection protects the motor from damage (a completely different function) and not the function being considered in the example.

Once the inherent risk has been determined, the review group *must* apply the RCM3 decision logic to determine the most appropriate risk management strategy to mitigate, reduce, or eliminate intolerable risks associated with the failure modes. For failure modes posing a tolerable risk, the review group (under the guidance of the facilitator) may decide to apply the decision logic to determine if a cost-effective way can be found to manage or reduce tolerable risks further, although this is not required as defined by the ISO 31000 standard.

An example of how the different levels of risk can be interpreted and ranked for specific action is illustrated in Figure 7.11.

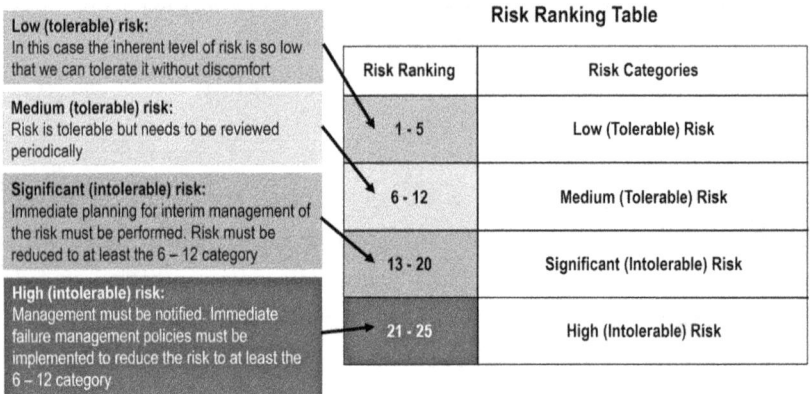

Low (tolerable) risk:
In this case the inherent level of risk is so low that we can tolerate it without discomfort

Medium (tolerable) risk:
Risk is tolerable but needs to be reviewed periodically

Significant (intolerable) risk:
Immediate planning for interim management of the risk must be performed. Risk must be reduced to at least the 6 – 12 category

High (intolerable) risk:
Management must be notified. Immediate failure management policies must be implemented to reduce the risk to at least the 6 – 12 category

Risk Ranking Table

Risk Ranking	Risk Categories
1 - 5	Low (Tolerable) Risk
6 - 12	Medium (Tolerable) Risk
13 - 20	Significant (Intolerable) Risk
21 - 25	High (Intolerable) Risk

FIGURE 7.11   Typical risk mitigation strategies based on risk ranking

7.3 Technically Feasible and Worth Doing

Every time a failure occurs, the organization that uses the asset is affected in some way. Some failures affect output, product quality, or customer service. Others threaten safety or the environment. Some increase operating costs, for instance by increasing energy consumption, while a few have an impact in four, five, or even all six of these areas. Still others may appear to have no effect at all if they occur on their own but may expose the organization to the risk of much more serious failures.

If any of these failures is not prevented, the time and effort that needs to be spent correcting them also affects the organization, because repairing failures consumes resources that might be better used elsewhere.

The nature and severity of these effects govern the consequences of the failure. In other words, they govern the extent to which the owners or users of the asset will believe that each failure matters. (Note that failure effect describes what happens when a failure occurs, while consequences severity describes how—and how much—it matters. Clearly, if we can reduce the effect of any failure in terms of frequency and/or severity, then it follows that we will also reduce the associated risk.)

If a failure matters very much (*intolerable risk*), then considerable efforts will be made to avoid, eliminate, or minimize the consequences. This is especially true if the failure could injure or kill someone or if it could have a serious effect on the environment. It is also true of failures that interfere with production or operations or that cause significant secondary damage.

On the other hand, if the failure only has minor consequences (*tolerable risk*), it is possible that no proactive action will be taken, and the failure will simply be corrected each time it occurs.

This focus on consequences means that RCM starts the task selection process by assessing the effects of each failure mode and classifying it into one of five broad categories of consequences. The second step is to find out if it is physically possible to perform a proactive task (proactive risk management strategy) that reduces, or enables action to be taken to reduce, the inherent risk of the failure mode to a tolerable level (to an extent that might be acceptable to the owner or user of the asset). If such a task is found, it is said to be technically feasible. The criteria governing technical feasibility are discussed in more detail in Chapter 8.

If a task is technically feasible, then the third step is to ask whether the task reduces the risk to a tolerable level (to an extent that justifies the direct and indirect costs of doing the task). Direct costs are the costs of labor or materials needed to do the task and to do any associated rectification work. Indirect costs include the cost of any downtime that may be needed to do the task. If the answer is yes, then the task is said to be worth doing.

> Proactive maintenance has much more to do with avoiding or mitigating the risk associated with the failure than it has to do with preventing the failures themselves.

If this is accepted, then it stands to reason that any proactive task is only worth doing if it deals successfully with the inherent risk posed by the failure that it is meant to prevent.

> A proactive task is worth doing if it deals successfully with the inherent risk posed by the failure that it is meant to prevent.

This, of course, presupposes that it is possible to anticipate or prevent the failure in the first place. Whether or not a proactive task is technically feasible depends on the technical characteristics of the task and of the failure that it is meant to prevent. Again, the criteria governing technical feasibility are discussed in more detail in Chapter 8.

If it is not possible to find a suitable proactive task, the nature of the failure consequences and associated risk will determine what other action is required. Any actions would be in the form of combinations of tasks, optimization of existing protective devices, and one-time changes (modifications and procedures).

The remainder of this chapter considers the criteria used to evaluate the risk associated with failures, and hence to decide whether any form of proactive risk management strategy is worth doing. These risks are divided in two criteria and into five categories. The risk categories are shown in Figure 7.12. The first criteria separates hidden risks from evident risks.

Hidden		Evident		
Hidden Operational Risk	Hidden Physical Risk	Evident Physical Risk	Evident Operational Risk	Tolerable Risk
If the effect of the multiple failure could result in an intolerable risk to operational capability	If the effect of the multiple failure could result in an intolerable risk of injuring or killing people or breaching an environmental standard or regulation	If the effect of the failure mode could result in an intolerable risk of injuring or killing people or breaching an environmental standard or regulation	If the effect of the failure mode could result in an intolerable risk of adversely effecting operational capability	When the effect of the failure can be tolerated OR if the failure mode only involves the direct cost of repair (and any secondary damage).

FIGURE 7.12   Risk categories

7.4. Hidden and Evident Functions

We have seen that every asset has more than one and sometimes dozens of functions. When most of these functions fail, it will inevitably become apparent to someone that the failure has occurred.

For instance, some failures cause warning lights to flash or alarms to sound, or both. Others cause machines to shut down or some other part of the process to be interrupted. Still others lead to product quality problems or increased use of energy, and yet others are accompanied by obvious physical effects such as loud noises, escaping steam, unusual smells, or pools of liquid on the floor.

For example, Chapter 3 showed three pumps, which are shown again in Figure 7.13. If a bearing on pump A seizes, pumping capability is lost. This failure on its own will inevitably become apparent to the operators, either as soon as it happens or when some downstream part of the process is interrupted. (The operators might not know immediately that the problem was caused by the bearing, but they would eventually and inevitably become aware of the fact that something unusual had happened.)

Stand-alone pump   Duty pump   Standby pump

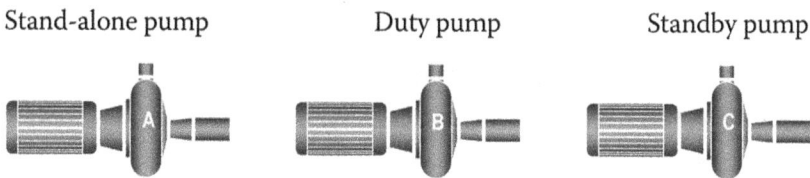

FIGURE 7.13   Three-pump example

Failures of this kind are classed as evident because someone will eventually find out about them when they occur on their own. This leads to the following definition of an evident function:

An evident function is one whose failure will eventually and inevitably become evident to the operating crew under normal circumstances.

When we consider evident failures, the question arises of what exactly is meant by "eventually" and "inevitably." In other words, is there a time limit in which the failure becomes evident? Is it immediately, or within the shift or day, or within the batch produced, and does it mean evident to some people only? According to the definition in the RCM logic, the loss of evident functions will become evident on their own in the fullness of time. There is no time limit, although it implies a reasonable period of time (not without limit) and it is not restricted to any one person, e.g., operators or maintainers.

However, some failures occur in such a way that nobody knows that the item is in a failed state unless or until some other failure also occurs.

For example, if pump C in Figure 7.13 failed, no one would be aware of the fact, because under normal circumstances pump B would still be working. In other words, the failure of pump C on its own has no direct impact unless or until pump B also fails (an abnormal circumstance).

Pump C exhibits one of the most important characteristics of a hidden function, which is that the failure of this pump on its own will not become evident to the operating crew under normal circumstances. In other words, it will not become evident unless pump B also fails. This leads to the following definition of a hidden function:

> A hidden function is one whose failure will not become evident to the operating crew under normal circumstances if it occurs on its own.

It is important to separate hidden functions from evident functions, because hidden functions need special handling. A further separation of hidden functions is required since hidden functions or protective devices provide protection against physical or operational risks. Where the multiple failure (failure of the protected function and the hidden function) leads to a physical risk, the hidden functions are treated differently from the hidden functions that result in operational consequences only.

These are discussed at length in Section 7.6 of this chapter. We will see later that these functions are associated with protective devices that are not fail-safe.

Since they can account for up to half the failure modes that could affect modern, complex equipment, hidden functions are recognized to play an important role in modern plants and facilities. Maintenance of these devices is largely overlooked, and it is now a known fact that major industrial incidents are an indirect effect of the failure of these devices. It is also a known fact that when protective devices are added, there is a tendency to "undermaintain" the protected function since it is believed that the protection will reduce or eliminate the risks associated with failure of the protected function. In RCM3 the focus will always be to increase the reliability of the protected function (if possible) before the protective device is addressed.

However, to place hidden functions in perspective, we first consider evident failures.

Categories of Evident Failures

Evident failures are classified into three categories in descending order of importance:

- **Safety and environmental (Physical Risk).** If the effect of the failure mode could result in an intolerable risk of injuring or killing people or breaching an environmental standard or regulation.
- **Operational (Economic Risk).** If the effect of the failure mode could result in an intolerable economic risk by adversely affecting operational capability (output, product quality, customer service, or operating costs in addition to the direct cost of repair, noncompliance, penalties, and legal proceedings).
- **Tolerable Risk.** Evident failures in this category pose a tolerable risk to safety, environment, and operational capability, so they involve mostly the direct cost of repair (and any secondary damage). Risks associated with these failures will be considered tolerable.

By ranking evident failures in this order, RCM ensures that the safety and environmental implications of every evident failure mode are considered. This unequivocally puts people and the environment ahead of production and operations.

This approach also means that the safety, environmental, and operational consequences of each failure are assessed in one exercise, which is much more cost-effective than considering them separately.

The next four sections of this chapter consider each of these categories in detail, starting with the evident categories and then moving on to the rather more complex issues surrounding hidden functions.

7.5 Safety and Environmental Consequences (Physical Risks)

Safety First

As we have seen, the first step in the risk evaluation process is to identify hidden functions so that they can be dealt with appropriately. All remaining failure modes—in other words, failures that are not classified as hidden—must by definition be evident. The above paragraphs explained that the RCM process considers the safety and environmental implications of each evident failure mode first. It does so for two reasons:

- A more and more firmly held belief among employers, employees, customers, and society in general that injuring or killing people in the course of business is simply not tolerable, and hence that everything possible should be done to minimize the possibility of any sort of safety-related incident or environmental excursion.
- The more pragmatic realization that the probabilities that are tolerated for safety-related incidents tend to be several orders of magnitude lower than those that are tolerated for failures that have operational consequences. As a result, in most of the cases where a proactive task is worth doing from the safety viewpoint, it is also likely to be more than adequate from the operational viewpoint.

At one level, safety refers to the safety of individuals in the workplace. Specifically, RCM asks whether anyone could get injured or killed either as a direct result of the failure mode itself or by other damage that may be caused by the failure.

> A failure mode poses an intolerable risk to safety if it causes
> a loss of function or other damage that could hurt or
> kill someone.

At another level, safety refers to the safety or well-being of society in general. Nowadays, failures that affect society tend to be classified as environmental issues. In fact, in many parts of the world, the point has been reached where either organizations conform to society's environmental expectations, or they will no longer be allowed to operate. So quite apart from any personal feelings that anyone may have on the issue, environmental probity is becoming a prerequisite for corporate survival.

Chapter 4 explained how society's expectations take the form of municipal, regional, and national environmental standards. Some organizations also have their own, sometimes even more stringent, corporate standards. A failure mode is said to have environmental consequences if it could lead to the breach of any of these standards.

> A failure mode poses an intolerable risk to environmental
> integrity if it causes a loss of function or other damage
> that could lead to the breach of an environmental
> standard or regulation.

Note that when considering whether a failure mode has safety or environmental consequences, we are considering whether one failure mode on its own could have the consequences. This is different from Section 7.6 of this chapter, in which we consider the failure of both elements of a protected system.

Safety and Proactive Maintenance

If the risk associated with a failure that could affect safety or the environment is intolerable, the RCM3 process stipulates that we must try to prevent or minimize it. The above discussion suggests that when a failure mode poses an intolerable risk to safety or environmental integrity, it is only worth doing a proactive task if it reduces the probability of the failure to a tolerably low level.

> For a failure mode that poses an intolerable risk to safety or environmental integrity, a proactive task is only worth doing if it reduces the probability of the failure to a tolerably low level.

If a single proactive task cannot be found that achieves this objective to the satisfaction of the group performing the analysis, we are dealing with a safety or environmental hazard that cannot be adequately anticipated or prevented. The next step is to seek a combination of tasks that reduce the risk to a tolerable level. Combinations of tasks are discussed later in this book in Chapter 8. This means that something must be changed in order to make the system safe. This "something" could be the asset itself, a process, or an operating procedure.

One-off changes of this sort are classified as "redesigns" and are usually undertaken with one of two objectives:

- To reduce the probability of the failure occurring to a tolerable level
- To change things so that the failure no longer poses an intolerable risk to safety or the environment

The question of a one-time change (redesign) is discussed in more detail in Chapter 8.

It is important to note that operating and maintenance procedures are considered as proactive risk mitigating measures and will be considered in the RCM3 process.

Note that when dealing with safety and environmental issues, RCM does not raise the question of economics. If it is not safe, we have an obligation either to prevent it from failing or to make it safe. This suggests that the decision process for failure modes that have safety or environmental consequences can be summarized as shown in Figure 7.14.

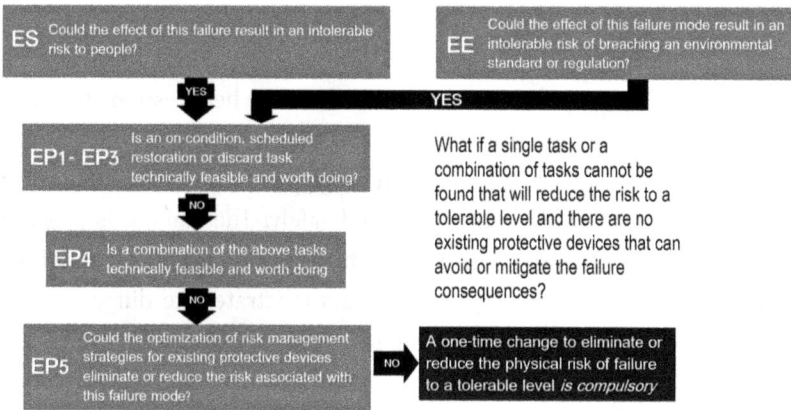

FIGURE 7.14   Identifying and developing a failure management strategy for a failure that affects safety or the environment

RCM, Safety Legislation, and Management Standards

A question often arises concerning the relationship between RCM and safety legislation (environmental legislation is dealt with directly).

Nowadays, most legislation governing safety merely demands that users are able to demonstrate that they are doing whatever is prudent to ensure that their assets are safe. This has led to rapidly increasing emphasis on the concept of an audit trail, which basically requires users of assets to be able to produce documentary evidence that there is a rational, defensible basis for their maintenance programs. In the vast majority of cases, RCM wholly satisfies this type of requirement.

However, some regulations demand that specific tasks should be done on specific types of equipment at specific intervals. If the RCM process suggests a different task and/or a different interval, it is wise to continue doing the task specified by the legislation and to discuss the suggested change with the appropriate regulatory authority. It is the author's experience that in many cases a regulation will specify the minimum requirements for doing maintenance and inspections—which may only be enough to satisfy the authority. In some cases, the regulations may be outdated and even obsolete—regulations are not updated or revised as rapidly as the changes that happen in technology and with our equipment.

With the release of the international standards for asset management and risk management, i.e., ISO 55000 (2014) and ISO 31000 (2009), the author is of the opinion that companies will be held responsible even more so in today's time if they do not do the right maintenance at the right time on the right equipment—in other words, if they fail to ensure that assets are operated and maintained safely. The standards provide guidance and a framework for organizations to organize their activities around physical asset management to demonstrate due diligence and responsibility. Ignorance of the standards—not knowing about the standards or what is in them—will not release asset owners of their responsibility and possible prosecution when incidents and accidents cause injury, death, and breach of environmental standards and regulations. The management's standards may actually be used by prosecutors to bring about justice.

7.6 Operational Consequences (Economic Risks)

How Failures Affect Operations

The primary function of most equipment in industry is connected in some way with the need to earn revenue or to support revenue-earning activities.

For example, the primary function of most of the assets used in manufacturing is to add value to materials, while customers pay directly for access to electricity, drinking water, telecommunications, and transport equipment (buses, trucks, trains, or aircraft).

Failures that affect the primary functions of these assets affect the revenue-earning capability of the organization. The magnitude of these effects depends on how heavily the equipment is loaded and the availability of alternatives. However, in nearly all cases the effects are greater—often much greater—than the cost of repairing the failures. This is also true of equipment in service industries such as entertainment, commerce, and even banking.

For example, if the lights fail at a ball game, fans tend to want their money back. The same applies if projectors fail at the movies. If the air conditioning fails in a shop or restaurant, customers walk out. Banks lose business if their ATMs fail.

In general, failures affect operations in five ways:

1. **They affect total output.** This occurs when equipment stops working altogether or when it works too slowly. It results either in increased production costs if the plant has to work extra time to catch up or in lost sales if the plant is already fully loaded.
2. **They affect product quality.** If a machine can no longer hold manufacturing tolerances or if a failure causes materials to deteriorate, the likely result is either scrap or expensive rework. In a more general sense, "quality" also covers concepts such as the precision of navigation systems, the accuracy of targeting systems, and so on.
3. **They affect customer service.** Failures affect customer service in many ways, ranging from the late delivery of orders to the late departure of passenger aircraft. Frequent or serious delays sometimes attract heavy penalties, but in most cases, they do not result in an immediate loss of revenue. However, chronic service problems eventually cause customers to lose confidence and take their business elsewhere.
4. **They will result in increased operating costs in addition to the direct cost of repair.** For instance, the failure might lead to the increased use of energy, or it might involve switching to a more expensive alternative process. It may also lead to penalties, fines, eroding customer confidence, and noncompliance.
5. **Legal and regulatory compliance.** Non-compliance may lead to financial penalties or even shutdown of operations (depending on the violation). Legal action may cost companies millions and although it may not affect production, it is still an economic risk that must be considered.

In nonprofit enterprises such as military undertakings, certain failures can also affect the ability of the organization to fulfill its primary function, sometimes with devastating results. We can sum it up easily in the old proverb:

"For want of a nail, a shoe was lost. For want of a shoe, a horse was lost. For want of a horse, the rider was lost. For want of a rider, the message was lost. For want of a message, a battle was lost. For want of a battle, a war was lost. All for want of a horseshoe nail."

While it may be difficult to cost out the results of losing a war, failures of this sort still have economic implications at a more mundane level. If they occur too often, it may be necessary to keep, say, two horses in order to ensure that one will be available to do the job—or sixty battle tanks instead of fifty—or six aircraft carriers instead of five. Redundancy on this scale can be very expensive indeed.

The severity of these consequences means that if an evident failure does not pose a threat to safety or the environment, the RCM process focuses next on the operational consequences of failure.

> If an evident failure does not pose a threat to safety or the environment, the RCM process focuses next on the operational consequences of failure.

A failure poses an intolerable operational risk if it has a direct adverse effect on operational capability, where operational capability is defined as:

- Lost time or downtime
- Lost production
- Lost sales
- Increased cost (increased energy cost or overtime)
- Litigation and lawsuits
- Eroding customer confidence and public perception
- Penalties and fines

As we have seen, these consequences tend to be economic in nature, so they are usually evaluated in economic terms. However, in certain more extreme cases (such as losing a war and eroding customer confidence) the "cost" may have to be evaluated on a more qualitative basis.

Not all failures may also have a direct economic impact. Insurance costs may rise as a result of a failure (higher risk—perceived or in reality); failures may result in penalties (from late delivery or contract breach); and failures may erode customer confidence, which in itself is not a direct economic consequence but over time will affect income. For example, a public utility may not suffer directly from a loss of income due to equipment failure and poor customer service but may find it difficult to push through a future rate increase when customers are not fully satisfied with the utility's performance.

Avoiding Operational Consequences

The overall economic effect of any failure mode that has operational consequences depends on the following factors:

- How much the failure costs each time it occurs, in terms of its effect on operational capability plus repair costs (for functions that are not protected)
- How often it happens
- Whether the function is protected with a protective device

In the previous section of this chapter, we did not pay much attention to how often failures are likely to occur. (Failure rates have little bearing on safety-related failures, because the objective in these cases is to avoid any failures on which to base a rate.) However, if the failure consequences are economic, the total cost is affected by how often the consequences are likely to occur. In other words, to assess the economic impact of these failures, we need to assess how much they are likely to cost over *a period of time.*

Consider, for example, the pump shown in Figure 3.1 and again in Figure 7.15. The pump is controlled by three switches: two float switches activate and turn the pump on when the level in tank Y drops to 120,000 gallons, and another turns the pump off when the level in tank Y reaches 240,000 gallons. A low-level alarm is located just below the 120,000-gallon level. If the tank runs dry, the downstream process must be shut down. This costs the organization using the pump $150,000 per hour.

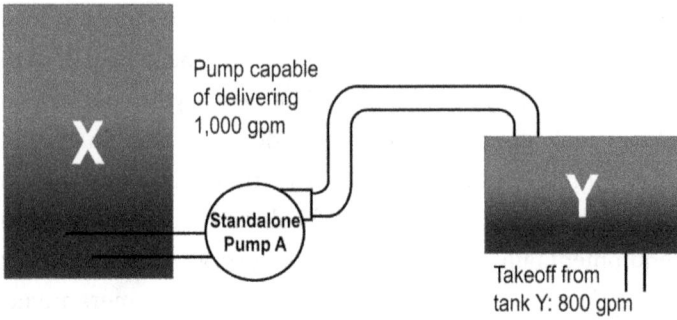

FIGURE 7.15   Stand-alone pump system

Assume that it has already been agreed that one failure mode that can affect this pump is "the bearing will seize due to normal wear." For the sake of simplicity, assume that the motor on this pump is equipped with an overload switch, but there is no trip alarm wired to the control room.

This failure mode and its effects might be described on an RCM3 Information Worksheet as shown in Figure 7.16.

Water is drawn out of the tank at a rate of 800 gallons per minute, so the tank runs dry in 2.5 hours (150 minutes) after the low-level alarm sounds. It takes 4 hours to replace the bearing (or swap out the pump), so the downstream process stops for 1.5 hours. So this failure costs:

$$1.5 \times \$150,000 = \$225,000$$

in lost production every 5 years, plus the cost of replacing the bearing.

FIGURE 7.16   Information Worksheet for bearing failure on stand-alone pump

Assume that it is technically feasible to check the bearing for audible noise once a week (the basis upon which we make this kind of judgment is discussed at length in the next chapter). If the bearing is found to be noisy, the operational consequences of failure can be avoided by ensuring that the tank is full before starting work on the bearing. This provides 5 hours of storage, so the bearing can now be replaced in 4 hours without interfering with the downstream process.

Assume also that the pump is located in an unmanned pumping station. It has been agreed that the check should be carried out by a maintenance craftsworker and that the total time needed to do each check is 20 minutes. Assume further that the total cost of employing the craftsworker is $60 per hour, in which case it costs $20 to perform each check. If the MTBF of the bearing is 5 years, the craftsworker will do about 250 checks per failure (assume a 50-week work year). In other words, the cost of the checks is:

$$250 \times \$20 = \$5,000$$

every 5 years, again plus the cost of replacing the bearing.

In this example, the scheduled task is clearly cost-effective relative to the cost of the operational consequences of the failure plus the cost of repair. This suggests that if a failure poses an intolerable operational risk, the basis for deciding whether a proactive task is worth doing is economic.

> For failure modes posing an intolerable operational risk, a proactive task is worth doing if, over a period of time, it costs less than the cost of the operational consequences plus the cost of repairing the failure that it is meant to prevent.

Conversely, if a cost-effective proactive task cannot be found, then it is not worth doing any scheduled maintenance to try to anticipate or prevent the failure mode under consideration. In some cases, the most cost-effective option at this point might simply be to decide to live with the failure.

However, if a proactive task cannot be found and the operational risk is still intolerable, it may be desirable to change the design of the asset (or to change the process) in order to reduce total costs by:

- Reducing the frequency (and hence the total cost) of the failure
- Reducing or eliminating the consequences of the failure
- Making a proactive task cost-effective

One-time changes (redesigns) are discussed in more detail in Chapter 9.

Note that in the case where a failure mode poses an intolerable safety and environmental risk, the objective is to reduce the probability of the failure to a very low level indeed. In the case of intolerable operational risks, the objective is to reduce the probability (or frequency) to an economically tolerable level. As mentioned at the start of Section 7.3 of this chapter, this frequency is likely to be several orders of magnitude greater than we would tolerate for most safety hazards, so the RCM process assumes that a proactive task that reduces the probability of a safety-related failure to a tolerable level will also deal with the operational risk associated with that failure.

To begin with, we again only consider the desirability of making changes after we have established whether it is possible to extract the desired performance from the asset as it is currently configured. However, in this case modifications also need to be cost-justified, whereas they were compulsory for failure modes with safety or environmental consequences.

In the light of these comments, the decision process for failures that poses an intolerable risk to operational capability, can be summarized as shown in Figure 7.17.

Note that this analysis is carried out for each individual failure mode, and not for the asset as a whole. This is because each proactive task is designed to prevent a specific failure mode, and so the economic feasibility of each task can only be compared with the costs of the failure mode that it is meant to prevent. In each case, it is a simple go-no-go decision. However, it must also be said that for certain failure modes this justification will not be necessary. Common sense will prove that certain PMs are always worth doing, e.g., lubrication and alignment.

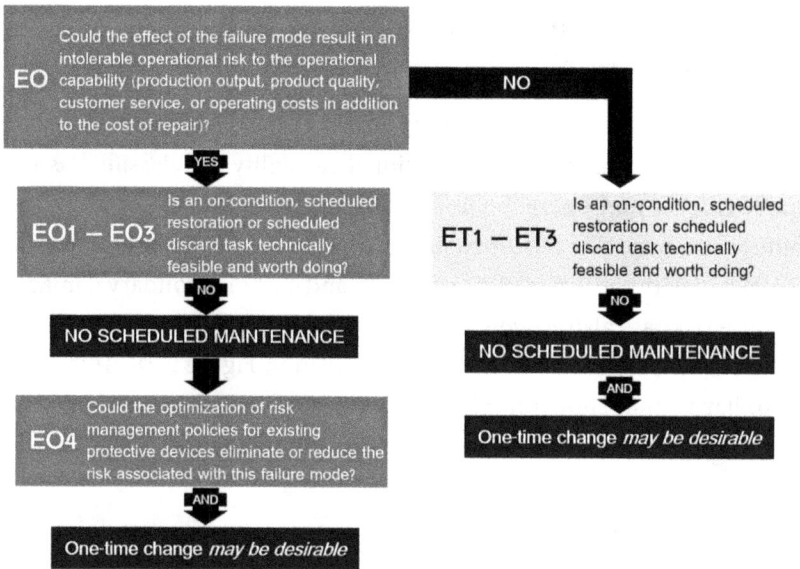

FIGURE 7.17 Identifying and developing a maintenance strategy for failure that has operational and nonoperational consequences

In practice, when assessing individual failure modes in this way, it is not always necessary to do a detailed cost-benefit study based on actual downtime costs and MTBFs as shown in the example on pages 218–219. This is because the economic desirability of proactive tasks is often intuitively obvious when assessing failure modes with operational consequences.

However, whether or not the economic consequences are evaluated formally or intuitively, this aspect of the RCM process must still be applied thoroughly. (In fact, this step is surprisingly often overlooked by people new to the process. Maintenance people in particular have a tendency to implement tasks on the basis of technical feasibility alone, which results in elegant but excessively costly maintenance programs.)

Finally, bear in mind that the operational consequences of any failure are heavily influenced by the context in which the asset is operating. This is yet another reason why care should be taken to ensure that the context is identical before taking a maintenance program developed for one asset and applying it to another. The key issues were discussed in Section 2.3 of Chapter 2 and in Chapter 3.

7.7 Tolerable Risk (Also Referred to as Non-operational Consequences)

The consequences of an evident failure that has a tolerable effect on safety, the environment, and operational capability are classified as tolerable risks or non-operational. The consequences associated with these failures are normally not impacting operations and the only impact is associated with the direct costs of repair (and cost of secondary damage), so these risks are also *economic*.

Consider, for example, the pumps shown in Figure 7.18. This setup is similar to that shown in Figure 7.15, except that there are now two pumps (both of which are identical to the pump in Figure 7.15).

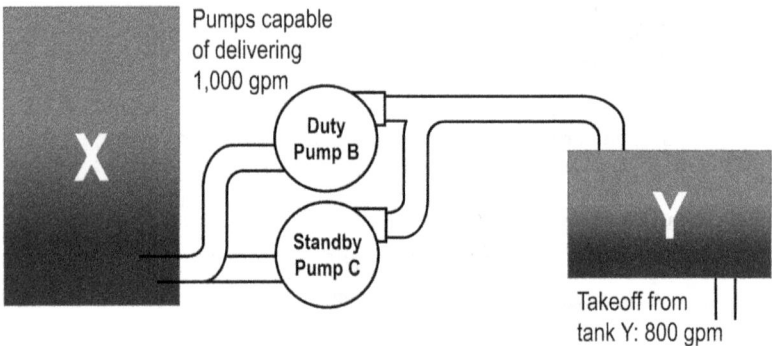

FIGURE 7.18   Pump system with standby system

The duty pump is switched on by one float switch when the level in tank Y drops to 120,000 gallons, and it is switched off by another when the level reaches 240,000 gallons. A third switch is located just below the low-level switch of the duty pump, and this switch is designed both to sound an alarm in the control room if the water level reaches it and to switch on the standby pump. If the tank runs dry, the downstream process must be shut down. This also costs the organization that uses the pump $150,000 per hour.

As before, assume that it has been agreed that one failure mode that can affect the duty pump is "the bearing will seize due to normal wear." Assume that the motor on the duty pump is also equipped with

RCM3™ Information Worksheet	Location ID			No	Compiled by		Date	Sheet	**Aladon**
© 2017 Aladon V0.0	Location Description			Ref.	Reviewed by		Date	of	The Risk & Reliability GLOBAL NETWORK

	Function		Failed State		Failure Mode Cause		Failure Mode Mechanism	Failure Effect		Inherent Risk			
										Pted	P	C	R
1	To transfer water at a minimum rate of 800 gpm from tank X to tank Y in the presence of a standby system	B	Transfers water at a minimum rate of 800 gpm from tank X to tank Y in the absence of a standby system	1	Duty pump bearing seized	Normal wear	**Local Failure Effect**	Over time, the bearing starts to wear. The bearing temperature and vibration increase and the bearing will eventually seize or collapse. The motor trips on electric overload and pumping stops. The pump bearing is failing on average once every 5 years.	2	3	9(M)
						Lack of lubrication						
						Axial thrust too high	**Next Higher Level Effect**	The water continues to drain from tank Y and the standby pump will automatically start when the low-level in the tank is reached. The standby pump will restore the level in tank Y. The duty pump will be repaired.				
						Incorrect installation	**End Effect**	It will take 8 hours to repair the duty pump (replace the bearing). A replacement bearing can be obtained within one day and repairs will be done on-site at a cost of $3,500.				
						Dirt ingress	**Potential Worst Case Effect**	If the standby pump fails to start, transfer of the water will stop and tank Y will drain empty. The low-level alarm will be generated. There is no more than 2.5 hours of water left in tank Y. The process will be down for 1.5 hours. One batch will be lost with a cost of $150,000.				
						Etc.						

FIGURE 7.19 Information Worksheet for bearing failure on duty pump (with standby system)

an overload switch, but again there is no trip alarm wired to the control room. This failure mode and its effects might be described on an RCM3 Information Worksheet, as shown in Figure 7.19.

In this example, the standby pump is switched on when the duty pump fails, so the tank does not run dry. So the only cost associated with this failure is . . . the cost of replacing the bearing.

Assume, however, that it is still technically feasible to check the bearing for audible noise once a week. If the bearing were found to be noisy, the operators would switch over manually to the standby pump and the bearing would be replaced.

Assume that these pumps are also located in an unmanned pumping station, and it has again been agreed that the check—which also takes 20 minutes—should be done by a maintenance craftsworker at a cost of $20 per check. Once again, the craftsworker will do about 150 checks per failure. In other words, the cost of the proactive maintenance program per failure is:

$$250 \times \$20 = \$5,000$$

plus the cost of replacing the bearing.

In this example, the cost of doing the scheduled task is now much greater than the cost of not doing it. As a result, it is not worth doing the proactive task even though the pump is technically identical to the pump described in Figure 7.15. This suggests that it is only worth trying to prevent a failure that has nonoperational consequences (tolerable risk) if, over a period of time, the cost of the preventive task is less than the cost of correcting the failure. If it is not, then scheduled maintenance is not worth doing.

> For failure modes posing a tolerable operational risk,
> a proactive task is worth doing if, over a period of time,
> it costs less than the cost of repairing the failures it is
> meant to prevent (plus any secondary damage).

If a proactive task is not worth doing, then in rare cases a modification might be justified for much the same reasons as those that apply to failures with operational consequences.

Another consideration about the previous examples is the following: When two pumps are considered, the cost of the second or standby pump must be offset to the cost of the downtime, which in our example amounts to $150,000. The standby pump will save the organization the downtime cost, but the pump (and the installation) costs money and will need to be maintained also. These decisions are best made during design when, through-life, costing can be calculated. Reliability-centered design is discussed in more detail later in this book.

Another consideration in RCM3 is the focus on the protected function. Although the redundancy or standby pump will reduce or eliminate the operational consequences, the standby pump will need maintenance. The maintenance required for the standby pump is discussed in length in the following sections. Making the protected function (duty pump) more reliable will enable the organization to rely on the protection (standby pump) less often. The maintenance intervals on the standby system will be reduced. It is therefore necessary to consider the maintenance requirements on both pumps to develop the most cost-effective PM program.

Further Points Concerning Tolerable Risks

Two more points need to be considered when reviewing failures with risks that are tolerable

Secondary Damage. Some failure modes cause considerable secondary damage if they are not anticipated or prevented, which adds to the cost of repairing them. A suitable proactive task could make it possible to prevent or anticipate the failure and avoid this damage. However, such a task is only justified if the cost of doing it is less than the cost of repairing the failure and the secondary damage.

For example, in Figure 7.18 the description of the failure effects suggests that the seizure of the bearing causes no secondary damage. If this is so, then the analysis is valid. However, if the unanticipated failure of the bearing also causes, say, the shaft to shear, then a proactive task that detects imminent bearing failure would enable the operators to shut down the pump before the shaft is damaged. In this case the cost of the unanticipated failure of the bearing is the cost of replacing the bearing and the shaft.

On the other hand, the cost of the proactive task (per bearing failure) is still $5,000 plus the cost of replacing the bearing.

Clearly, the task is worth doing if it costs more than $5,000 to replace the shaft. If it costs less than $5,000, then this task is still not worth doing.

Protected Functions. It is only valid to say that a failure will have non-operational consequences or a tolerable risk because a standby or redundant component is available and if it is reasonable to assume that the protective device will be functional when the failure occurs. This, of course, means that a suitable maintenance program must be applied to the protective device (the standby pump in the example given above). This issue is discussed at length in the next section of this chapter.

If the consequences of the multiple failure of a protected system are particularly serious, it may be worth trying to prevent the failure of the protected function as well as the protective device in order to reduce the probability of the multiple failure to a tolerable level. (As explained on page 237, if the multiple failure has safety consequences, it may be wise to assess the consequences as if the protection were not present at all and then to revalidate the protection as part of the task selection process.)

7.8 Risks Associated with Hidden Failures

Hidden Failures and Protective Devices

Chapter 4 (Section 4.3) mentioned that the growth in the number of ways in which equipment can fail has led to corresponding growth in the variety and severity of failure consequences that fall into the evident categories. Section 4.3 also mentioned that protective devices are being used increasingly in an attempt to eliminate (or at least reduce) these consequences, and it explained how these devices work in one of five ways:

- To alert operators to abnormal conditions
- To shut down the equipment in the event of a failure
- To eliminate or relieve abnormal conditions that follow a failure and that might otherwise cause much more serious damage
- To take over from a function that has failed
- To prevent dangerous situations from arising in the first place (such as guards, railings, safety signs, etc.)

In essence, the function of these devices is to ensure that the consequences of the failure of the protected function are much less serious than they would be if there were no protection. So any protective device is, in fact, part of a system with at least two components:

- The protective device
- The protected function

For example, pump C in Figure 7.18 can be regarded as a protective device, because it "protects" the pumping function if pump B should fail. Pump B is, of course, the protected function.

The existence of such a system creates two sets of failure possibilities, depending on whether the protective device is fail-safe or not. We consider the implications of each set, starting with devices that are fail-safe.

Fail-Safe Protective Devices. In this context, "fail-safe" means that the failure of the device on its own will become evident to the operating crew under normal circumstances.

In the context of this book, a "fail-safe" device is one whose failure on its own will become evident to the operating crew under normal circumstances.

This means that in a system that includes a fail-safe protective device, there are three failure possibilities in any period.

The first possibility is that *neither device fails*. In this case everything proceeds normally.

The second possibility is that the *protected function fails before the protective device*. In this case the protective device carries out its intended function, and depending on the nature of the protection, the consequences of failure of the protected function are reduced or eliminated.

The third possibility is that the *protective device fails before the protected function*. This would be evident, because if it were not, the device would not be fail-safe in the sense defined above. If normal good practice is followed, the chance of the protected device failing while the protective device is in a failed state can be almost eliminated, either by shutting down the protected function or by providing alternative protection while the failed protective device is being rectified.

For instance, an operator could be asked to keep an eye on a pressure gauge—and his finger by a stop button—while a pressure switch is being replaced.

This means that the consequences of the failure of a fail-safe protective device usually fall into the "operational" or "non-operational" categories. This sequence of events is summarized in Figure 7.20.

Protective Devices That Are Not Fail-Safe. In a system that contains a protective device that is not fail-safe, the fact that the device is unable to fulfill its intended function is not evident under normal circumstances. This creates four failure possibilities in any given period, two of which are the same as those that apply to a fail-safe device. The first is where neither device fails, in which case everything proceeds normally as before.

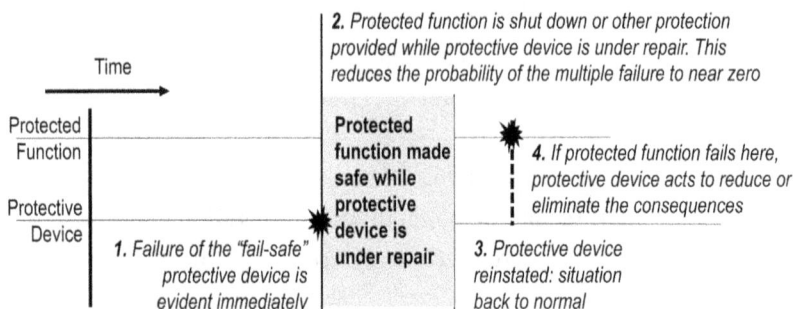

FIGURE 7.20   Failure of a "fail-safe" protective device

The second possibility is that the protected function fails at a time when the protective device is still functional. In this case the protective device also carries out its intended function, so the consequences of the failure of the protected function are again reduced or eliminated altogether.

For instance, consider a pressure relief valve (the protective device) mounted on a pressure vessel (the protected function). If the pressure rises above acceptable limits, the valve relieves and so reduces or eliminates the consequences of the overpressurization. Similarly, if pump B in Figure 7.18 fails, pump C takes over.

The third possibility is that the *protective device fails while the protected function* is still working. In this case, the failure has no direct consequences. In fact, no one even knows that the protective device is in a failed state.

For example, if the pressure relief valve was jammed shut, no one would be aware of the fact as long as the pressure in the vessel remained within normal operating limits. Similarly, if pump C were to fail somehow while pump B was still working, no one would be aware of the fact unless or until pump B also failed.

The above discussion suggests that hidden functions can be identified by asking the following question: Will the loss of function caused by this failure mode on its own become evident to the operating crew under normal circumstances? If the answer to this question is no, the failure mode is hidden. If the answer is yes, it is evident. Note that in this context, "on its own" means that nothing else has failed.

Note also that we assume at this point in the analysis that no attempts are being made to check whether the hidden function is still working. This is because such checks are a form of scheduled maintenance, and the whole purpose of the analysis is to find out whether such maintenance is necessary.

The fourth possibility during any one cycle is that the protective device fails, and then the protected function fails while the protective device is in a failed state. This situation is known as a multiple failure. (This is a real possibility simply because the failure of the protective device is not evident, and so no one would be aware of the need to take corrective or alternative action to avoid the multiple failure.) See Figure 7.21.

A multiple failure only occurs if a protected function fails while the protective device is in a failed state.

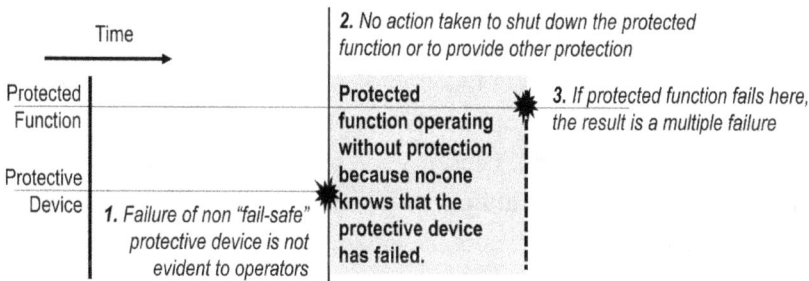

FIGURE 7.21   Failure of protective device whose function is hidden

In the case of the relief valve, if the pressure in the vessel rises excessively while the valve is jammed, the vessel will probably explode (unless someone acts very quickly or unless there is other protection in the system). If pump B fails while pump C is in a failed state, the result will be a total loss of pumping.

Given that failure prevention is mainly about avoiding the consequences of failure, this example also suggests that when we develop maintenance programs for hidden functions, our objective is actually to prevent—or at least to reduce the probability of—the associated multiple failure.

> The objective of a maintenance program for a hidden function is to prevent—or at least to reduce the probability of—the associated multiple failure.

How hard we try to prevent the hidden failure depends on the consequences of the multiple failure. We have discussed earlier that hidden functions are separated between hidden physical and hidden economic-type risks. If the consequences of the multiple failure are very serious indeed (hidden physical risk), we would go to great lengths to preserve the integrity of the hidden function. If the consequences of the multiple failure are purely economic, then how much it costs would influence how hard we would try to prevent the hidden failure.

For example, pumps B and C might be pumping cooling water to a nuclear reactor. In this case, if the reactor could not be shut down fast enough, the ultimate consequences of the multiple failure could be a meltdown, with catastrophic safety, environmental, and operational consequences.

On the other hand, the two pumps might be pumping water into a tank that has enough capacity to supply a downstream process for 2 hours. In this case, the consequence of the multiple failure would be that production stops after 2 hours, and then only if neither of the pumps could be repaired before the tank ran dry. Further analysis might suggest that, at worst, this multiple failure might cost the organization $200,000 in lost production.

In the first of these examples, the consequences of the multiple failure are very serious, and so, as we mentioned above, we would go to great lengths to maintain the integrity of the hidden function. In the second case, the consequences of the multiple failure are purely economic, and how strenuous our efforts to prevent the hidden failure would depend on how much it costs.

Further examples of hidden failures and the multiple failures that could follow if they are not detected are:

- **Vibration switches.** A vibration switch designed to shut down a large fan might be configured in such a way that its failure is hidden. However, this only matters if the fan vibration rises above tolerable limits (a second failure), causing the fan bearings and possibly the fan itself to disintegrate (the consequences of the multiple failure).

- **Ultimate-level switches.** Ultimate-level switches are designed to activate an alarm or shut down equipment if a primary-level switch fails to operate. In other words, if an ultimate low-level switch jams, there are no consequences unless the primary switch also fails (the second failure), in which case the vessel or tank would run dry (the consequences of the multiple failure).

- **Fire hoses.** The failure of a fire hose has no direct consequences. It only matters if there is a fire (a second failure), when the failed hose may result in the place burning down and people being killed (the consequence of the multiple failure).

Other typical hidden functions include emergency medical equipment, most types of fire detection, fire warning and firefighting equipment, emergency stop buttons and trip wires, secondary containment structures, pressure and temperature switches, overload or overspeed protection devices, a standby plant, redundant structural components, overcurrent circuit breakers and fuses, and emergency power supply systems.

The Required Availability of Hidden Functions

So far, this part of this chapter has defined hidden failures and described the relationship between protective devices and hidden functions. The next question involves a closer look at the performance we require from hidden functions.

One of the most important conclusions that has been drawn so far is that the only direct consequence of a hidden failure is increased exposure to the risk of a multiple failure. Since it is the latter that we most wish to avoid, a key element of the performance required from a hidden function must be connected with the associated multiple failure.

We have seen that where a system is protected by a device that is not fail-safe, a multiple failure only occurs if the protected device fails while the protective device is in a failed state, as illustrated in Figure 7.21.

So the probability of a multiple failure in any period must be given by the probability that the protected function will fail while the protective device is in a failed state during the same period. This can be calculated as follows:

Probability of a multiple failure = probability of failure of the protected function × average availability of the protective device

The tolerable probability of the multiple failure is determined by the users of the system, as discussed in the next part of this chapter and again in Appendix IV. The probability of failure of the protected function is usually given (by its MTBF). So if these two variants are known, the allowed unavailability of the protective device can be expressed as follows:

$$\text{Allowed unavailability the protective device} = \frac{\text{probability of a multiple failure}}{\text{probability of failure of the protective device}}$$

So a crucial element of the performance required from any hidden function is the availability required to reduce the probability of the associated multiple failure to a tolerable level. The above discussion suggests that this availability is determined in the following three stages:

- First establish what probability the organization is prepared to tolerate for the multiple failure.
- Then determine the probability that the protected function will fail in the period under consideration (this is also known as the demand rate).
- Finally determine what availability the hidden function must achieve to reduce the probability of the multiple failure to the required level.

When calculating the risks associated with protected systems, there is sometimes a tendency to regard the probability of failure of the protected and protective devices as fixed. This leads to the belief that the only way to change the probability of a multiple failure is to change the hardware (in other words, to modify the system), perhaps by adding more protection or by replacing existing components with ones that are thought to be more reliable.

In fact, this belief is incorrect, *because it is usually possible to vary both the probability of failure of the protected function and (especially) the unavailability of the protective device* by adopting suitable maintenance and operating policies. As a result, *it is also possible to reduce the probability of the multiple failure to almost any desired level within reason by adopting such policies.* (Zero is, of course, an unattainable ideal.)

Calculating the Probability of a Multiple Failure

The probability that a protected function will fail in any period is the inverse of its mean time between failures, as illustrated in Figure 7.22*a*.

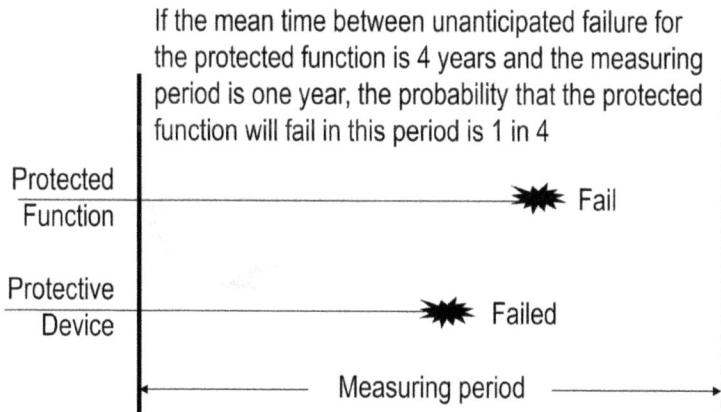

FIGURE 7.22A   Probability and protected functions

The probability that the protective device will be in a failed state at any time is given by the percentage of time that it is in a failed state. This is, of course, measured by its unavailability (also known as downtime or fractional dead time), as shown in Figure 7.22*b*.

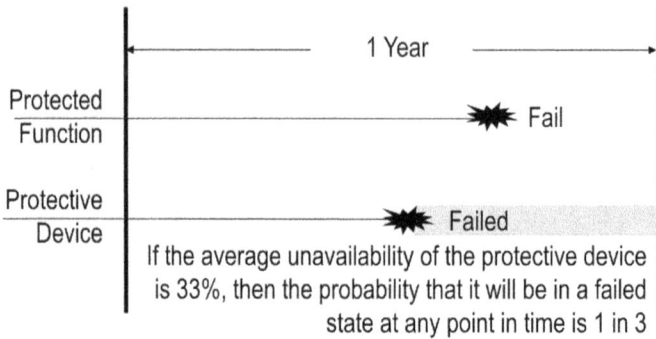

FIGURE 7.22B   Probability and protective devices

The probability of the multiple failure is calculated by multiplying the probability of failure of the protected function by the average unavailability of the protective device. For the case described in Figures 7.22*a* and 7.22*b*, the probability of a multiple failure would be as indicated in Figure 7.22*c*.

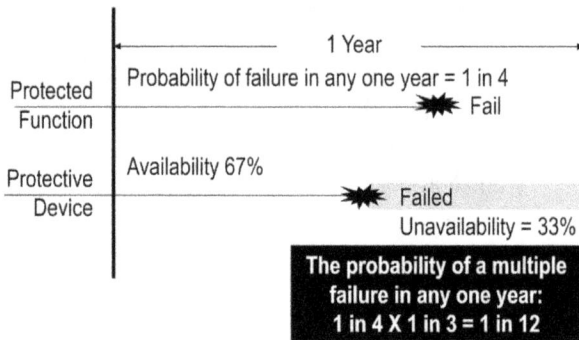

FIGURE 7.22C   Probability of a multiple failure

For example, the consequences of both pumps in Figure 7.18 being in a failed state may be such that the users are prepared to tolerate a probability of multiple failure of less than 1 in 1,000 in any one year (or 10^{-3}). Assume that it has also been estimated that if the duty pump is suitably maintained, the mean time between unanticipated failures of the duty pump can be increased to 10 years which corresponds to a probability of failure in any one year of 1 in 10, or 10^{-1}.

So to reduce the probability of the multiple failure to less than 10^{-3}, the unavailability of the standby pump must not be allowed to exceed 10^{-2}, or 1%. In other words, it must be maintained in such a way that its availability exceeds 99%. This is illustrated in Figure 7.23.

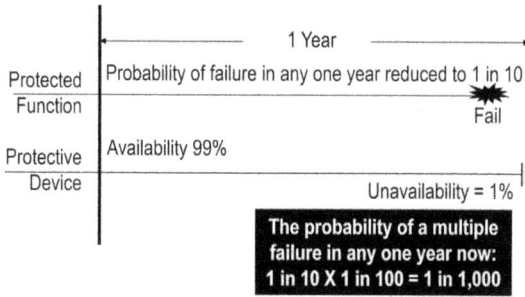

FIGURE 7.23   Desired availability of a protected device

In practice, the probability that is considered to be tolerable for any multiple failure depends on its consequences. In the vast majority of cases, the assessment has to be made by the users of the asset. These consequences vary hugely from system to system, so what is deemed to be tolerable varies equally widely. To illustrate this point, Figure 7.24 suggests four possible such assessments for four different systems.

Failure of Protected Function	Failure of Protective Device	Multiple Failure Consequence	Acceptable Rate of Multiple Failure
Spelling mistake in internal memo or office communication	Spell-checker in a word processing program unable to detect errors	Spelling mistake undetected	10 per month
10-hp motor on pump B becomes overloaded	Trip switch jams in closed position	Motor burns out: $1,500 to rewind	1 in 50 years
Duty pump (pump B) fails	Standby pump (pump C) fails	Total loss of pumping capability: $100,000 in lost production	1 in 100 years
Boiler overpressurizes	Relief valves jam shut	Boiler blows up: 10 people die	1 in 1,000,000 years

FIGURE 7.24   Multiple failure rates

As before, these levels of tolerability are not meant to be prescriptive and do not necessarily reflect the views of the author. The examples above are meant to demonstrate that in any protected system, someone must decide what is acceptable before it is possible to decide on the level of protection needed, and they are also meant to show that this assessment will differ for different systems. Part 3 of this chapter suggested that if the multiple failure could affect safety, the "someone" in the previous sentence should be a group that includes representatives of the likely victims together with their managers. This is also true of multiple failures that have economic consequences.

For instance, in the case of the spelling error, the "likely victim" is the author of the correspondence. In most organizations, the consequences are likely to be no more than mild embarrassment (if anyone even notices the error). In the case of the electric motor, the person most likely to be held accountable (in other words, the "likely victim") will be either the manager responsible for the maintenance budget or the maintenance manager. In the case of loss of pumping, the larger sums involved mean that higher levels of management should become involved in setting tolerability criteria.

Figure 7.24 also suggests that the probabilities that any organization might be prepared to tolerate for failures that have economic consequences tend to decrease as the magnitude of the consequences increases. This further suggests that it should be possible for any organization to develop a schedule of tolerable "standard" economic risks that could, in turn, be used to help develop maintenance programs designed to deliver those risks. This might take the form shown in Figure 7.25.

Yet again, please note that these levels of tolerability are not meant to be prescriptive and are not meant to be any kind of proposed universal standard. The economic risks that any organization is prepared to tolerate are quite literally that organization's business.

Figures 7.6 and 7.25 suggest that it might be possible to produce a schedule of risk that combines safety risks and economic risks in one continuum. How this might be done is discussed in Appendix IV.

In some cases, it may be unnecessary—indeed it is sometimes impossible to perform a rigorous quantitative analysis of the probability

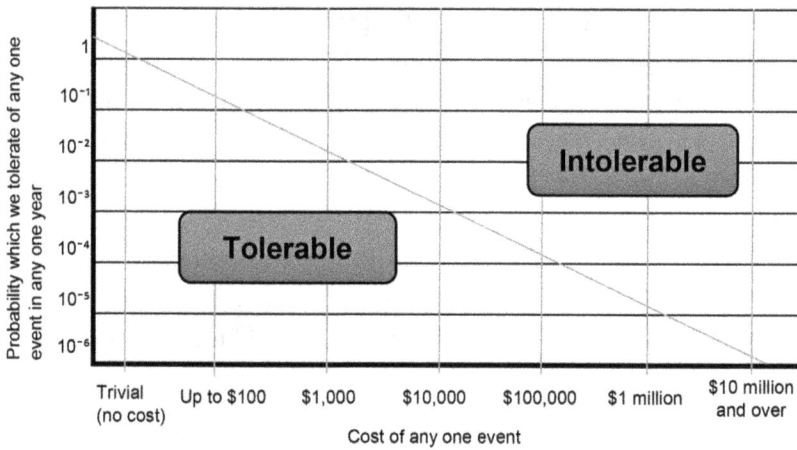

FIGURE 7.25 Tolerability of economic risk

of multiple failure in the manner described above. In such cases, it may be enough to make a judgment about the required availability of the protective device based on a qualitative assessment of the reliability of the protected function and the possible consequences of the multiple failure. This approach is discussed further in Chapter 8. However, if the multiple failure is particularly serious, then a rigorous analysis should be performed.

The following paragraphs consider in more detail how it is possible to influence:

- The rate at which protected functions fail
- The availability of protected devices

Routine Maintenance and Hidden Functions Posing a Physical Risk

Many of the functions in modern plants and facilities are protected by one or more protective devices. However, we have discussed how many of these devices fail hidden (not fail-safe), and because they are not fail-safe, the failure could result in catastrophic consequences when the protected function fails while the protective device is also in a failed state. In order to reduce or eliminate the consequences of the multiple failure, the author suggests increasing the reliability of the protected function first, therefore reducing the dependency on the protective device(s). Furthermore, it is much easier to manage evident failures. In a system that incorporates a non-fail-safe protective device, the probability of a multiple failure can be reduced as follows:

- Reduce the rate of failure of the protected function by:
  - Doing some sort of proactive maintenance
  - Changing the way in which the protected function is operated
  - Changing the design of the protected function
- Increase the availability of the protective device by:
  - Doing some sort of proactive maintenance
  - Checking periodically if the protective device has failed
  - Modifying the protective device

Prevent the Failure of the Protected Function. We have seen that the probability of a multiple failure is partly based on the rate of failure of the protected function. This could almost certainly be reduced by improving the maintenance or operation of the protected device, or even (as a last resort) by changing its design. For failures posing a physical risk, RCM will always consider proactive risk management strategies for making the evident function (protected function) more reliable and less likely to rely on the protection.

Specifically, if the failures of a protected function can be anticipated or prevented, the mean time between (unanticipated) failures of this function would be increased. This, in turn, would reduce the probability of the multiple failure.

For example, one way to prevent the simultaneous failure of pumps B and C is to try to prevent unanticipated failures of pump B. By reducing the number of these failures, the mean time between failures of pump B would be increased, and so the probability of the multiple failure would be correspondingly reduced, as shown in Figure 7.23.

However, bear in mind that the reason for installing a protective device is that the protected function is vulnerable to unanticipated failures with serious consequences.

Also, if no action is taken to prevent the failure of the protective device, it will inevitably fail at some stage and hence cease to provide any protection. *After this point, the probability of the multiple failure is equal to the probability of the protected function failing on its own.*

This situation must be intolerable, or a protective device would not have been installed to begin with. This suggests that we must at least try to find a practical way of preventing the failure of protective devices that are not fail-safe.

Prevent the Hidden Failure. In order to prevent a multiple failure, we must try to ensure that the hidden function is not in a failed state if and when the protected function fails. If a proactive task could be found that was good enough to ensure 100% availability of the protective device, then a multiple failure is theoretically almost impossible.

For example, if a proactive task could be found that could ensure 100% availability of pump C while it is in the standby state, then we can be sure that C would always take over if B failed. (In this case a multiple failure is only possible if the users operate pump C while pump B is being repaired or replaced. However, even then the risk of the multiple failure is low, because B should be repaired quickly and so the amount of time the organization is at risk is fairly short. Whether or not the organization is prepared to take the risk of running pump C while pump B is down depends on the consequences of the multiple failure and on whether it is possible to arrange other forms of protection, as discussed earlier.)

In practice, it is most unlikely that any proactive task would cause any function, hidden or otherwise, to achieve an availability of 100% indefinitely. What it must do, however, is deliver the availability needed to reduce the probability of the multiple failure to a tolerable level.

For example, assume that a proactive task is found that enables pump C to achieve an availability of 99%. If the mean time between unanticipated failures of pump B is 10 years, then the probability of the multiple failure would be 10^{-3} (1 in 1,000) in any one year, as discussed earlier. If the availability of pump C could be increased to 99.9%, then the probability of the multiple failure would be reduced to 10^{-4} (1 in 10,000), and so on.

So for a hidden failure, a proactive task is only worth doing if it secures the availability needed to reduce the probability of the multiple failure to a tolerable level.

> For hidden failures, a proactive task is worth doing if it secures the availability needed to reduce the probability of a multiple failure to a tolerable level.

The ways in which failures can be prevented are discussed in Chapters 8 and 10. However, these chapters also explain that it is often impossible to find a proactive task that secures the required availability. This applies especially to the type of equipment that suffers from hidden failures. So, if we cannot find a way to *prevent* a hidden failure, we must find some other way of improving the availability of the hidden function.

Detect the Hidden Failure. If it is not possible to find a suitable way of *preventing* a hidden failure, it is still possible to reduce the risk of the multiple failure by checking the hidden function periodically to find out if it is still working. If this check (called a "failure-finding task") is carried out at suitable intervals and if the function is rectified as soon as it is found to be faulty, it is still possible to secure high levels of availability. Scheduled failure-finding is discussed in detail in Chapter 8.

Combine Tasks. In the event a suitable failure-finding task cannot be found that would reduce the probability of a multiple failure to a tolerable level and the multiple failure has safety and environmental consequences, a combination of tasks will be considered. For a combination of tasks to be considered, both the tasks must be technically feasible and worth doing but not worth doing on their own.

Consider a One-Time Change. In a very small number of cases, either it is impossible to find any kind of routine task that secures the desired level of availability, or it is impractical to do it at the required frequency, and a combination of tasks will also not reduce the risk to a tolerable level. However, something must still be done to reduce the risk of the multiple failure to a tolerable level, so in these cases, it is usually necessary to go back to the drawing board and reconsider the design. If the multiple failure could affect safety or the environment, a one-time change (redesign) is compulsory. If the multiple failure only has economic consequences, the need for redesign is assessed on economic grounds.

Ways in which redesign can be used to reduce the risk or to change the consequences of a multiple failure are discussed in Chapter 8.

Routine Maintenance and Hidden Functions Posing Operational Risk

Similarly, hidden functions posing an operational risk should also be managed appropriately in order to reduce the economic impact of the multiple failure. In the case where the multiple failure only has economic consequences, we have to distinguish between two things:

- First, consider the cost of the multiple failure:
  - How much the multiple failure will cost when it occurs
  - How often the multiple failure would occur (based on the MTBF of the protected function and failure rate of the protective device)
- Second, consider the cost of doing the failure-finding task:
  - How much the failure-finding task will cost
  - How often the task must be done

The most economic task interval is where the sum of the annualized cost of the multiple failure and the cost of the failure-finding task is a minimum, as can be seen in Figure 7.26. If the most economic failure-finding task will not reduce the probability of the multiple failure to a tolerable level, the need for a one-time change (redesign) is assessed on economic grounds.

Hidden economic risk

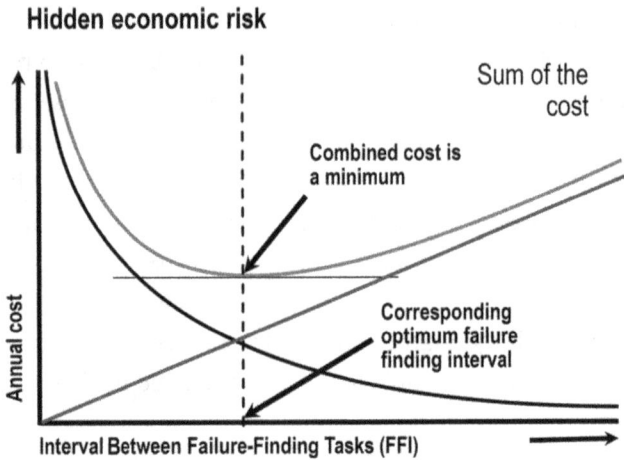

FIGURE 7.26 Failure-finding tasks for hidden functions with economic consequences

Hidden Functions: The Decision Process

All the points made so far about the development of a maintenance strategy for hidden functions can be summarized as shown in Figures 7.27 and 7.28.

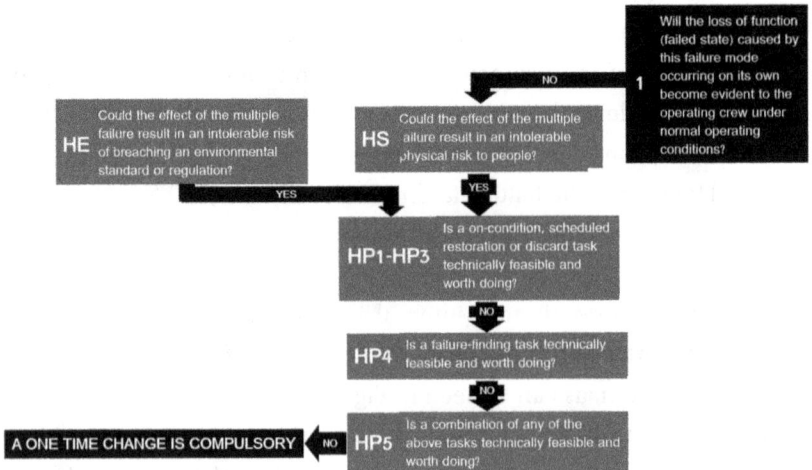

FIGURE 7.27 Decision diagram for hidden functions with safety and environmental consequences

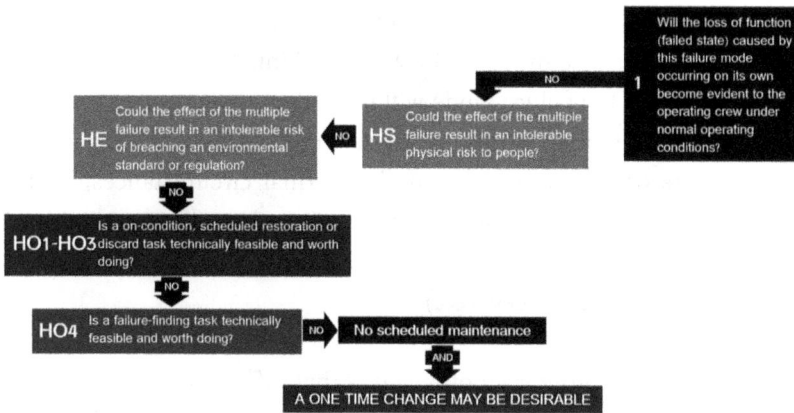

FIGURE 7.28   Decision diagram tasks for hidden functions with economic consequences

Further Points about Hidden Functions

Six issues need special care when asking the first question in Figure 10.1:

- Failed states and failure modes
- The question of time
- The primary and secondary functions of protective devices
- The operating crew
- Normal circumstances
- Fail-safe devices

These are all discussed in more detail in the following paragraphs.

Failed States and Failure Modes.   At this stage in the RCM process, every failure mode that is reasonably likely to cause each failed state will already have been identified on the RCM3 Information Worksheet. This has two key implications:

- First, we are not asking what failures could occur. All we are trying to establish is whether each failure mode that has already been identified as a possibility would be hidden or evident if it did occur.

- Second, we are not asking whether the operating crew can diagnose the failure mode itself. We are asking if the loss of function caused by the failure mode will be evident under normal circumstances. (In other words, we are asking if the failure mode has any effects or symptoms that, under normal circumstances, would lead the observer to believe that the item is no longer capable of fulfilling its intended function—or at least that something out of the ordinary had occurred.)

For example, consider a motor vehicle that suffers from a blocked fuel line. The average driver (in other words, the average "operator") would not be able to diagnose this failure mode without expert assistance, so there might be a temptation to call this a hidden failure. However, the loss of the function caused by this failure mode is evident, because the car stops working.

The Question of Time. There is often a temptation to describe a failure as "hidden" if a considerable period of time elapses between the moment the failure occurs and the moment it is discovered. In fact, this is not the case. If the loss of function eventually becomes apparent to the operators, and it does so as a direct and inevitable result of this failure on its own, then the failure is treated as evident, no matter how much time elapses between the failure in question and its discovery.

For example, a tank fed by pump A in Figure 7.15 may take weeks to empty, so the failure of this pump might not be apparent as soon as it occurs. This might lead to the temptation to describe the failure as hidden. However, this is not so because the tank runs dry as a direct and inevitable result of the failure of pump A on its own. Therefore, the fact that pump A is in a failed state will inevitably become evident to the operating crew.

Conversely, the failure of pump C in Figure 7.18 will only become evident if pump B also fails (unless someone makes a point of checking pump C from time to time). If pump B were to be operated and maintained in such a way that it is never necessary to switch on pump C, it is possible that the failure of pump C on its own would never be discovered.

This example demonstrates that time is not an issue when considering hidden failures. We are simply asking whether anyone will eventually become aware of the fact that the failure has occurred on its own, and not if anyone will become aware of it when it occurs.

Primary and Secondary Functions of Protective Devices. Thus far we have focused on the primary function of protective devices, which is to be capable of fulfilling the function they are designed to fulfill when called upon to do so. As we have seen, this is usually after the protected function has failed. However, an important secondary function of many of these devices is that they should not work when nothing is wrong.

For instance, the primary function of a pressure switch might be listed as follows:

- To be capable of transmitting a signal when pressure falls below 250 psi

The implied secondary function of this switch is:

- To be incapable of transmitting a signal when pressure is above 250 psi

The failure of the first function is hidden, but the failure of the second is evident, because if it occurs, the switch transmits a spurious shutdown signal and the machine stops. If this is likely to occur in practice, it should be listed as a failure mode of the function that is interrupted (usually the primary function of the machine). As a result, there is usually no need to list the implied second function separately, but the failure mode would be listed under the relevant function if it is reasonably likely to occur.

The Operating Crew. When asking whether a failure is evident, the term "operating crew" refers to anyone who has occasion to observe the equipment or what it is doing at any time in the course of normal daily activities, and who can be relied upon to report that it has failed.

Failures can be observed by people with many different points of view. These people include operators, drivers, quality inspectors, craftspeople, supervisors, and even the tenants of buildings. However, whether any of these people can be relied upon to detect and report a failure depends on four critical elements:

- The observer must be in a position either to detect the failure mode itself or to detect the loss of function caused by the failure mode. This may be a physical location or access to equipment or information (including management information) that will draw attention to the fact that something is wrong.
- The observer must be able to recognize the condition as a failure.
- The observer must understand and accept that it is part of his or her job to report failures.
- The observer must have access to a procedure for reporting failures.

Normal Circumstances. Careful analysis often reveals that many of the duties performed by operators are actually maintenance tasks. It is wise to start from a zero base when considering these tasks, because it may transpire that either the tasks or their frequencies need to be radically revised. In other words, when asking if a failure will become evident to the operating crew under "normal" circumstances, the word "normal" has the following meanings:

- That nothing is being done to prevent the failure. If a proactive task is currently successfully preventing the failure, it could be argued that the failure is "hidden" because it does not occur. However, in Chapter 6 it was pointed out that failure modes and effects should be listed, and the rest of the RCM process should be applied as if no proactive tasks are being done, because one of the main purposes of the exercise is to review whether we should be doing any such tasks in the first place.

- That no specific task is being done to detect the failure. A surprising number of tasks that already form part of an operator's normal duties are, in fact, routines designed to check if hidden functions are working. For example, pressing a button on a control panel every day to check if all the alarm lights on the panel are working is a failure-finding task.

  We shall see later that failure-finding tasks are covered by the RCM task selection process, so once again it should be assumed at this stage in the analysis that this task is not being done (even though the task is currently genuinely part of the operator's normal duties). This is because the RCM process might reveal a more effective task, or it might reveal the need to do the same task at a higher or lower frequency.

(Quite apart from the question of maintenance tasks, there is often considerable doubt about what the "normal" duties of the operating crew actually are. This occurs most often where standard operating procedures either are poorly documented or do not exist. In these cases, the RCM review process does much to help clarify what these duties should be and can do much to help lay the foundations of a full set of operating procedures. This applies especially to high-technology plants.)

Fail-Safe Devices. It often happens that a protective circuit is said to be fail-safe when it is not. This usually occurs when only part of a circuit is considered instead of the circuit as a whole.

An example is again provided by a pressure switch, this time attached to a hydrostatic bearing. The switch was meant to shut down the machine if the oil pressure in the bearing fell below a certain level. It emerged during discussion that if the electrical signal from the switch to the control panel was interrupted, the machine would shut down, so the failure of the switch was initially judged to be evident.

However, further discussion revealed that a diaphragm inside the switch could deteriorate with age, so the switch could become incapable of sensing changes in the pressure. This failure was hidden, and the maintenance program for the switch was developed accordingly.

To avoid this problem, take care to include the sensors and the actuators in the analysis of any control loop, as well as the electric circuit itself.

Protective devices that are not fail-safe are classified as hidden functions. There is no consequence if the protective device fails on its own (other than increasing the risk of another failure). Failure of a protective device on demand resulting in a multiple failure that poses a threat to the safety of people, or damage to the environment, is treated as an "intolerable" risk. Therefore, risk management policies for protective devices should ensure the required availability of these devices to satisfy the users and society as a whole.

Layers of Protection

Much evaluation work, including a hazard and risk assessment, has to be performed by the customer to identify the overall risk reduction requirements and to allocate these to independent protection layers (IPLs). No single safety measure can eliminate risk and protect a plant and its personnel against harm or mitigate the spread of harm if a hazardous incident occurs. For this reason, safety exists in protective layers: a sequence of mechanical devices, process controls, shutdown systems, and external response measures that prevent or mitigate a hazardous event. If one protection layer fails, successive layers will be available to take the process to a safe state. If one of the protection layers is a safety instrumented function (SIF), the risk reduction allocated to it determines its safety integrity level (SIL). As the number of protection layers and their reliabilities increase, the safety of the process increases. Figure 7.29 illustrates layers of protection

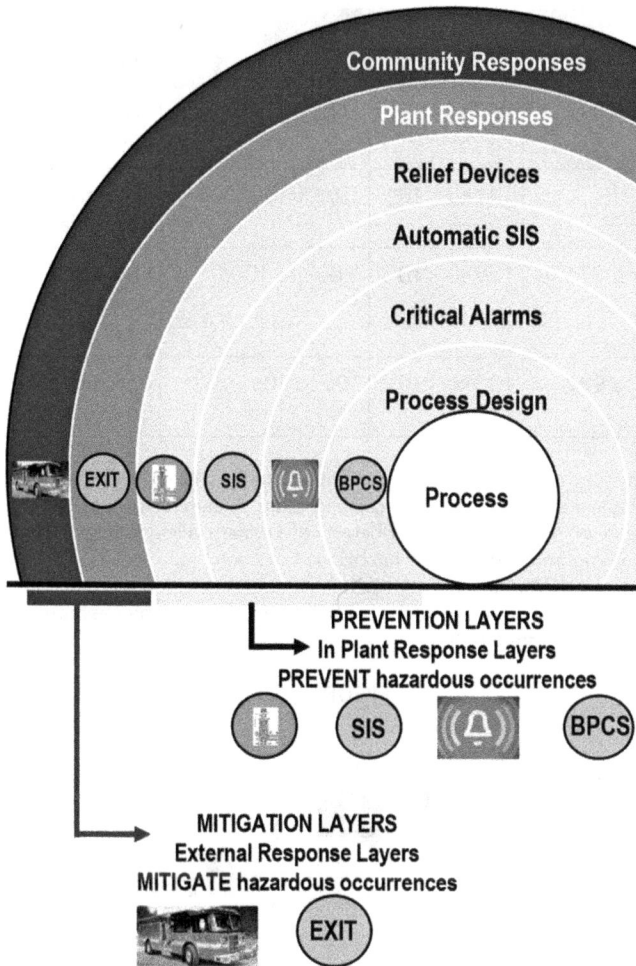

FIGURE 7.29 Layers of protection

7.9 Conclusion

This chapter has demonstrated how the RCM process provides a comprehensive strategic framework for managing failures. As summarized in Figure 7.31, this framework:

- Classifies all failures on the basis of their consequences and associated risk. In doing so, it separates hidden failures from evident

SIL	Availability*	PFD$_{avg}$†	Risk Reduction	Qualitative Consequence
4‡	>99.99%	10^{-5} to 10^{-4}	100,000 to 10,000	Potential for fatalities in the community
3	99.9%	10^{-4} to $<10^{-3}$	10,000 to 1,000	Potential for multiple on-site fatalities
2	99 to 99.9%	10^{-3} to $<10^{-2}$	1,000 to 100	Potential for major on-site injuries or a fatality
1	90 to 99%	10^{-2} to $<10^{-1}$	100 to 10	Potential for minor on-site injuries

*Availability: The probability that equipment will perform its task.
†PFD$_{avg}$: The average PFD used in calculating safety system reliability. (PFD—the probability of failure on demand—is the probability of a system failing to respond to a demand for action arising from a potentially hazardous condition.)
‡Both IEC and ANSI/ISA standards utilize similar tables covering the same range of PFD values. ANSI/ISA, however, does not show a SIL 4. *No standard process controls have yet been defined and tested for SIL 4.*

FIGURE 7.30

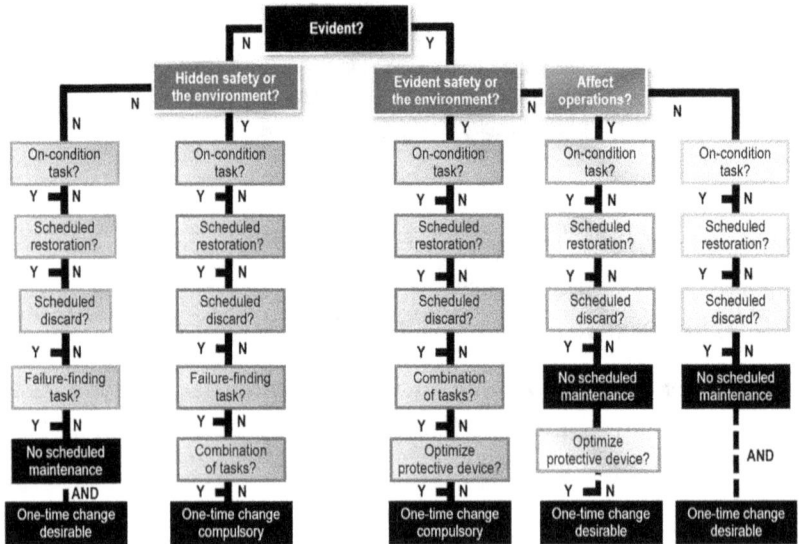

FIGURE 7.31   Summary of the RCM3 decision diagram

failures, separates hidden functions with safety and environmental consequences (physical risks) from hidden functions with economic consequences, and then ranks the consequences of the evident failures in descending order of importance, safety being the highest priority.

- Provides a basis for deciding whether proactive maintenance is worth doing in each case.
- Suggests what action should be taken if a suitable proactive task cannot be found.

The different types of proactive tasks and default actions are discussed in the next four chapters, together with an integrated approach to risk evaluation and task selection.

CHAPTER 8

Risk Management Strategies— Technical Feasibility

8.1 Proactive Strategies and Technical Feasibility

We have seen in Chapter 2 that risk is divided into the following two categories:

- **Intolerable risk.** The converse of acceptable risk as defined by the organization's risk management policy. If safety is defined by freedom from unacceptable risk, an unacceptable risk is therefore not safe.
- **Tolerable risk.** A level of risk deemed acceptable by the user and society in order that some particular benefit or functionality can be obtained, but in the knowledge that the risk has been evaluated and is being managed.

In order to answer the seventh and eighth questions of the eight questions that make up the RCM3 process, each failure mode must be evaluated to determine the most cost-effective risk management strategy that is both worth doing and technically feasible. The seventh and eighth questions are:

- What *must* be done to reduce intolerable risk to a tolerable level (using proactive risk management strategies)?
- What *can* be done to reduce tolerable risk in a cost-effective way?

This chapter focuses on the risk management strategies for *intolerable risks* and the technical feasibility of these strategies in the RCM3 process. The risk management strategies for intolerable risks will be discussed first. The discussion deals with the criteria used to decide what risk management strategies are technically feasible. As well, the discussion looks in more detail at how we decide whether specific categories of tasks (risk management strategies) are worth doing, as discussed in previous chapters. (The next chapter reviews actions dealing with tolerable risks.)

Proactive risk management strategies are actions undertaken to reduce or eliminate risks before they occur. We have seen that in order to reduce or mitigate a risk, the analyst has three options:

- Reduce the probability of the event happening.
- Reduce the consequence severity of the event.
- Or both . . .

In terms of physical asset management, it is possible to reduce the probability of the event happening through preventing the item from getting into a failed state. This is possible through addressing design integrity (defect elimination) and proactive maintenance (failure management strategies). Proactive failure management strategies embrace what is traditionally known as "predictive" and "preventive" maintenance, although RCM uses the terms "scheduled restoration," "scheduled discard," and "on-condition maintenance." Risk management strategies related to proactive maintenance will be discussed first.

8.2 Treating Intolerable Risks—Proactive Failure Management Strategies

When we ask whether a proactive task is technically feasible, we are simply asking whether it is possible for the task to prevent or anticipate the failure in question. This has nothing to do with economics—economics are part of the "worth-doing" evaluation process that has already been considered at length in the previous chapter. Instead, technical feasibility depends on the technical characteristics of the failure mode and of the task itself

A task or engineering solution is technically feasible if it makes it physically possible to reduce, or enable action to be taken to reduce, the risk associated with the failure mode to an extent that might be tolerable to the owner or user of the asset.

Two issues dominate proactive task selection from the technical viewpoint:

- The relationship between the age of the item under consideration and how likely it is to fail
- What happens once a failure has started to occur

RCM3 takes account of the fact that the conditional probability of some failure modes will increase with age (or exposure to stress), that the conditional probability of others will not change with age, and that the conditional probability of others will actually decrease with age.

Figure 8.1 summarizes the proactive tasks that will be discussed in this chapter.

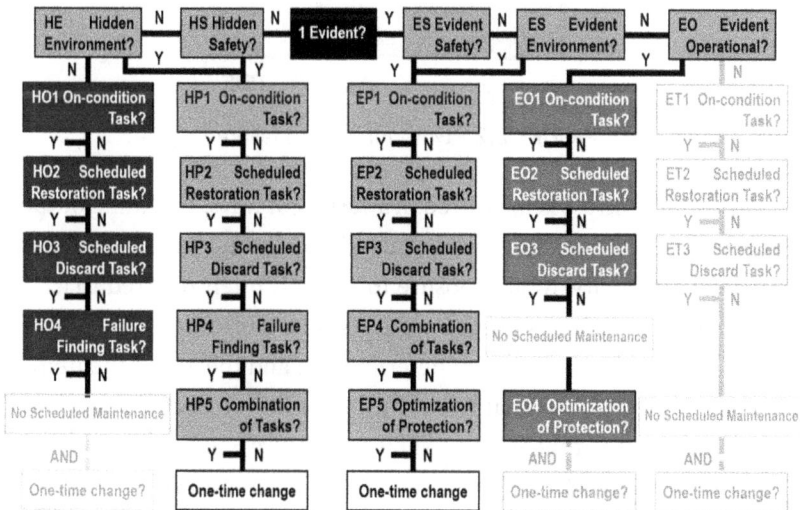

FIGURE 8.1   RCM3 decision diagram and proactive risk management strategies for intolerable risks

8.3 Age and Deterioration

Any physical asset that fulfills a function that brings it into contact with the real world will be subjected to a variety of stresses. These stresses cause the asset to deteriorate by lowering its resistance to stress. Eventually this resistance drops to the point at which the asset can no longer deliver the desired performance—in other words, it fails. This process is illustrated in Figure 8.2.

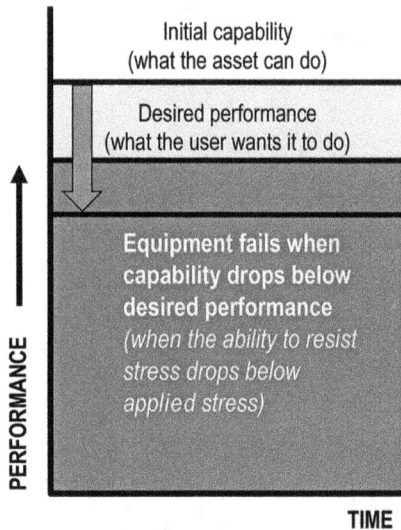

FIGURE 8.2   Deterioration to failure

Exposure to stress is measured in a variety of ways, including output, distance traveled, operating cycles, calendar time, and running time. For the sake of simplicity, total exposure to stress is often referred to as the *age* of the item. Common sense suggests that there should be a direct relationship between the rate of deterioration and the age of the item. If this is so, then it follows that the point where failure occurs should also depend on the age of the item, as shown in Figure 8.3.

However, Figure 8.2 is based on two key assumptions:

- Deterioration is directly proportional to the applied stress.
- The stress is applied consistently.

If this were true of all assets, we would be able to predict equipment life with great precision. The classical view of preventive maintenance suggests that this can be done—all we need is enough information about failures.

In reality, however, the situation is much less clear-cut. This chapter first looks at the real world by considering a situation where there is a clear relationship between age and failure, and then the chapter moves on to a more general view of reality.

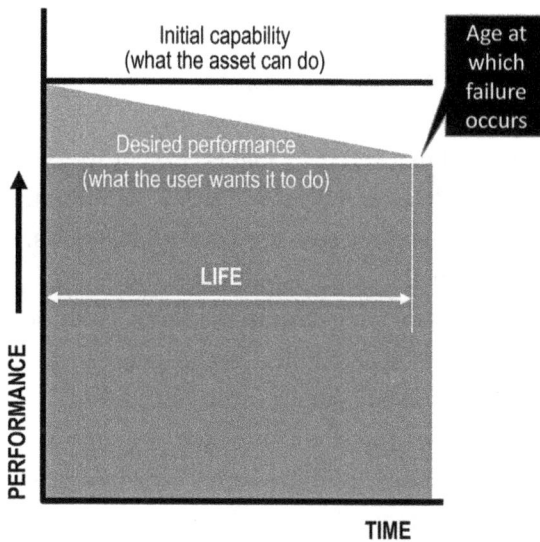

FIGURE 8.3   Absolute predictability

Age-Related Failures

Even parts that seem to be identical vary slightly in their initial resistance to failure. The rate at which this resistance declines with age also varies. Furthermore, no two parts are subject to exactly the same stresses throughout their lives. Even when these variations are quite small, they can have a disproportionate effect on the age at which the part fails. This is illustrated in Figure 8.4, which shows what happens to two components that are put into service with similar resistance to failure.

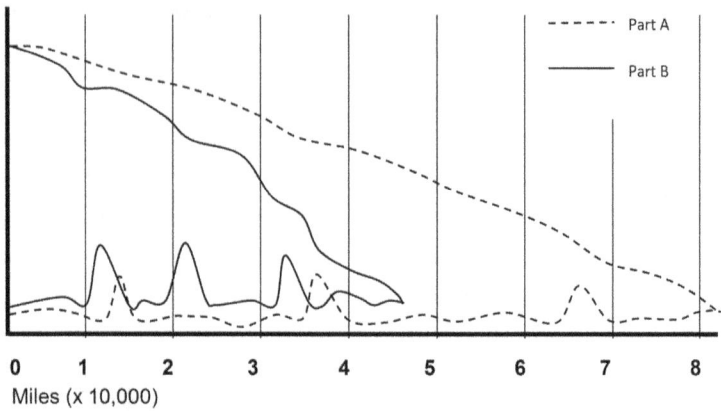

FIGURE 8.4 A realistic view of age-related failures

Part B is exposed to a generally higher level of stress throughout its life than part A, so it deteriorates more quickly. Deterioration also accelerates in response to the three stress peaks at 13,000 miles, 22,000 miles, and 33,000 miles. On the other hand, for some reason part A seems to deteriorate at a steady pace despite the three stress peaks at 15,000 miles, 35,000 miles, and 65,000 miles. So one component fails at 46,000 miles and the other at 82,000 miles.

This example shows that the failure age of identical parts working under apparently identical conditions varies widely. In practice, although some parts last much longer than others, the failures of a large number of parts that deteriorate in this fashion would tend to congregate around some average life, as shown in Figure 8.5.

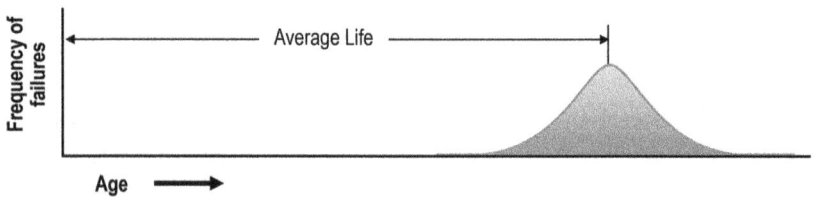

FIGURE 8.5 Frequency of failure and "average life"

So even when resistance to failure does decline with age, the point at which failure occurs is often much less predictable than common sense suggests. It can be shown that the failure frequency curve depicted in Figure 8.5 can be drawn as a conditional probability of failure curve, as shown in Figure 8.6. (The term "useful life" defines the age at which there is a rapid increase in the conditional probability of failure. It is used to distinguish this age from the average life shown in Figure 8.5.)

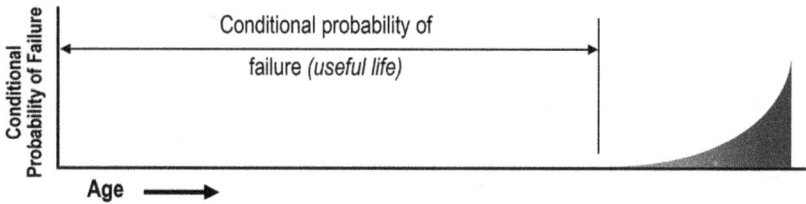

FIGURE 8.6   Conditional probability of failure and "useful life"

If large numbers of apparently identical age-related failure modes are analyzed in this fashion, it is not unusual to find a number that occurs prematurely. The result of such premature failures is a conditional probability curve, as shown in Figure 8.7. This is the same as failure pattern B.

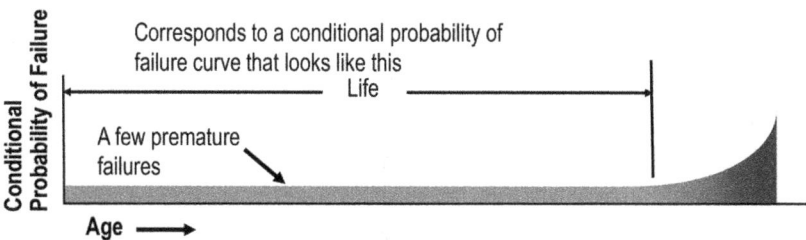

FIGURE 8.7   The effect of premature failures

Even this is actually a somewhat simplistic view of age-related failures, because there are, in fact, three sets of ways in which the probability of failure can increase as an item gets older. These are shown in Figure 8.8.

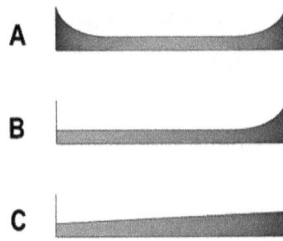

FIGURE 8.8   Failures that are age-related

The patterns in Figure 8.8 were introduced in Chapter 2 and are discussed at much greater length in Chapter 12. The characteristic shared by patterns A and B is that they both display a point at which there is a rapid increase in the conditional probability of failure. Pattern C shows a steady increase in the probability of failure, but no distinct wear-out zone. The next three sections of this chapter consider the implications of these failure patterns from the viewpoint of preventive maintenance.

8.4   Age-Related Failures and Preventive Maintenance

For centuries—certainly since machines have come into widespread use—people have tended to believe that most equipment will behave as shown in Figures 8.5 to 8.7. In other words, most people still tend to assume that similar items performing a similar duty will perform reliably for a period, perhaps with a small number of random early failures, and then most of the items will wear out at about the same time.

In general, age-related failure patterns apply to items that are very simple, or to complex items that suffer from a dominant failure mode. In practice, they are commonly found under conditions of direct wear (most often where equipment comes into direct contact with the product). They are also associated with fatigue, corrosion, oxidation, and evaporation.

> Wear-out characteristics most often occur where equipment comes into direct contact with the product. Age-related failures also tend to be associated with fatigue, oxidation, corrosion, abrasion, and evaporation.

Examples of points where equipment comes into contact with the product include furnace refractories, pump impellers, valve seats, seals, machine tooling, screw conveyors, crusher and hopper liners, the inner surfaces of pipelines, dies, and so on.

Fatigue affects items—especially metallic items—that are subjected to reasonably high-frequency cyclic loads. The rate and extent to which oxidation and corrosion affect any item depend, of course, on the item's chemical composition, the extent to which it is protected, and the environment in which it is operating. Evaporation affects solvents and the lighter fractions of petrochemical products.

Under certain circumstances, two preventive options that are available for reducing the incidence of failure modes like these are scheduled restoration tasks and scheduled discard tasks. These two categories of tasks are considered in more detail in the next section of this chapter.

8.5 Scheduled Restoration and Scheduled Discard Tasks

Failure modes that conform to pattern A or B in Figure 8.8 become more likely to occur after the end of the useful life, as shown in Figure 8.7. If an item or component is one of those that survives to the end of this life, it is possible to remove it from service before it enters the wear-out zone and to take some sort of action either to prevent it from failing or at least to reduce the consequences of the failure. Sometimes, this action entails doing something to restore the initial capability of the item or component that has been removed. If this is done at fixed intervals without attempting to assess the condition of the item or component concerned before subjecting it to the restoration process, the action is known as "scheduled restoration."

Specifically, scheduled restoration entails restoring the initial capability of an existing item to specified standards, remanufacturing a single component, or overhauling an entire assembly at or before a specified age limit, regardless of its apparent condition at the time.

Scheduled restoration tasks also used to be known as "scheduled rework tasks." They include overhauls or turnarounds that are performed at preset intervals in order to prevent specific age-related failure modes.

In the case of some failure modes that are age-related, it is simply impossible to restore anything like the initial capability of the affected item or component once it has reached the end of its useful life. In these cases, initial capability can only be restored by discarding the item and replacing it with a new one. In other cases, scheduled restoration of the existing item may be technically possible, but it is much more cost-effective to replace it with a new one. In both cases, if the item or component is replaced with a new one at fixed intervals without attempting to assess the condition of the old one beforehand, the task is known as "scheduled discard."

> Scheduled discard tasks entail discarding an item or component at or before a specified age limit, regardless of its condition at the time.

Note that the terms "scheduled restoration" and "scheduled discard" can often be applied to exactly the same task, and which term is appropriate is a function of the level at which the analysis is being performed.

For instance, if a pump impeller wears out at a predictable rate and so is replaced with a new one at fixed intervals, the replacement task could be described as scheduled discard of the impeller or scheduled restoration of the pump.

This is why we tend to consider scheduled restoration and scheduled discard together. However, the distinction does become important when considering a failure mode that could be prevented by either of the two tasks when considered at the same level of analysis.

For instance, a certain type of electric motor may be known to suffer from failure of the windings after a predictable amount of time in service. In this case, it may be possible to restore initial capability by rewiring the motor (scheduled restoration) or by replacing it with a new one (scheduled discard).

For this reason, the remainder of this section considers the common features of scheduled restoration and scheduled discard together, but it also takes care to highlight key differences.

The Frequency of Scheduled Restoration and Scheduled Discard Tasks

If the failure mode under consideration conforms to pattern A or B, it is possible to identify the age at which wear-out begins. The scheduled restoration task is done at intervals slightly less than this age. In other words:

> The frequency of a scheduled restoration and scheduled discard task is governed by the age at which the item or component shows a rapid increase in the conditional probability of failure.

In the case of pattern C, at least four different restoration intervals need to be analyzed to determine the optimum interval (if one exists at all). In practice, the frequency of a scheduled restoration task can only be determined satisfactorily on the basis of reliable historical data. This is seldom available when assets first go into service, so it is usually impossible to specify scheduled restoration tasks in prior-to-service maintenance programs. However, items subject to very expensive failure modes should be put into age-exploration programs as soon as possible to find out if they would benefit from scheduled restoration tasks.

In general, there is a particularly widely held belief that all items "have a life," and that overhauling the item or installing a new part before this "life" is reached will automatically make it safe. This is not always true, so RCM takes special care to focus on safety when considering scheduled restoration and scheduled discard tasks.

In fact, RCM recognizes two different types of life limits when dealing with these tasks. Life limits of the first type apply to tasks meant to avoid failures that have safety consequences and are called "safe-life limits." Life limits of the second type apply to tasks that are intended to prevent failures that do not have safety consequences and are called "economic-life limits."

Safe-Life Limits. Safe-life limits only apply to failures that have safety or environmental consequences, so the associated tasks must reduce the probability of a failure occurring before the end of the life limit to a tolerable level. (A method of deciding what is tolerable is discussed in Chapter 7, Section 7.1, and Appendix III of this book. Probabilities as low as 10^{-6} and even 10^{-9} are sometimes used in this context.) This means that safe-life limits cannot apply to items that conform to pattern A, because start-up failures mean that a significant number of items must fail prematurely. In fact, they cannot apply to any failure mode where there is a significant probability of the failure occurring when the item enters service.

Ideally, safe-life limits should be determined before the item is put into service. The item should be tested in a simulated operating environment to determine what life is actually achieved, and a conservative fraction of this life should be used as the safe-life limit. This is illustrated in Figure 8.9.

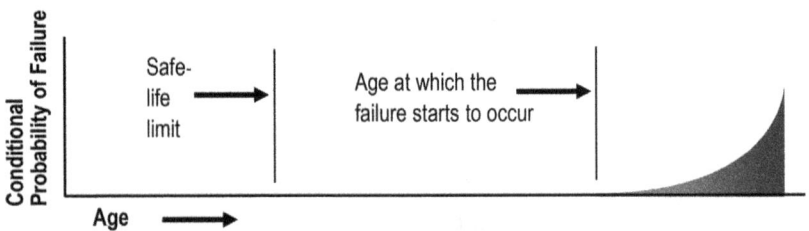

FIGURE 8.9   Safe-life limits

There is never a perfect correlation between a test environment and the operating environment. Testing a long-lived part to failure is also costly and obviously takes a long time, so there is usually not enough test data for survival curves to be drawn with confidence. In these cases safe-life limits are sometimes established by dividing the average by an arbitrary factor as large as 3 or 4. This implies that the conditional probability of failure at the life limit should essentially be zero.

Economic-Life Limits. Operating experience sometimes suggests that scheduled restoration or scheduled discard is desirable on economic grounds. The associated life limit is known as an economic-life limit. It is usually equal to the useful life, rather than a fraction of this life. The economics of scheduled restoration and scheduled discard are discussed in more detail later in this section.

The Technical Feasibility of Scheduled Restoration

The above comments indicate that for a scheduled restoration task to be technically feasible, the criteria that must be satisfied are that:

- There must be a point at which there is an increase in the conditional probability of failure (in other words, the item must have a life).
- We must be reasonably sure what the life is.

As well, most of the items must survive to this age. If too many items fail before reaching it, the net result is an increase in unanticipated failures. Not only could this have unacceptable consequences, but it means that the associated restoration tasks are done out of sequence. This, in turn, disrupts the entire schedule planning process.

Note that if the failure has safety or environmental consequences, *all* the items must survive to the age at which the scheduled restoration task is to be done, because we cannot risk failures that might hurt people or damage the environment. (In this situation, "all" means that a large enough percentage—preferably 100%—of the items must last longer than the safe-life limit. The safe-life limit is usually some fraction of the useful life.) In this context, the comments about safe-life limits that were made in the previous part of this chapter apply equally to scheduled restoration tasks.

Finally, scheduled restoration must restore the original resistance to failure of the asset, or at least something close enough to the original condition (initial capability) to ensure that the item continues to be able to fulfill its intended function for a reasonable period of time that should ideally be equal to the original useful life limit.

For example, no one in his or her right mind would try to overhaul a domestic light bulb, simply because it is not possible to restore it to its original condition (regardless of the economics of the matter). On the other hand, it could be argued that retreading a truck tire restores the tread to something approaching its original condition.

These points lead to the following general conclusions about the technical feasibility of scheduled restoration:

Scheduled restoration tasks are technically feasible if:

- There is an identifiable age at which the item shows a rapid increase in the conditional probability of failure.
- Most of the items survive to that age (all the items if the failure has safety or environmental consequences).
- They restore the original resistance to failure of the item.

The Technical Feasibility of Scheduled Discard Tasks

The above comments indicate that scheduled discard tasks are technically feasible under the following circumstances:

Scheduled discard tasks are technically feasible if:

- There is an identifiable age at which the item shows a rapid increase in the conditional probability of failure.
- Most of the items survive to that age (all the items if the failure has safety or environmental consequences).

There is no need to ask if the task will restore the original condition because the item is replaced with a new one. The RCM3 process will always favor *scheduled restoration* tasks ahead of *scheduled discard* tasks unless the discard task is clearly more effective.

For example, it may be possible to overhaul a small gearbox that drives an electric actuated valve, but it will probably be less expensive to replace the gearbox with an in-kind replacement. This may also be true for most electric motors below a certain size.

The Effectiveness of Scheduled Restoration and Scheduled Discard Tasks

Even if it is technically feasible, scheduled restoration and scheduled discard might still not be worth doing, because other tasks may deal with the failure consequences even more effectively, as explained in Chapter 7.

If a more effective task cannot be found, there is often a temptation to select scheduled restoration or scheduled discard purely on the grounds of technical feasibility. An age limit applied to an item that behaves as shown in Figure 8.7 means that some items will receive attention before they need it while others might fail early. But the net effect may be an overall reduction in the number of unanticipated failures. However, even then, scheduled restoration or discard might not be worth doing, because, as mentioned earlier, a reduction in the number of failures is not sufficient if the failure has *safety or environmental consequences*. This is so, because to be *worth doing*, the task should reduce the probability of failures that have these consequences to a vanishingly low level (effectively zero).

On the other hand, if the consequences are economic, we need to be sure that over a period of time, the cost of doing the scheduled restoration task or scheduled discard task will be less than the cost of allowing the failure to occur. In other words, the only justification for an economic-life limit is cost-effectiveness. This is so because scheduled restoration increases the number of jobs passing through the workshop, while scheduled discard increases the consumption of the items or components that are subject to discard. Why this is so is illustrated by the example that follows.

When considering failures that have operational consequences, bear in mind also that scheduled restoration and scheduled discard may themselves also affect operations. However, in most cases, this effect is likely to be less than the consequences of the failure because:

- The task would normally be done when it is likely to have the least effect on operations (usually during a so-called production window).
- The task is likely to take less time than it would to repair the failure because it is possible to plan more thoroughly for the scheduled task.

Figure 8.10 shows an age-related failure mode where the useful life is 12 months, while the average life is 18 months. In a period of 3 years, the failure occurs twice if no preventive maintenance is done, while the preventive task would be done three times. In other words, the preventive task has to be done 50% more often than the corrective task that would have to be performed if the failure were allowed to occur on its own.

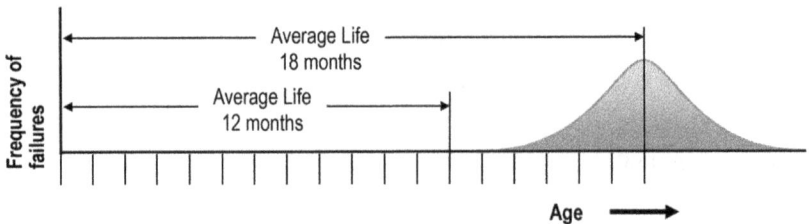

FIGURE 8.10 Useful life and average life

If each failure costs $2,000 in lost production and repair, failures would cost $4,000 over a 3-year period. If the total cost of each preventive task is $1,100, these tasks would cost $3,300 over the same period. So in this case, the task is cost-effective.

On the other hand, if the average life was 24 months and all other figures remained the same, failures only occur 1.5 times every 3 years and would cost $3,000 over this period. The scheduled task still costs $3,300 over the same period, so it would not be cost-effective.

If there are no operational consequences, scheduled restoration and scheduled discard are only justified if they cost substantially less than the cost of repair (which may be the case if the failure causes extensive secondary damage).

All of this means that, in general, an economic-life limit is worth applying if it avoids or reduces the operational consequences of an unanticipated failure, and/or if the failure that it prevents causes significant secondary damage. Clearly, we must know the failure pattern before we can assess the cost-effectiveness of scheduled discard tasks.

For new assets, this means that a failure mode that has major economic consequences should also be put into an age-exploration program to find out if a life limit is applicable. However, there is seldom enough evidence to include scheduled restoration or scheduled discard in an initial scheduled maintenance program. In practice, the frequency of these tasks can only be determined satisfactorily on the basis of reliable historical data. This is seldom available when assets first go into service, so it is usually impossible to specify scheduled restoration or scheduled discard in prior-to-service maintenance programs. (For example, scheduled restoration tasks were only assigned to seven components in the initial program developed for the Douglas DC 10.) However, items subject to very expensive failure modes should be put into age-exploration programs as soon as possible to find out if they would benefit from these tasks.

8.6 Failures That Are Not Age-Related

One of the most challenging developments in modern maintenance management has been the discovery that very few failure modes actually conform to any of the failure patterns shown in Figure 8.7. As discussed in the following paragraphs, this is due primarily to a combination of variations in applied stress and increasing complexity.

Variable Stress

Contrary to the assumptions listed in Section 8.3 of this chapter, deterioration is not always proportional to applied stress, and stress is not always applied consistently. For instance, we saw earlier that many failures are caused by increases in applied stress, which are caused, in turn, by incorrect operation, incorrect assembly, or external damage.

Examples of such increases in stress given in Chapter 6 included operating errors (starting up a machine too quickly, accidentally putting it into reverse while it is going forward, feeding material into a process too quickly), assembly errors (overtorqueing bolts, misfitting parts), and external damage (lightning, the "thousand-year flood," and so on).

In all these cases, there is little or no relationship between the length of time the asset has been in service and the likelihood that the failure will occur. This is shown in Figure 8.11. (Ideally, preventing failures of this sort is a matter of preventing whatever causes the increase in stress levels, rather than a matter of doing anything to the asset.)

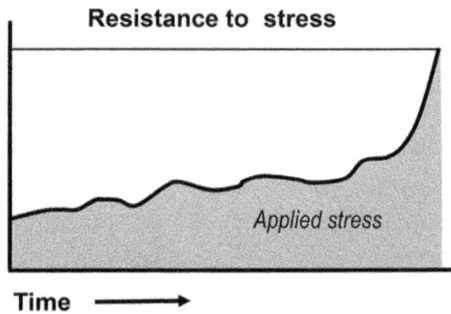

Resistance to stress

Applied stress

Time ⟶

FIGURE 8.11   Increase in stress levels

In Figure 8.12, the stress peak permanently reduces resistance to failure, but it does not actually cause the item to fail (an earthquake cracks a structure but does not cause it to fall down). The reduced failure resistance makes the part vulnerable to the next peak, which may or may not occur before the part is replaced for another reason.

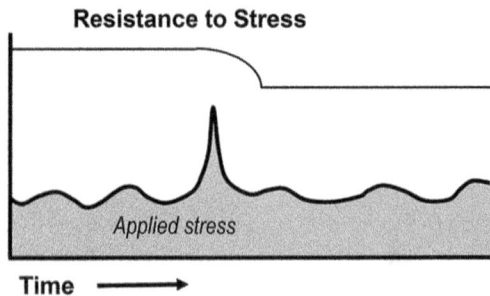

Resistance to Stress

Applied stress

Time ⟶

FIGURE 8.12   A stress peak permanently reduces resistance to failure

In Figure 8.13, the stress peak only temporarily reduces failure resistance (as in the case of thermoplastic materials that soften when temperature rises and harden again when it drops).

Resistance to Stress

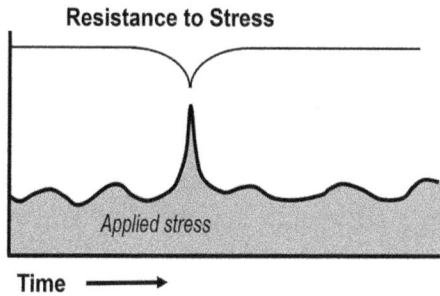

Applied stress

Time ➝

FIGURE 8.13   A stress peak only temporarily reduces failure resistance

Finally, in Figure 8.14 a stress peak accelerates the decline of failure resistance and eventually greatly shortens the life of the component. When this happens, the cause and effect relationship can be very difficult to establish, because the failure could occur months or even years after the stress peak.

Resistance to Stress

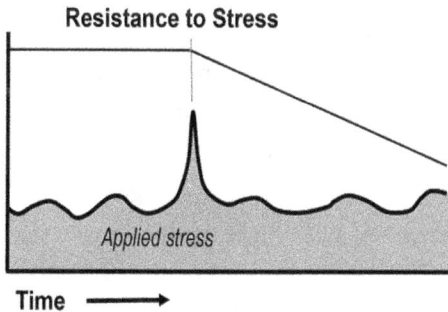

Applied stress

Time ➝

FIGURE 8.14   A stress peak accelerates the decline of failure resistance

This often happens if a part is damaged during installation (which might happen if a ball bearing is misaligned), if it is damaged prior to installation (the bearing is dropped on the floor in the parts store), or if it is mistreated in service (dirt gets into the bearing). In these cases, failure prevention is ideally a matter of ensuring that maintenance and installation work are done correctly and that parts are looked after properly in storage.

In all four of these examples, when the items enter service, it is not possible to predict when the failures will occur. For this reason, such failures are described as "statistically random."

Complexity

The failure processes depicted in Figure 8.8 apply to fairly simple mechanisms. In the case of complex items, the situation becomes even less predictable. Items are made more complex to improve their performance (by incorporating new or additional technology or by automation) or to make them safer (using protective devices).

For example, Nowlan and Heap (1978) cite developments in the field of civil aviation. In the 1930s, an air trip was a slow, somewhat risky affair, undertaken in reasonably favorable weather conditions in an aircraft with a range of a few hundred miles and space for about 20 passengers. The aircraft had one or two reciprocating engines, fixed landing gear, fixed pitch propellers, and no wing flaps.

Today an air trip is much faster and very much safer. It is undertaken in almost any weather conditions in an aircraft with a range of thousands of miles and space for hundreds of passengers. The aircraft has several jet engines, anti-icing equipment, retractable landing gear, movable high-lift devices, pressure and temperature control systems for the cabin, extensive navigation and communications equipment, complex instrumentation, and complex ancillary support systems.

In other words, better performance and greater safety are achieved at the cost of greater complexity. This is true in most branches of industry.

Greater complexity means balancing the lightness and compactness needed for high performance, on the one hand, with the size and mass needed for durability, on the other. This combination of complexity and compromise:

- Increases the number of components that can fail and also increases the number of interfaces or connections between components. This, in turn, increases the number and variety of failures that can occur. For example, a great many mechanical failures involve welds or bolts, while a significant proportion of electrical and electronic failures involve the connections between components. The more such connections there are, the more such failures there will be.

• Reduces the margin between the initial capability of each component and the desired performance (in other words, the "can" is closer to the "want"), which reduces scope for deterioration before failure occurs.

These two developments, in turn, suggest that complex items are more likely to suffer from random failures than are simple items.

Patterns D, E, and F

The combination of variable stress and erratic response to stress coupled with the increasing complexity means that in practice, more and more failure modes conform to the failure patterns shown in Figure 8.15. The most important feature of patterns D, E, and F is that after the initial period, there is little or no relationship between reliability and operating age. In these cases, unless there is a dominant age-related failure mode, age limits do little or nothing to reduce the probability of failure.

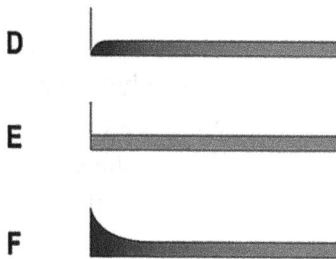

FIGURE 8.15 Failures that are not age-related

(In fact, scheduled overhauls can actually increase overall failure rates by introducing start-up failure into otherwise stable systems. This is borne out by the high and rising number of nasty accidents around the world that have occurred either while maintenance is under way or immediately after a maintenance intervention. It is also borne out by the machine operator who says that "every time maintenance works on it over the weekend, it takes us until Wednesday to get it going again.")

From the maintenance management viewpoint, the main conclusion to be drawn from these failure patterns is that the idea of a wear-out age simply does not apply to random failures, so the idea of fixed-interval replacement or overhaul prior to such an age cannot apply.

As mentioned in Chapter 2 of this book, an awareness of these facts has led some people to abandon the idea of preventive maintenance altogether. Although this can be the right thing to do for failures with minor consequences, when the failure consequences are serious, something must be done to prevent the failures or at least to avoid the consequences.

The continuing need to prevent certain types of failure, and the growing inability of classical techniques to do so, is behind the growth of new types of failure management. Foremost among these are the techniques known as predictive or on-condition maintenance. These techniques are discussed at length in the rest of this chapter.

8.7 Potential Failures and On-Condition Maintenance

We have seen that there is often little or no relationship between how long an asset has been in service and how likely it is to fail. However, although many failure modes are not age-related, most of them give some sort of warning that they are in the process of occurring or are about to occur. If evidence can be found that something is in the final stages of failure, it may be possible to take action to prevent it from failing completely and/or to avoid the consequences.

Figure 8.16 illustrates what happens in the final stages of failure. It is called the P-F curve, because it shows how a failure starts, deteriorates to the point at which it can be detected (point P), and then, if it is not detected and corrected, continues to deteriorate—usually at an accelerating rate—until it reaches the point of functional failure (F).

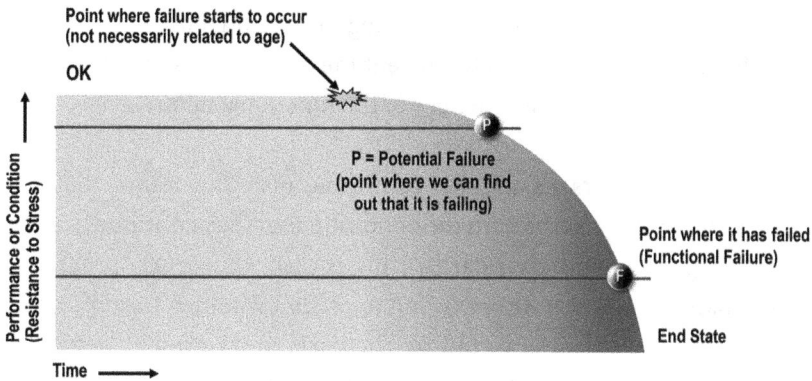

FIGURE 8.16   P-F curve

The point in the failure process at which it is possible to detect whether the failure is occurring or is about to occur is known as a potential failure.

A potential failure is an identifiable condition that indicates that a functional failure is either about to occur or in the process of occurring.

In practice, there are thousands of ways of finding out if failures are in the process of occurring.

Examples of potential failures include hot spots showing deterioration of furnace refractories or electrical insulation, vibrations indicating imminent bearing failure, cracks show- ing metal fatigue, particles in gearbox oil showing imminent gear failure, excessive tread wear on tires, etc.

If a potential failure is detected between point P and point F in Figure 8.16, it may be possible to take action to prevent or to avoid the consequences of the functional failure. (Whether or not it is possible to take meaningful action depends on how quickly the failure occurs, as discussed in Section 8.8 of this chapter.) Tasks designed to detect potential failures are known as "on-condition tasks."

> On-condition tasks entail checking for potential failures, so
> that action can be taken to prevent the functional failure or to
> avoid the consequences of the functional failure.

On-condition tasks are so called because the items that are inspected are left in service on the condition that they continue to meet specified performance standards. This is also known as "predictive maintenance" (because we are trying to predict whether—and possibly when—the item is going to fail on the basis of its present behavior) or "condition-based maintenance" (because the need for corrective or consequence-avoiding action is based on an assessment of the condition of the item).

8.8 The P-F Interval

In addition to the potential failure itself, we need to consider the amount of time (or the number of stress cycles) that elapses between the point at which a potential failure occurs—in other words, the point at which it becomes *detectable*—and the point where it deteriorates into a functional failure. As shown in Figure 8.17, this interval is known as the "P-F interval."

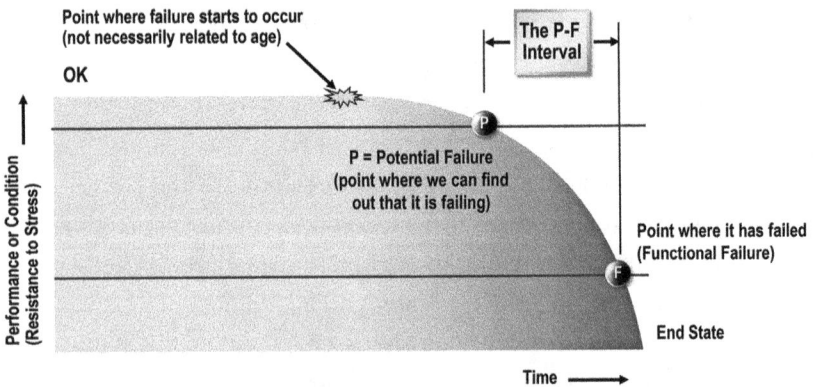

FIGURE 8.17   The P-F interval

> The P-F interval is the interval between the occurrence of the
> potential failure and its decay into the functional failure.

A potential failure is an identifiable condition that indicates that a functional failure is either about to occur or in the process of occurring. The P-F interval tells us how often on-condition tasks must be done. If we want to detect the potential failure before it becomes a functional failure, the interval between checks must be less than the P-F interval.

> On-condition tasks must be carried out at intervals less than
> the P-F interval.

The P-F interval is also known as the "warning period," the "lead time to failure," or the "failure development period." It can be measured in any units that provide an indication of exposure to stress (running time, units of output, stop-start cycles, etc.), but for practical reasons, it is most often measured in terms of elapsed time. For different failure modes, it varies from fractions of a second to several decades.

Note that if an on-condition task is done at intervals that are longer than the P-F interval, there is a chance that we will miss the failure altogether. On the other hand, if we do the task at too small a percentage of the P-F interval, we will waste resources on the checking process.

For instance, if the P-F interval for a given failure mode is 2 weeks, the failure will be detected if the item is checked once a week. Conversely, if it is checked once a month, it is possible to miss the whole failure process. On the other hand, if the P-F interval is 3 months, it is a waste of effort to check the item every day.

In practice it is usually sufficient to select a task frequency equal to half the P-F interval. This ensures that the inspection will detect the potential failure before the functional failure occurs, while (in most cases) providing a reasonable amount of time to do something about it. This leads to the concept of the net P-F interval.

The Net P-F Interval

The net P-F interval is the minimum interval likely to elapse between the discovery of a potential failure and the occurrence of the functional failure.

Figure 8.18 shows that if the item is inspected monthly, the net P-F interval is 8 months. On the other hand, if it is inspected at six monthly intervals, as shown in Figure 8.19, the net P-F interval is 3 months. So in the first case the minimum amount of time available to do something about the failure is 5 months longer than in the second, but the inspection task has to be done six times more often.

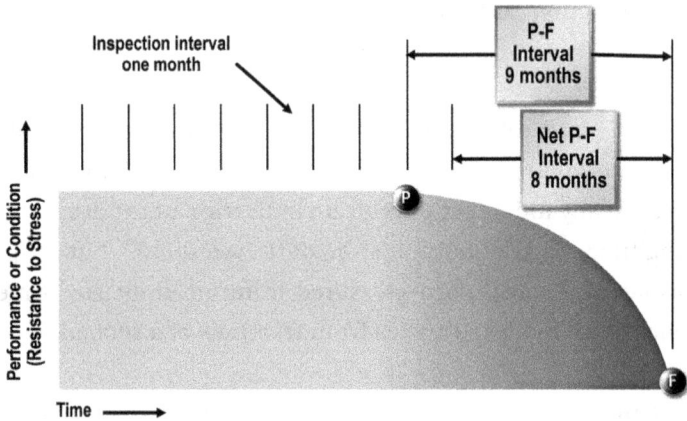

FIGURE 8.18   Net P-F interval (1)

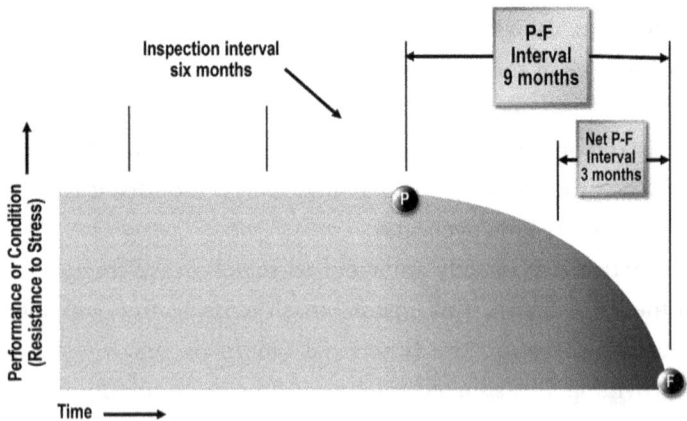

FIGURE 8.19   Net P-F interval (2)

The net P-F interval governs the amount of time available to take whatever action is needed to reduce or eliminate the consequences of the failure. Depending on the operating context of the asset, warning of incipient failure enables the users of an asset to reduce or avoid consequences in a number of ways:

- **Downtime.** Corrective action can be planned at a time that does not disrupt operations. The opportunity to plan the corrective action properly also means that it is likely to be done more quickly.
- **Repair costs.** Users may be able to take action to eliminate the secondary damage that would be caused by unanticipated failures. This would reduce the downtime and the repair costs associated with the failure.
- **Safety.** Warning of failure provides time either to shut down a plant before the situation becomes dangerous or to move people who might otherwise be injured out of harm's way. For instance, if a crack in a wall is discovered in good time, it may be possible to shore up the foundations and so prevent the wall from deteriorating so much that it falls down. It is highly likely that we would have to vacate the premises while this work is done, but at least we avoid the safety consequences that would arise if the wall fell down.

For an on-condition task to be technically feasible, the net P-F interval must be *longer* than the time required to take action to avoid or reduce the consequences of the failure. If the net P-F interval is too short for any sensible action to be taken, then the on-condition task is clearly not technically feasible.

In practice, the time required varies widely. In some cases, it may be a matter of hours (say, until the end of an operating cycle or the end of a shift) or even minutes (to shut down a machine or evacuate a building). In other cases, it can be weeks or even months (say, until a major shutdown). In general, longer P-F intervals are desirable for two reasons:

- It is possible to do whatever is necessary to avoid the consequences of the failure (including planning the corrective action) in a more considered and hence more controlled fashion.
- Fewer on-condition inspections are required.

This explains why so much energy is being devoted to finding potential failure conditions and associated on-condition techniques that give the longest possible P-F intervals. However, it is possible to make use of very short P-F intervals in certain cases.

For example, failures that affect the balance of large fans cause serious problems very quickly, so online vibration sensors are used to shut the fans down when such failures occur. In this case, the P-F interval is very short, so monitoring is continuous. Note that once again, the monitoring device is being used to avoid the consequences of the failure.

P-F Interval Consistency

The P-F curves illustrated so far in this chapter indicate that the P-F interval for any given failure is constant. In fact, this is not the case—some actually vary over a wide range of values, as shown in Figure 8.20.

For example, when discussing the P-F interval associated with a change in noise levels, someone might say, "This thing rattles away for any time from 2 weeks to 3 months before it collapses." In another case, tests might show that any time from 6 months to 5 years elapses from the moment a crack becomes detectable at a particular point in a structure until the moment the structure fails.

Clearly, in these cases a task interval should be selected that is substantially less than the shortest of the likely P-F intervals. In this way, we can always be reasonably certain of detecting the potential failure before it becomes a functional failure. If the net P-F interval associated with this minimum interval is long enough for suitable action to be taken to deal with the consequences of the failure, then the on-condition task is technically feasible.

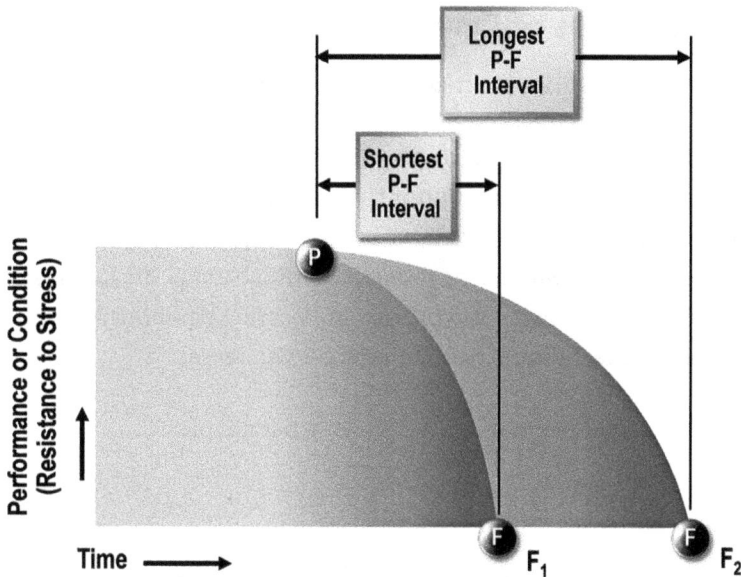

FIGURE 8.20   Inconsistent P-F intervals

On the other hand, if the P-F interval is very inconsistent, then it is not possible to establish a meaningful task interval, and the task should again be abandoned in favor of some other way of dealing with the failure.

8.9 Technical Feasibility of On-Condition Tasks

In the light of the above discussion, the criteria that any on-condition task must satisfy to be technically feasible can be summarized as follows:

Scheduled on-condition tasks are technically feasible if:

- It is possible to define a clear potential failure condition.
- The P-F interval is reasonably consistent.
- It is practical to monitor the item at intervals less than the P-F interval.
- The net P-F interval is long enough to be of some use (in other words, long enough for action to be taken to reduce or eliminate the consequences of the functional failure).

8.10 Categories of On-Condition Techniques

The four major categories of on-condition techniques are:

- *Condition monitoring techniques*, which involve the use of specialized equipment to monitor the condition of other equipment
- Techniques based on variations in *product quality*
- *Primary effects monitoring techniques*, which entail the intelligent use of existing gauges and process monitoring equipment
- Inspection techniques based on the *human senses*

These are each reviewed in the following paragraphs.

Condition Monitoring

The most sensitive on-condition maintenance techniques usually involve the use of some type of equipment to detect potential failures. In other words, equipment is used to monitor the condition of other equipment. These techniques are known as "condition monitoring" to distinguish them from other types of on-condition maintenance.

Condition monitoring embraces several hundred different techniques, so a detailed study of the subject is well beyond the scope of this chapter. All the techniques are designed to detect failure effects (or more precisely, potential failure effects, such as changes in vibration characteristics, changes in temperature, particles in lubricating oil, leaks, and so on). The numerous different techniques of condition monitoring can be classified under the following headings:

- Dynamic effects
- Particle effects
- Chemical effects
- Physical effects
- Temperature effects
- Electrical effects

These techniques can be seen as highly sensitive versions of the human senses. Many of them are now very sensitive indeed, and a few give several months' (if not several years') warning of failure. However, a major limitation of nearly every condition monitoring device is that it monitors only one condition. For instance, a vibration analyzer only monitors vibration and cannot detect chemicals or temperature changes. So greater sensitivity is bought at the price of the versatility inherent in the human senses.

The P-F intervals associated with different monitoring techniques vary from a few minutes to several months. Different techniques also pinpoint failures with different degrees of precision. Both of these factors must be considered when assessing the feasibility of any technique.

In general, condition monitoring techniques can be spectacularly effective when they are appropriate, but when they are inappropriate, they can be a very expensive and sometimes bitterly disappointing waste of time. As a result, the criteria for assessing whether on-condition tasks are technically feasible and worth doing should be applied especially rigorously to condition monitoring techniques.

In modern developments, technology is now becoming so inexpensive that organizations *overmonitor asset health* using devices capable of transmitting information across wireless networks. Decisions are now made using information and no longer symptoms. Information can be available anywhere it is needed, and devices are now capable of "talking to one another" and learning the health characteristics through *cognitive learning*. The Industrial Internet of Things (IIoT) is now the topic of many conferences and white papers and will not be covered by this book.

Product Quality Variation

In some industries, an important source of data about potential failures is the quality management function. Often the emergence of a defect in an article produced by a machine is directly related to a failure mode in the machine itself. Many of these defects emerge gradually, and so provide timely evidence of potential failures. If the data-gathering and -evaluation procedures exist already, it costs very little to use them to provide warning of equipment failure.

One popular technique that can often be used in this way is statistical process control (SPC). SPC entails measuring some attribute of a product, such as a dimension, filling level, or packing weight, and using the measurements to draw conclusions about the stability of the process.

Figure 5.3 in Chapter 5 showed how such measurements might appear for a process that is in control and in specification. Figures 5.4 and 5.5 in Chapter 5 showed two ways in which a process could be out of control and out of specification (in other words, failed). In a great many cases, the transition from being in control to failed takes place gradually. SPC charts frequently track this transition.

For instance, there are many ways in which SPC charts could be used to give warning of potential failures, as illustrated in Figure 8.21.

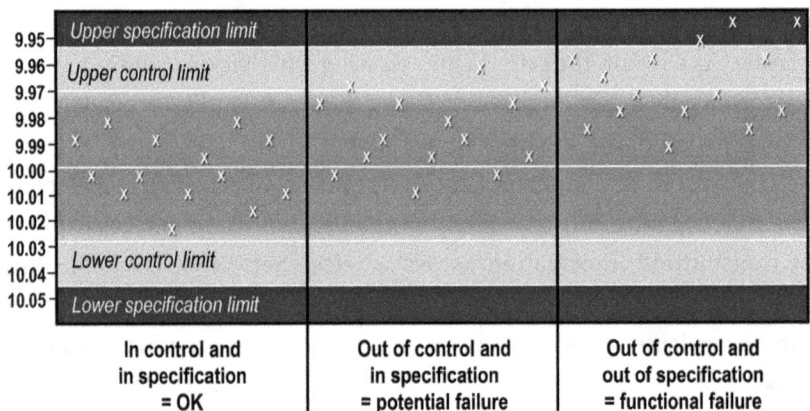

| | In control and in specification = OK | Out of control and in specification = potential failure | Out of control and out of specification = functional failure |

FIGURE 8.21   On-condition maintenance and SPC

Figure 8.21 shows a typical SPC chart on which the readings are in control to start with. A failure mode occurs that causes the measurements to start drifting in one direction.

For example, as a grinding wheel wears, the diameter of successive workpieces increases until the wheel is adjusted or replaced.

In zone 2 in Figure 8.21, the process is out of control but still within specification. [It is possible to identify very gradual shifts of this sort using a "cusum chart."] This shift in the mean is a clearly identifiable condition that indicates that a functional failure is about to occur. In other words, it is a potential failure. If nothing is done to rectify the situation, the process eventually begins to produce out-of-spec products, as shown in zone 3 in Figure 8.21.

This example describes only one of many ways in which SPC can be used to measure and manage process variability. A full description of all the techniques is well beyond the scope of this book. However, the key point to note at this stage is that if deviations on charts like these can be related directly to specific failure modes, then the charts are sources of on-condition data that can make a valuable contribution to overall proactive maintenance efforts.

Primary Effects Monitoring

Primary effects (speed, flow rate, pressure, temperature, power, current, etc.) are yet another source of information about equipment condition. The effects can be monitored by a person reading a gauge and perhaps recording the reading manually, by a computer as part of a process control system, or even by a traditional chart recorder.

The records of these effects or their derivatives are compared with reference information, and so these records provide evidence of a potential failure. However, in the case of the first option in particular, take care to ensure that:

- The person taking the reading knows what the reading should be when all is well, what reading corresponds to a potential failure, and what corresponds to functional failure.
- The readings are taken at a frequency that is less than the P-F interval (in other words, the frequency should be less than the time it takes the pointer on the dial to move from the potential failure level to the functional failure level when the failure mode in question is occurring).
- The gauge itself is maintained in such a way that it is sufficiently accurate for this purpose.

The process of taking readings can be greatly simplified if gauges are marked up (or even colored), as shown in Figure 8.22. In this case, all the operator—or anyone else—needs to do is look at the gauge and report if the pointer is in the potential failure (yellow?) zone, or take more drastic action if it is in the functional failure (red?) zone. However, the gauge must still be monitored at intervals that are less than the P-F interval. (For obvious reasons, this suggestion only applies to gauges that are measuring a steady state. Also take care to ensure that gauges marked up in this way are not taken off and remounted in the wrong place.)

FIGURE 8.22   Primary effects monitoring

The Human Senses

Perhaps the best known on-condition inspection techniques are those based on the human senses (look, listen, feel, smell, and taste). The main disadvantages of using these senses to detect potential failures are:

- By the time it is possible to detect most failures using the human senses, the process of deterioration is already quite far advanced. This means that the P-F intervals are usually short, so the checks must be done more frequently than most and response has to be rapid.
- The process is subjective, so it is difficult to develop precise inspection criteria, and the observations depend very much on the experience and even the state of mind of the observer

- The process is limited to access and restricts certain areas of being monitored by human senses, e.g.. confined spaces, hazardous chemicals, and unfriendly process materials.
- The process is inconsistent, and baseline data and condition variances may vary from individual to individual.

However, the advantages of using these senses are:

- The average human being is highly versatile and can detect a wide variety of failure conditions, whereas any one condition monitoring technique can only be used to monitor one type of potential failure.
- It can be very cost-effective if the monitoring is done by people who are at or near the assets anyway in the course of their normal duties.
- A human is able to exercise judgment about the severity of the potential failure and hence about the most appropriate action to be taken, whereas a condition monitoring device can only take readings and send a signal.
- It brings the operations and maintenance people "back" to the equipment and machines instead of operating remotely in control rooms only.

Selecting the Right Category

Many failure modes are preceded by more than one—often several—different potential failures, so more than one category of on-condition task might be appropriate. Each of these will have a different P-F interval, and each will require different types and levels of skill.

For example, consider a ball bearing whose failure is described as "bearing seizes due to normal wear and tear." Figure 8.23 shows how this failure could be preceded by a variety of potential failures, each of which could be detected by a different on-condition task. This does not mean that all ball bearings will exhibit these potential failures, nor will they necessarily have the same P-F intervals. The extent to which any technique is technically feasible and worth doing depends very much on the operating context of the bearing. For instance:

- The bearing may be buried so deep in the machine that it is impossible to monitor its vibration characteristics.
- It is only possible to detect particles in the oil if the bearing is operating in a totally enclosed oil-lubricated system.
- Background noise levels may be so high that it is impossible to detect the noise made by a failing bearing.
- It may not be possible to reach the bearing housing to feel how hot it is.

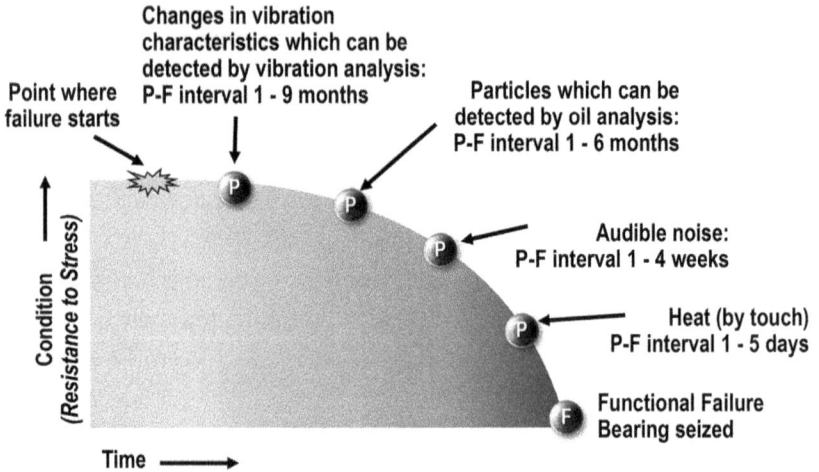

FIGURE 8.23   Different potential failures that can precede one failure mode

This means that no one single category of tasks will always be more cost-effective than any other. It is important to bear this in mind, because there is a tendency in some quarters to present condition monitoring in particular as "the answer" to all our maintenance problems.

In fact, if RCM is correctly applied to typical modem, complex industrial systems, it is not unusual to find that condition monitoring as defined in this part of this chapter is technically feasible for no more than 20% of failure modes, and it is worth doing in less than half these cases. (All four categories of on-condition maintenance together are usually suitable for about 25–35% of failure modes.) This is not meant to imply that condition monitoring should not be used—where it is good, it is very, very good—but that we must also remember to develop suitable strategies for managing the other 80–90% of our failure modes. In other words, condition monitoring is only part of the answer—and a fairly small part at that.

So to avoid unnecessary bias in task selection, we need to:

- Consider all the warnings that are reasonably likely to precede each failure mode, together with the full range of on-condition tasks that could be used to detect those warnings.
- Apply the RCM task selection criteria rigorously to determine which (if any) of the tasks is likely to be the most cost-effective way of anticipating the failure mode under consideration.

As with so much else in maintenance, the right choice ultimately depends on the operating context of the asset.

8.11 On-Condition Tasks: Some of the Pitfalls

When considering the technical feasibility of on-condition maintenance, two issues need special care. They concern the distinction between potential and functional failures and the distinction between potential failure and age. These issues are discussed in more detail below.

Potential and Functional Failures

In practice, confusion often arises over the distinction between potential and functional failures. This happens because certain conditions can correctly be regarded as potential failures in one context and as functional failures in another. This is especially common in the case of leaks.

For example, a minor leak in a flanged joint on a pipeline might be regarded as a potential failure if the pipeline is carrying water. In this case, the on-condition task would be "Check pipe joints for leaks."

The task frequency is based on the amount of time it takes for an "acceptable" minor leak to become an "unacceptable" major leak, and suitable corrective action would be initiated whenever a minor leak was discovered.

However, if the same pipeline were carrying a toxic substance like cyanide, any leak at all would be regarded as a functional failure. In this case it is not feasible to ask anyone to check for leaks, so some other method would need to be found to manage the failure. This would almost certainly entail some sort of modification.

This example reemphasizes how important it is to agree on what is meant by a functional failure before considering what should be done to prevent it.

The P-F Interval and Operating Age

When applying these principles for the first time, people often have difficulty in distinguishing between the life of a component and the P-F interval. This leads them to base on-condition task frequencies on the real or imagined life of the item. If it exists at all, this life is usually many times greater than the P-F interval, so the task achieves little or nothing. In reality, we measure the life of a component forward from the moment it enters service. The P-F interval is measured back from the functional failure, so the two concepts are often completely unrelated. The distinction is important, because failures that are not related to age (in other words, random failures) are as likely to be preceded by a warning as those that are not.

For example, Figure 8.24 depicts a component that conforms to a random failure pattern (pattern E). One of the components failed after 5 years, a second after 6 months, and a third after 2 years. In each case, the functional failure was preceded by a potential failure with a P-F interval of 4 months.

Figure 8.24 shows that in order to detect the potential failure, we need to do an inspection task every 2 months. Because the failures occur on a random basis, we don't know when the next one is going to happen,

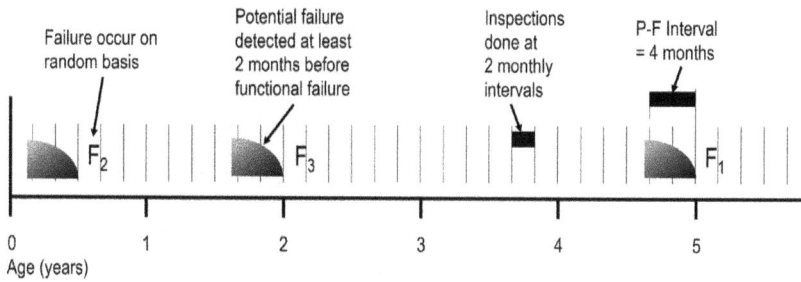

FIGURE 8.24   Random failures and the P-F interval

so the cycle of inspections must begin as soon as the item is put into service. In other words, the timing of the inspections has nothing to do with the age or life of the component.

However, this does not mean that on-condition tasks apply only to items that fail on a random basis. They can also be applied to items that suffer age-related failures, as discussed later in this chapter.

8.12   Nonlinear and Linear P-F Curves

Section 8.7 of this chapter explained that the final stages of deterioration can be described by the P-F curve. In this part of this chapter, we consider this curve in more detail, starting with a look at nonlinear P-F curves and then going on to consider linear P-F curves.

The Final Stages of Deterioration

Figure 8.16 suggests that deterioration usually accelerates in the final stages. To see why this is so, let us consider in more detail what happens when a ball bearing fails due to "normal wear and tear."

Figure 8.25 illustrates a typical vertically loaded ball bearing that is rotating clockwise. The most heavily and frequently loaded part of the bearing will be the bottom of the outer race. As the bearing rotates, the inner surface of the outer race moves up and down as each ball passes over it. These cyclic movements are tiny, but they are sufficient to cause subsurface fatigue cracks to develop, as shown in Figure 8.25.

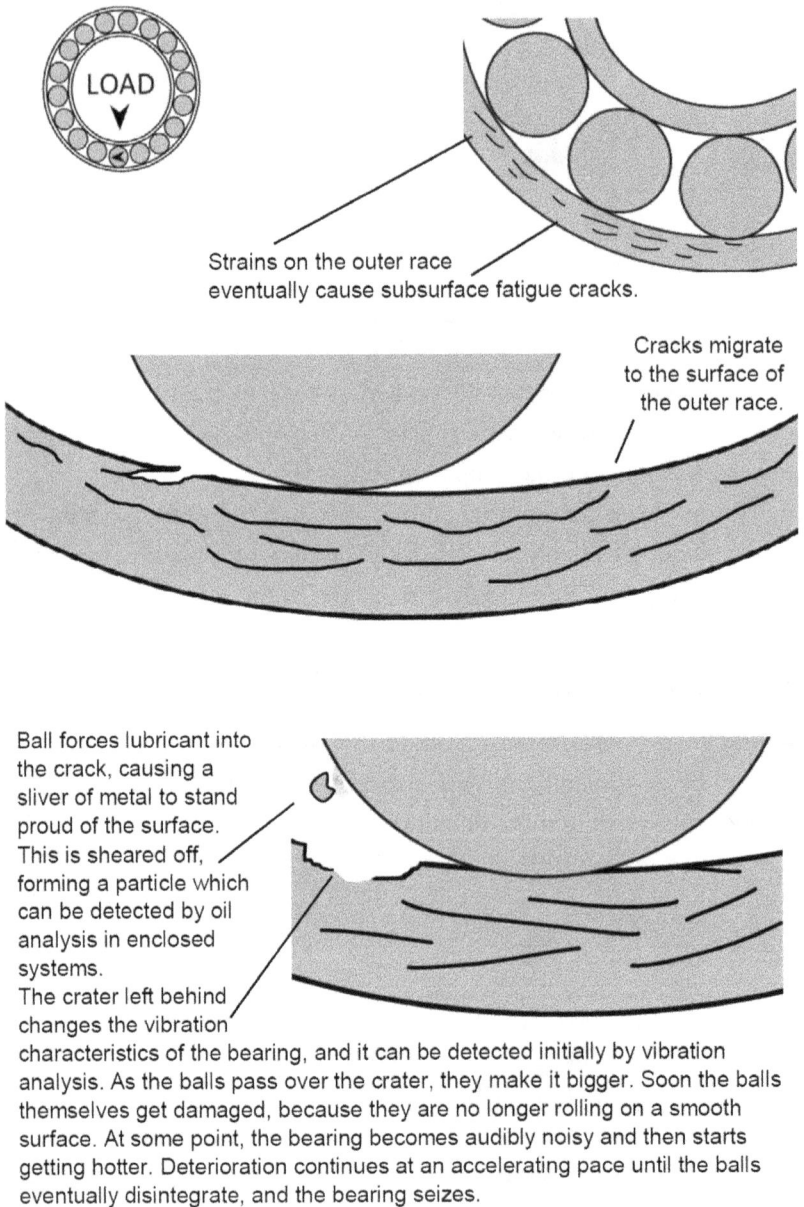

Strains on the outer race eventually cause subsurface fatigue cracks.

Cracks migrate to the surface of the outer race.

Ball forces lubricant into the crack, causing a sliver of metal to stand proud of the surface. This is sheared off, forming a particle which can be detected by oil analysis in enclosed systems.

The crater left behind changes the vibration characteristics of the bearing, and it can be detected initially by vibration analysis. As the balls pass over the crater, they make it bigger. Soon the balls themselves get damaged, because they are no longer rolling on a smooth surface. At some point, the bearing becomes audibly noisy and then starts getting hotter. Deterioration continues at an accelerating pace until the balls eventually disintegrate, and the bearing seizes.

FIGURE 8.25   Random failures of rolling element bearing due to normal wear

Figure 8.25 also explains how these cracks eventually give rise to detectable symptoms of deterioration. These are, of course, potential failures, and the associated P-F intervals are shown in

Figure 8.23. This example raises several further points about potential failures:

- In the example, the deterioration process accelerates. This suggests that if a quantitative technique such as vibration analysis is used to detect potential failures, we cannot predict when failure will occur by drawing a straight line based on just two observations. This, in turn, leads to the notion that after an initial deviation is observed, additional vibration readings should be taken at progressively shorter intervals until some further point is reached, at which action should be taken. In practice, this can only be done if the P-F interval is long enough to allow time for the additional readings. It also does not escape the fact that the initial readings need to be taken at a frequency that is known to be less than the P-F interval.

  In fact, if the shape of the P-F curve is fairly well known and the P-F interval is reasonably consistent, it should not be necessary to take additional readings after the first sign of deviation is discovered. This suggests that the process of deterioration should only be tracked by taking additional readings if the P-F curve is poorly understood or if the P-F interval is highly inconsistent.
- Different failure modes can often exhibit similar symptoms. For example, the symptoms described in Figure 8.25 are based on failure due to normal wear and tear. Very similar symptoms would be exhibited in the final stages of the failure of a bearing where the failure process has been initiated by dirt, lack of lubrication, or brinelling.

  In practice, the precise root cause of many failures can only be identified using sophisticated instruments. For instance, it might be possible to determine the root cause of the failure of a bearing by using a ferrograph to separate particles from the lubricating oil and examining the particles under an electron microscope.

  However, if two different failures have the same symptoms and if the P-F interval is broadly similar for each set of symptoms—as it probably would be in the case of the bearing examples—the distinction between root causes is irrelevant from the failure detection viewpoint. (The distinction does, of course, become relevant if we are seeking to eliminate the root cause of the failure.)

- Failure only becomes detectable when the fatigue cracks migrate to the surface and the surface starts breaking up. The point at which this happens in the life of any one bearing depends on the speed of rotation of the bearing, the magnitude of the load, the extent to which the outer race itself rotates, whether the bearing surface is damaged prior to or during installation, how hot the bearing gets in service, the alignment of the shaft relative to the housing, the materials used to manufacture the bearing, how well it was made, etc. Effectively, this combination of variables makes it impossible to predict how many operating cycles must elapse before the cracks reach the surface, and hence when the bearing will start exhibiting the symptoms mentioned in Figure 8.24. [For those interested in pursuing this subject further, chaos theory—in particular, the "butterfly effect"—shows how tiny differences between the initial conditions that apply to any dynamic system lead to dramatic differences after the passage of time. This may explain why minute variations between the initial conditions of two rolling element bearings can lead to huge differences between the ages at which they fail.

Deterioration accelerates in the final stages of most failures. For instance, deterioration is likely to accelerate when bolts start to loosen, when filter elements get blinded, when V-belts slacken and start slipping, when electrical contactors overheat, when seals start to fail, when rotors become unbalanced, and so on. But it does not accelerate in every case.

Linear P-F Curves

If an item deteriorates in a more or less linear fashion over its entire life, it stands to reason that the final stages of deterioration will also be more or less linear. A close look at Figures 8.3 and 8.4 suggests that this is likely to be true of age-related failures.

For example, consider tire wear. The surface of a tire is likely to wear in a more or less linear fashion until the tread depth reaches the legal minimum. If this minimum is 1/8 inch, it is possible to specify a depth of tread greater than 1/8 inch, which provides adequate warning that functional failure is imminent. This is, of course, the potential failure level.

If the potential failure is set at 3/8 inch, then the P-F interval is the distance the tire could be expected to travel while its tread depth wears down from 3/8 inch to 1/8 inch, as illustrated in Figure 8.26.

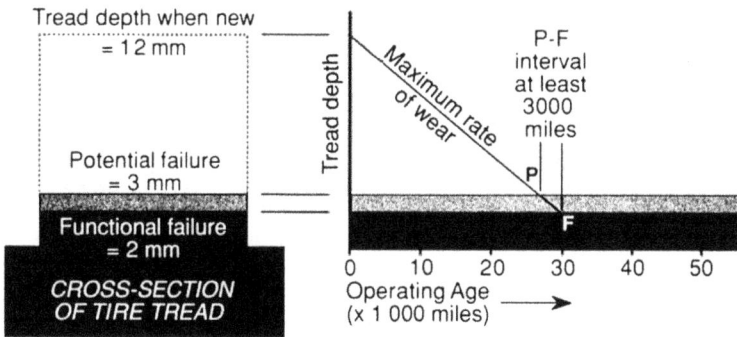

FIGURE 8.26   A linear P-F curve

Figure 8.26 also suggests that if the tire enters service with a tread depth of, say, 1 inch, it should be possible to predict the P-F interval based on the total distance usually covered before the tire has to be retreaded. For instance, if the tires last at least 30,000 miles before they have to be retreaded, it is reasonable to conclude that the tread wears at a maximum rate of 1/16 inch for every 3,000 miles traveled. This amounts to a P-F interval of 3,000 miles. The associated on-condition task would call for the driver to "Check tread depth every 1,500 miles and report tires whose tread depth is less than 3/8 inch."

Not only will this task ensure that wear is detected before it exceeds the legal limit, but it also allows plenty of time—1,500 miles in this case—for the vehicle operators to plan to remove the tire before it reaches the limit.

In general, linear deterioration between P and F is only likely to be encountered where the failure mechanisms are intrinsically age-related. (This is true except in the case of fatigue, which is a somewhat more complex case. This failure process is discussed in more detail later.)

Note that the P-F interval and the associated task frequency can only be deduced in this way if deterioration is linear. As we have seen, the P-F interval cannot be determined in this way if deterioration accelerates between P and F.

A further note about linear failures concerns the point at which one should start to look for potential failures.

For example, Figure 8.26 suggests that it would be a waste of time to measure the overall depth of the tire tread at 10,000 or 20,000 miles, because we know that it only approaches the potential failure point at 30,000 miles. So perhaps we should only start measuring the tread depth of each tire after it has passed the point where we know tread depth will be approaching 3/8 inch—in other words, when the tire has been in service for more than, say, 25,000 miles.

However, if we want to ensure that this checking regime is adopted in practice, consider how the checks for a four-wheeled truck would have to be planned if the actual history of a set of tires is as follows: "Tread depth of LF tire dropped below 3/8 inch, and tire replaced at depot; 6-inch nail caused tire to blow out at 13,000 miles—replaced with new spare."

If we are seriously going to try to ensure that the driver only checks each tire after it passes 25,000 miles in service, we must devise a system which tells her to:

- Start checking the LF tire only when the truck reaches 126,000 miles.
- Check the LF and RR tires when the truck reaches 141,500 miles.
- Check again at 143,000 miles.
- Check the RR and RF tires at 144,500 miles, but not the LF tire.

Clearly this is nonsense, because the cost of administering such a planning system would be far greater than the cost of asking the driver to check the tread depth of every tire on the vehicle every 1,500 miles. In other words, in this example the cost of fine-tuning the planning system would be far greater than the cost of doing the tasks. So we would simply ask the driver to check the tread depth of every tire at 1,500-mile intervals, rather than direct her attention to specific tires.

However, if the process of deterioration is linear and the task itself is very expensive, then it might be worth ensuring that we only start checking for potential failures when it is really necessary.

For instance, if an on-condition task entails shutting down and opening up a large turbine to check the turbine discs for cracks, and we are certain that deterioration only becomes detectable after the turbine has been in service for a certain length of time (in other words, the failure is age-related), then we should only start taking the turbine out of service to check for the cracks after it has passed the age at which there is a reasonable likelihood that detectable cracks will start to emerge. Thereafter, the frequency of checking is based on the rate at which a detectable crack is likely to deteriorate into a failure.

For the record, the age at which cracks are likely to start becoming detectable is known as the crack initiation life, whereas the time (or number of stress cycles) that elapses from the moment a crack becomes detectable until it grows so large that the item fails is known as the crack propagation life.

In cases like these, the cost of doing the task would be much greater than the cost of the associated planning systems, so it is worth ensuring that we only start doing the tasks when it is really necessary. However, if it is felt that this fine-tuning is worthwhile, bear in mind that the planning process has to employ two completely different time frames:

- The first time frame is used to decide when we should start doing the on-condition tasks. This is the operating age at which potential failures are likely to start becoming detectable.
- The second time frame governs how often we should do the tasks after this age has been reached. This time frame is, of course, the P-F interval.

For example, it might be felt that the turbine disc is unlikely to develop any detectable cracks until it has been in service for at least 50,000 hours, but that it takes a minimum of 10,000 hours for a detectable crack to deteriorate into disc failure. This suggests that we don't need to start checking for cracks until the item has been in service for 50,000 hours, but thereafter it must be checked at intervals of less than 10,000 hours.

Planning with this degree of sophistication requires a very detailed understanding of the failure mode under consideration, together with highly sophisticated planning systems. In practice, few failure modes are this well understood. When they are, even fewer organizations possess planning systems that can switch from one time frame to another, as described above, so this issue needs to be approached with care.

In closing this discussion, it must be stressed that all the curves—P-F and age-related—that have been drawn in this part of this chapter have been drawn for one failure mode at a time.

For instance, in the example concerning tires, the failure process was "normal wear." Different failure modes (such as flat spots worn on the tires due to emergency braking, or damage to the carcass caused by hitting objects) would lead to different conclusions because both the technical characteristics and the consequences of these failure modes are different.

It is one matter to speculate on the nature of P-F curves in general, but it is quite another to determine the magnitude of the P-F interval in practice. This issue is considered in the next section of this chapter.

8.13 How to Determine the P-F Interval

It is usually a fairly simple matter to determine the P-F interval for age-related failure modes whose final stages of deterioration are linear. It is done by applying logic similar to that used in the tire example above.

On the other hand, the P-F interval can be surprisingly difficult to determine in the case of random failures where deterioration accelerates. The main problem with random failures is that we don't know when the next one is going to occur, so we don't know when the next failure mode is going to start on its way down the P-F curve. So if we don't even know where the P-F curve is going to start, how can we go about finding out how long it is? The following paragraphs review five possibilities, only the fourth and fifth of which have any merit.

Continuous Observation

In theory, it is possible to determine the P-F interval by continuously observing an item that is in service until a potential failure occurs, noting when that happens, and then continuing to observe the item until it fails completely. (Note that we cannot chart a full P-F curve by observing the item intermittently, because when we eventually discover that it is failing, we still wouldn't know precisely when the failure process started. What is more, if the P-F interval is shorter than the intermittent observation period, we might miss the P-F curve altogether, in which case we would have to start all over again with a new item.)

Clearly this approach is impractical, first because continuous observation is very expensive—especially if we were to try to establish every P-F interval in this way. Second, waiting until the functional failure occurs means that the item actually has to fail. This might end up with us saying to the boss after the compressor blew up: "Oh, we knew it was failing, but we just wanted to see how long it would take before it finally went so that we could determine the P-F interval."

Start with a Short Interval and Gradually Extend It

The impracticality of the above approach leads some people to suggest that P-F intervals can be established by starting the checks at some quite short but arbitrary interval (say, 10 days), and then waiting until "we find out what the interval should be," perhaps by gradually extending the interval. Unfortunately, this is again the point at which the functional failure occurs, so we would still end up blowing up the compressor.

This approach is, of course, potentially very dangerous, because there is also no guarantee that the initial arbitrary interval, no matter how short, will be shorter than the P-F interval to begin with (unless serious consideration is given to the failure process itself).

Arbitrary Intervals

The difficulties associated with the two approaches described above lead some people to suggest—quite seriously—that some arbitrary "reasonably short" interval should be selected for all on-condition tasks. This arbitrary approach is the least satisfactory (and the most dangerous) way to set on-condition task frequencies, because there is again no guarantee that the "reasonably short" arbitrary interval will be shorter than the P-F interval. On the other hand, the true P-F interval may be much longer than the arbitrary interval, in which case the task ends up being done much more often than necessary.

For instance, if a daily task really only needs to be done once a month, that task is costing 30 times as much as it should.

Research

The best way to establish a precise P-F interval is to simulate the failure in such a way that there are no serious consequences when it eventually does occur. For example, this is done when aircraft components are tested to failure on the ground rather than in the air. Not only does this provide data about the life of the components, as discussed in Chapter 6, but it also enables the observers to study at leisure how failures develop and how quickly this happens. However, laboratory testing is expensive, and it takes time to yield results even when it is accelerated. So it is only worth doing in cases where a fairly large number of components are at risk—such as an aircraft fleet—and the failures have very serious consequences.

A Rational Approach

The above paragraphs indicate that in most cases, it is either impossible, impractical, or too expensive to try to determine P-F intervals on an empirical basis. On the other hand, it is even more unwise simply to take

a shot in the dark. Despite these problems, P-F intervals can still be estimated with surprising accuracy on the basis of judgment and experience.

The first trick is to ask the right question. It is essential that anyone
who is trying to determine a P-F interval understand that we are asking how quickly the item fails. In other words, we are asking how much
time (or how many stress cycles) elapses from the moment the potential
failure becomes detectable until the moment it reaches the functionally
failed state. We are not asking how often it fails or how long it lasts.

The second trick is to ask the right people—people who have an intimate knowledge of the asset, the ways in which it fails, and the symptoms of each failure. For most equipment, this usually means the people
who operate it, the craftspeople who maintain it, and their first-line
supervisors. If the detection process requires specialized instruments
such as condition monitoring equipment, then appropriate specialists
should also take part in the analysis.

In practice, John Moubray explained that an effective way to crys-
tallize thinking about P-F intervals is to provide a number of mental
"coat hooks" on which people can hang their thoughts. For instance, one
could ask, "Do you think that the P-F interval is likely to be on the order
of days, weeks, or months?" If the answer is, say, weeks, the next step is
to ask, "One, two, four, or eight weeks?"

If everyone in the group achieves consensus, then the P-F interval
has been established, and the analysts go on to consider other task selection criteria such as the consistency of the P-F interval and whether the
net interval is long enough to avoid the failure consequences.

If the group cannot achieve consensus, then it is not possible to provide a positive answer to the question, "What is the P-F interval?" When
this happens, the associated on-condition task must be abandoned as a
way of detecting the failure mode under consideration, and the failure
must be dealt with in some other way.

The third trick is to concentrate on one failure mode at a time. In other
words, if the failure mode is wear, then the analysts should concentrate
on the characteristics of wear and should not discuss, say, corrosion or
fatigue (unless the symptoms of the other failure modes are almost identical and the rate of deterioration is also very similar).

Finally, it must be clearly understood by everyone taking part in such an analysis that the objective is to arrive at an on-condition task interval that is less than the P-F interval, but not so much less that resources will be squandered on the checking process.

The effectiveness of such a group is redoubled if management expresses an appreciation of the fact that the group is made up of human beings and that humans are not infallible.

However, the analysts must also be aware that if the failure has safety consequences, the price of getting it badly wrong could (literally) be fatal for themselves or their colleagues, so they need to take special care in this area.

8.14 When On-Condition Tasks Are Worth Doing

On-condition tasks must satisfy the following criteria to be worth doing:

- If a failure is *hidden*, it has no direct consequences. So an on-condition task intended to prevent a hidden failure should reduce the risk of the multiple failure to an acceptably low level. In practice, because the function is hidden, many of the potential failures that normally affect evident functions would also be hidden. What is more, much of this type of equipment suffers from random failures with very short or nonexistent P-F intervals, so it is fairly unusual to find an on-condition task that is technically feasible and worth doing for a hidden function. But this does not mean that one should not be sought.
- If the failure poses an intolerable risk to safety or the environment, an on-condition task is only worth doing if it can be relied on to give enough warning of the failure to ensure that action can be taken in time to avoid the safety or environmental consequences.
- If the failure does not involve safety, the task must be cost-effective, so over a period of time, the cost of doing the on-condition task must be less than the cost of not doing it. The question of cost-effectiveness applies to failures with operational and nonoperational consequences, as follows:

- Operational consequences are usually expensive, so an on-condition task that reduces the rate at which the operational consequences occur is likely to be cost-effective. This is because the cost of inspection is usually low. This was illustrated in the example in Section 7.6, Chapter 7.
- The only cost of a functional failure that has nonoperational consequences is the cost of repair. Sometimes this is almost the same as the cost of correcting the potential failure that precedes it. In such cases, even though an on-condition task may be technically feasible, it would not be cost-effective, because over a period of time, the cost of the inspections plus the cost of correcting the potential failures would be greater than the cost of repairing the functional failure. However, an on-condition task may be justified if the functional failure costs a lot more to repair than the potential failure, especially if the former causes secondary damage.

8.15 Selecting Proactive Risk Management Strategies (Proactive Tasks)

It is seldom difficult to decide whether a proactive task is *technically feasible*. The characteristics of the failure govern this decision, and they are usually clear enough to make the decision a simple yes-no affair.

Deciding whether they are *worth doing* usually needs more judgment. For instance, Figure 8.23 indicates that it may be technically feasible for two or more tasks of the same category to prevent the same failure mode. They may even be so closely matched in terms of cost-effectiveness that which one is chosen becomes a matter of personal preference.

The situation is complicated further when tasks from two *different* categories are both technically feasible for the same failure mode.

For example, most countries nowadays specify a minimum legal tread depth for tires (usually about 1/8 inch). Tires that are worn below this depth must be either replaced or retreaded. In practice, truck tires—especially tires on similar vehicles in a single fleet working the same routes—show a fairly close relationship between age and failure. Retreading restores nearly all the original failure resistance, so the tires could be scheduled for restoration after they have covered a set distance. This means that all the tires in the truck fleet would be retreaded after they had covered the specified mileage, whether or not they needed it.

Figure 8.5, repeated as Figure 8.27, could have been drawn for just such a fleet. This shows that in terms of normal wear, all tires last between 30,000 and 50,000 miles. If a scheduled restoration policy were to be adopted on the basis of this information, there is a rapid increase in the conditional probability of this failure mode at 30,000 miles and none of these failures occur before this age, so all the tires would be retreaded at 30,000 miles. However, if this policy were adopted, many tires would be retreaded long before it was really necessary. In some cases, tires that could have lasted as much as 50,000 miles would be retreaded at 30,000 miles, so they could lose up to 20,000 miles of useful life.

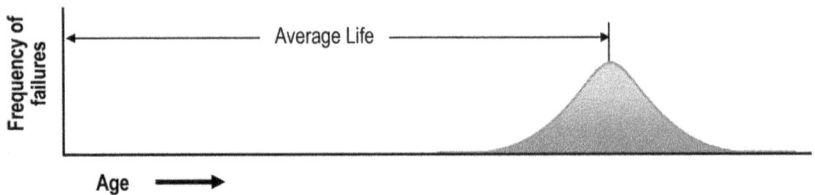

FIGURE 8.27   Frequency of failure and "average life"

On the other hand, as discussed in Section 8.6, it is possible to define a potential failure condition for tires related to tread depth. Checking tread depth is quick and easy, so it is a simple matter to check the tires every 1,500 miles and to arrange for them to be retreaded only when they need it. In this way, the tires would average 40,000 miles between retreads (due to normal wear) without endangering the drivers, instead of 30,000 miles if the scheduled restoration task is done as described above—an increase in useful tire life of 33%. So, in this case on-condition tasks are much more cost-effective than scheduled restoration.

This example suggests the following basic order of preference for selecting proactive tasks.

On-Condition Tasks

On-condition tasks are considered first in the task selection process, for the following reasons:

- They can nearly always be performed without moving the asset from its installed position and usually while it is in operation, so they seldom interfere with the production process. They are also easy to organize.
- They identify specific potential failure conditions, and so corrective action can be clearly defined before work starts. This reduces the amount of repair work to be done and enables it to be done more quickly.
- By identifying equipment on the point of potential failure, they enable it to realize almost all of its useful life (as illustrated by the tire example).

Scheduled Restoration Tasks

If a suitable on-condition task cannot be found for a particular failure, the next choice is a scheduled restoration task or scheduled discard task. It too must be technically feasible, so the failures must be concentrated about an average age. If they are, scheduled restoration prior to this age can reduce the incidence of functional failures. This may be cost-effective if the failures have major economic consequences, or if the cost of doing the scheduled restoration task is significantly lower than the cost of repairing the functional failure. The disadvantages of scheduled restoration and scheduled discard tasks are:

- They can only be done when items are stopped and (usually) sent to the workshop, so the tasks nearly always affect production in some way.

- The age limit applies to all items, so many items or components that might have survived to higher ages will be removed.
- Restoration tasks involve shop work, so they generate a much higher workload than on-condition tasks.

These disadvantages mean that when both categories are technically feasible, on-condition tasks are nearly always more cost-effective than scheduled restoration or discard, so the former are considered first.

As mentioned in Chapter 2, scheduled restoration and scheduled discard are usually considered together because they have so much in common. When they are encountered in practice, it is usually obvious whether the failure mode concerned should be dealt with by scheduled discard or scheduled restoration. However, in the case of some failure modes, both categories of tasks can satisfy the criteria for technical feasibility. In these cases, the most cost-effective of the two should be selected.

In general, however, scheduled restoration is usually considered before scheduled discard, because it is inherently more conservative to restore things instead of throwing them away.

Scheduled discard is usually the least cost-effective of the three proactive tasks, but where it is technically feasible, it does have a few desirable features.

Safe-life limits may be able to prevent certain critical failures, while an economic-life limit can reduce the frequency of functional failures that have major economic consequences. However, these tasks suffer from all the same disadvantages as scheduled restoration tasks.

8.16 Combinations of Tasks

For a very small number of evident failure modes that have safety or environmental consequences, or for hidden failures where the multiple failure has safety and environmental consequences, a task cannot be found that on its own reduces the risk of failure to a tolerably low level, and for which a suitable modification does not readily suggest itself.

In these cases, it is sometimes possible to find a combination of tasks (usually from two different task categories, such as an on-condition task and a scheduled discard task), which reduces the risk of the failure to a tolerable level. Each task is carried out at the frequency appropriate for that task. Figure 8.28 illustrates what a combination of tasks may look like. However, it must be stressed that situations in which this is necessary are very rare, and care should be taken not to employ such tasks on a "belt and braces" basis.

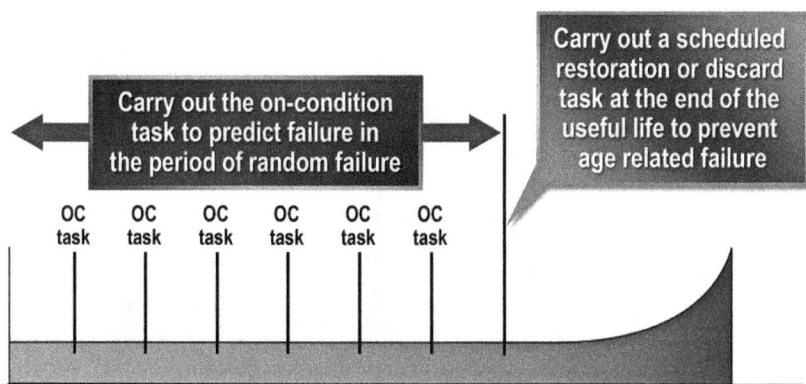

FIGURE 8.28 Combination of tasks

The Task Selection Process

The task selection process is summarized in Figure 8.29. This basic order of preference is valid for the vast majority of failure modes, but it does not apply in every single case. If a lower-order task is clearly going to be a more cost-effective method of managing a failure than a higher-order task, then the lower-order task should be selected. Although a combination of tasks is also classified as a *proactive task*, it is not shown in Figure 8.28. It is imperative that a combination of tasks will only be considered for failures (evident or hidden) that impact safety and/or the environment. It is further true that both tasks considered as the combination of tasks will be technically feasible.

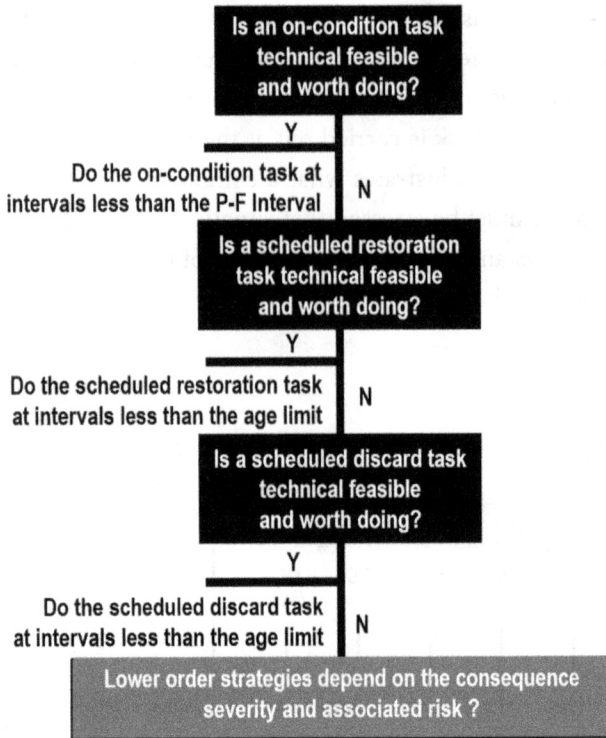

FIGURE 8.29 The task selection process

Example: Combination of tasks where the failure is hidden with physical risk (impact on environment), see Figure 8.30.

A water utility uses lime slurry to treat drinking water for pH control and to prohibit rust in the pipes and steel tanks. Lime powder is mixed with water to form the slurry, and the slurry is pumped into the drinking water clear wells through a diffuser. From the clear wells, drinking water is distributed to reservoirs and residential homes.

Raw lime in powder form is delivered via tanker trucks and offloaded into silos. The tankers are "pressurized" with a blower system, and the lime is pushed from the trucks into the silos. The level in the silo is monitored through an ultrasonic level sensor. The level sensor not only monitors the level but will also generate an alarm in the event of a low and high level, and it will shut the blower down and generate an alarm in the event of a high-high level. For protection, the level sensor is installed inside the silo in a tube.

Four functions were identified for the level sensor: (1) to monitor the level of lime inside the silo, (2) to generate an alarm when the lime level is low, (3) to generate an alarm when the lime level is high, and (4) to shut down the blower and generate an alarm when the lime has reached the high-high level. An RCM analysis revealed a failure of lime dust building up around the inside-bottom edge of the tube and blocking the ultrasonic sensor from detecting the level of lime in the silo. Another failure that was identified was the sensor failing in such a way that it was unable to detect the level or send an incorrect reading, causing the silos to be overfilled. The roof of the silo is fitted with a "relief valve" that would open in the event of overpressurization and release the pressure to the atmosphere. If this should happen, lime powder will escape the silo and enter the atmosphere. The release of lime to the atmosphere has to be reported to the local authorities, and it is considered a permit violation with environmental impact.

The review group felt that a failure-finding task alone would not secure the availability of the sensor to generate the high-high-level alarm, but a combination of tasks would be the only way to prevent the sensor from failing. Cleaning of the tube will be required at a set interval, while the sensor is being functionally tested at some other interval.

FIGURE 8.30

Previously it has been mentioned that if a proactive task (routine maintenance task), or combinations thereof, cannot be found that is both technically feasible and worth doing for failure modes posing an intolerable risk, we must consider alternative strategies, i.e., optimization of protective systems and one-time changes that would reduce the risk to a tolerable level. The location of the alternative strategies (other than routine maintenance) in the RCM3 decision framework is shown in Figure 8.31. At this point, we are still dealing with proactive risk management strategies and answering the seventh question in the RCM3 process:

What *must* be done to reduce intolerable risks to a tolerable level (using proactive risk management strategies)?

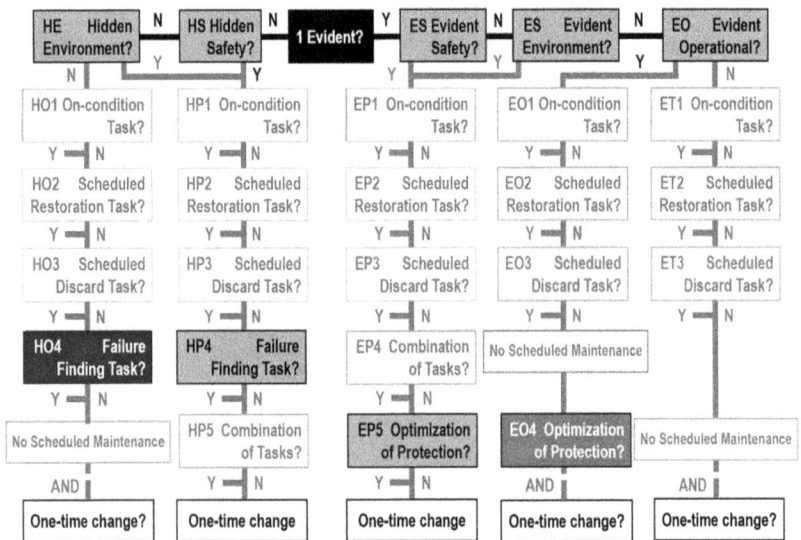

FIGURE 8.31   Proactive risk management strategies—engineering solutions

First, we focus on a review of failure-finding and optimizing protective systems; then we consider one-time changes and finally run-to-failure and routine tasks that fall outside the RCM3 decision framework such as walk-around checks. The latter will apply to the eighth question in the RCM3 process, although routine tasks may also be selected to satisfy the last step:

What *can* be done to manage or reduce tolerable risks in a cost-effective way?

For many years the belief was (and still is) that there are only three types of maintenance actions: predictive maintenance (check if something is failing), preventive maintenance (overhaul or replace at fixed intervals regardless of condition), and corrective maintenance (fix something that is failing or has failed). The tasks of failure-finding and functional checks were simply not considered.

Third-Generation Maintenance theory defined maintenance strategy development in two distinct categories: proactive maintenance (tasks done before the failure happens) and default actions (tasks done when no effective proactive task can be found). This implied that failure-finding, one-time changes (redesigns), and run-to-failure were only selected as default tasks—the focus during the third generation was very much on maintenance.

If Third-Generation Maintenance thinking is still followed today, many risks will go untreated and many failures will end up with recommendations for one-time changes or redesigns, because all failures were simply treated to avoid or minimize consequences, regardless of severity and probability of occurrence. Many more failure consequences will be managed through reliance on protective devices to reduce the consequences associated with the failure. It is the author's experience that the default actions as described were never properly implemented, and with growing complexity, reliance on the integrity of protective systems became even more important.

Fourth-Generation Maintenance requires a different approach, and RCM3 addresses the requirements through distinct differentiation between functions that are protected (through some protective device or system) and functions that are not. This distinction allows true optimization and risk treatment by optimizing the protective systems and not simply relying on them as a default. Protective devices directly related to the failure (as described in our discussion of the next-higher-level effect in Chapter 6) are optimized with consideration of the reliability of the protected system and risk associated with the failure.

Managing the integrity of the protective devices and systems requires a more structured approach than relying on experience and tribal knowledge only, and so we therefore refer to these actions as one of the *engineering solutions*. The other is the one-time change, i.e., redesign and defect elimination, which will be discussed at a later stage.

John Moubray mentioned in his book *RCMII* that, at the time of writing, many existing maintenance programs provided for fewer than one-third of protective devices to receive any attention at all (and then usually at inappropriate intervals). The people who operated and maintained the plant covered by these programs were aware that another third of these devices existed but paid them no attention, while it was not unusual to find that no one even knew that the final third even existed. This lack of awareness and attention meant that most of the protective devices in industry—our last line of protection when things go wrong—were maintained poorly or not at all.

This has not seemed to have changed much in the past 20+ years since *RCMII* was written. Even though some protective systems are tested, it is still done at the incorrect interval (normally to suit some production outage), or not enough information is available to develop proper strategies and the devices are not tested as a whole (tests are simulated). This remains a challenge, but if industry is serious about safety and environmental integrity, then this must be given a top priority as a matter of urgency. As John Moubray mentioned in his book, more and more maintenance professionals would become aware of the importance of this neglected area of maintenance, and it was likely to become a bigger maintenance strategy issue in the future than what predictive maintenance had been. We saw this prediction come true, and the next section will explore some of the old and new thinking around this issue.

8.17 Optimization of Protective Devices

Existing protective devices may have been installed to reduce the consequences and risk associated with failures that matter. To successfully manage protective systems, the risk associated with these devices must be addressed proactively as part of the primary risk management strategy. These are no longer treated as default actions and include the following two scenarios:

- **Evident failures.** If a routine maintenance task cannot be found that reduces the risk of a failure to a tolerable level and the item is protected by one or more protective devices, the question arises whether the protection will do so when the item fails. This assumes that the protection will take over from a function that has failed. Please note, this is different from the multiple failure we discussed earlier where the protection is also in a failed state when the protected function fails.

  If the optimization of the protective device will reduce the risk to a tolerable level by reducing the consequence severity, a periodic *failure-finding task* for the protective device will be recommended. If a suitable failure-finding task cannot be found, then a one-time change will have to be considered for failures posing an intolerable risk to safety and the environment (the item may have to be redesigned), and no scheduled maintenance and a one-time change may be recommended for failures posing an economic risk.

- **Hidden failures.** If a routine task cannot be found that reduces the risk of the *multiple failure* associated with a hidden function to a tolerably low level, then a periodic *failure-finding task* must be performed. If a suitable failure-finding task cannot be found for failures posing an intolerable risk to safety and environment, then the secondary decision is to seek a combination of tasks; and if a combination of tasks cannot be found, a one-time change will have to be considered (the item may have to be redesigned).

For multiple failures posing an intolerable economic risk, if the failure-finding task will not reduce the risk to a tolerable level, no scheduled maintenance will be selected, and a one-time change may be considered.

8.18 Failure-Finding Tasks

Multiple Failures and Failure-Finding

A multiple failure occurs if a protected function fails while a protective device is in a failed state. This phenomenon was illustrated in Figure 7.24 in Chapter 7. Figure 7.25 showed that the probability of a multiple failure can be calculated as follows:

$$\begin{aligned} \text{Probability of a multiple failure} = \text{probability of failure} \\ \text{of the protected function} \times \text{average availability} \quad (1) \\ \text{of the protective device} \end{aligned}$$

This led to the conclusion that the probability of a multiple failure can be reduced by reducing the unavailability of the protective device—in other words, by increasing its availability. Chapter 7 went on to explain that the best way to do this is to prevent the protective device from getting into a failed state by applying some sort of proactive maintenance.

Chapter 7 explained how to decide whether any sort of proactive maintenance is worth doing. However, when the criteria described in Chapter 7 are applied to hidden functions, it transpires that fewer than 5% of these functions are susceptible to any form of predictive or preventive maintenance.

Nonetheless, although proactive maintenance is often inappropriate, it is still essential to do something to reduce the probability of the multiple failure to the required level. This can be done by checking periodically whether the hidden function is still working.

For example, we cannot prevent the failure of a brake light bulb. So, if there is no warning circuit to show that a bulb has failed, the only way to reduce the possibility that a burned-out bulb will fail to warn other drivers of our intentions is to check if it is still working and replace it if it has failed.

Such checks are known as failure-finding tasks.

Scheduled failure-finding entails checking a hidden function at regular intervals to find out if it has failed.

Technical Aspects of Failure-Finding

This section looks at key technical aspects of failure-finding, describes how to determine failure-finding intervals, defines the formal technical feasibility criteria for failure-finding, and considers what should be done if a suitable failure-finding task cannot be found.

The objective of failure-finding is to satisfy ourselves that a protective device will provide the required protection if it is called upon to do so. In other words, we are checking whether it still works as it should. (This is why failure-finding tasks are also known as functional checks.) The following subsections consider some of the key issues in this area.

Check the Entire Protective System. A failure-finding task must be sure of detecting all the failure modes that are reasonably likely to cause the protective device to fail. This is especially true of complex devices such as electric circuits. In these cases, the function of the entire system should be checked from sensor to actuator. Ideally, this should be done by simulating the conditions that the circuit should respond to and by checking if the actuator gives the right response. Do not simulate tests; it is possible that not all components will be tested during simulated tests.

For example, a pressure switch may be designed to shut down a machine if the lubricating oil pressure drops below a certain level. Wherever possible, switches like this should be checked by dropping the oil pressure to the required level and checking whether the machine shuts down.

Similarly, a fire detection circuit should be checked from a smoke detector to a fire alarm by blowing smoke at the detector and checking if the alarm sounds.

Do Not Disturb. Dismantling anything always creates the possibility that it will be put back together incorrectly. If this happens to a hidden function, the fact that it is hidden means that no one will know it has been left in a failed state until the next check (or until it is needed). For this reason, we should always look for ways of checking the functions of protective devices without disconnecting or otherwise disturbing them.

This having been said, some devices simply have to be dismantled or removed altogether to check if they are working properly. In these cases, great care must be taken to do the task in such a way that the devices will still work when they are returned to service. (The mathematical implications of the fact that a failure-finding task might induce a failure are considered later in this chapter.)

It Must Be Physically Possible to Check the Function. In a very small but still significant number of cases, it is impossible to carry out a failure-finding task of any sort. These are:

- Where it is impossible to gain access to the protective device in order to check it (this is almost always a result of thoughtless design).
- When the function of the device cannot be checked without destroying it (as in the case of fusible devices and rupture discs). In most such cases, other technologies are available (such as circuit breakers instead of fuses). However, in one or two cases, our only options are to find some other way of managing the risks associated with untestable protection until something better comes along or to abandon the processes concerned.

Minimize Risk While the Task Is Being Done. It should be possible to carry out a failure-finding task without significantly increasing the risk of the multiple failure. An example is overspeeding something in order to check whether the overspeed protection mechanism works.

If a protective device must be disabled in order to carry out a failure-finding task, or if such a device is checked and found to be in a failed state, then alternative protection should be provided, or the protected function should be shut down until the original protection is restored. This issue is discussed in more detail later.

Failure-finding should not be carried out on systems where it is called for but would simply be too dangerous. (If society is serious about safety and environmental integrity, it is debatable whether such systems should be allowed to exist at all.)

The Frequency Must Be Practical. It must be practical to do the failure-finding task at the required intervals. However, before we can decide whether a required interval is practical, we need to determine what interval is actually required. This is considered next and discussed in more detail.

8.19 Failure-Finding Task Intervals

This section describes how to determine the frequency of failure-finding tasks. It starts by explaining that this frequency depends on two variables: the desired availability (which depends on the associated risk) and the frequency of failure of the protective device (as given by its mean time between failures). It goes on to describe how we establish *desired availability* and then explores different methods for determining or calculating failure-finding intervals under different circumstances.

Failure-Finding Intervals, Availability, and Reliability

We have seen that predictive and preventive maintenance task intervals are each based on just one variable (P-F interval and useful life, respectively). The following paragraphs will show that not one but two variables—availability and reliability—are used to set failure-finding intervals.

Figure 8.32 shows a situation in which 10 motorbikes have been in service for 4 years. This means that the total service life of the fleet of bikes in this period is

$$10 \text{ bikes} \times 4 \text{ years} = 40 \text{ years}$$

The brake light on each motorbike has been checked once a year for 4 years. (This example assumes that no attempt is made to check the lights

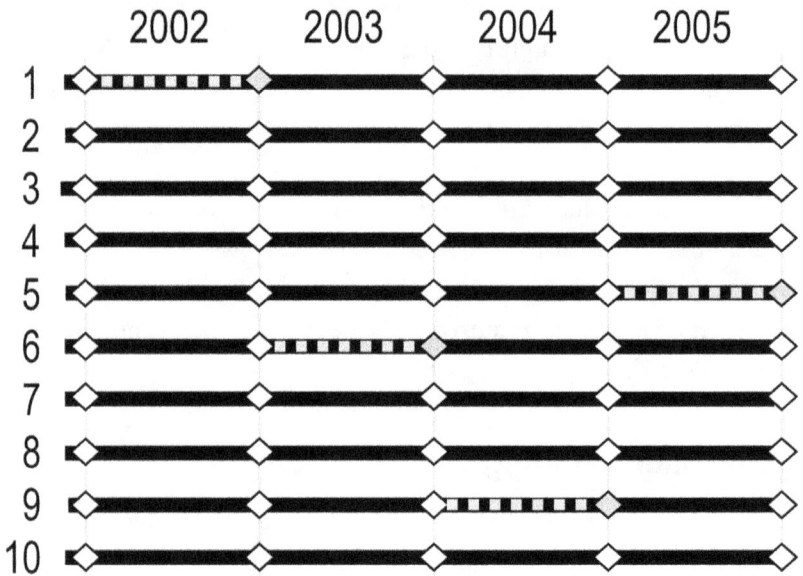

FIGURE 8.32 Brake light failures

between the annual checks.) Over the 4-year period, the lights have been found to be in a failed state on four occasions, as shown in Figure 8.30. So the mean time between failures (MTBF) of the brake lights is

40 years in service ÷ 4 failures = 10 years

In this case, the failure-finding interval of 1 year is equal to 10% of the MTBF of 10 years. However, we don't know exactly when each failed light ceased to function. One might have failed the day after the last check, another the day before the current check, and the rest at some time in between. All we know for sure is that each of the four lights failed some time during the year preceding the check. So in the absence of any better information, we assume that, on average, each failed light failed half way through the year. In other words, on average, each of the failed lights was out of service for half a year. This means that over the 4-year period, our failed lights were in a failed state for a total of

4 failed lights × 0.5 year each in a failed state = 2 years

So on the basis of the above information, it seems that we can expect an average unavailability from our brake lights of

2 years in a failed state ÷ 40 years in service = 0.05 (5%)

This corresponds to an availability of 95%.

The above example suggests that there is a linear correlation between the unavailability (5%), the failure-finding interval (1 year), and the reliability of the protective device as given by its MTBF (10 years), as follows:

$$\text{Unavailability} = 0.5 \times \text{failure-finding interval} \div \text{MTBF}$$
$$\text{of the protective device}$$

It can be shown that this linear relationship is valid for all unavailabilities of less than 5%, provided that the protective device conforms to an exponential survival distribution (failure pattern E or random failure). See Cox and Tait (1991) or Andrews and Moss (1993).

Excluding Task Time and Repair Time. The "unavailability" of the protective device in the above formula does not include any unavailability incurred while the failure-finding task is being performed, nor does it include any unavailability caused by the need to repair the device if it is found to be failed. This is so for two reasons:

- The unavailability required to carry out the failure-finding task and to effect any repairs is likely to be very small relative to the unrevealed unavailability between tasks, to the extent that it will usually be negligible on purely mathematical grounds.
- Both the failure-finding task and any repairs that might be needed should be carried out under tightly controlled conditions. These conditions should greatly reduce—if not completely eliminate—the chance of a multiple failure while the intervention is under way. This entails either shutting down the protected system or arranging alternative protection until the system has been fully

restored. If this is done properly, the unavailability resulting from the (controlled) intervention can be ignored in any assessments of the probability of a multiple failure.

In the RCM decision process, the latter point is covered by the criteria for assessing whether a failure-finding task is worth doing. If there is a significant increase in the likelihood of a multiple failure while the task is under way, the answer to the question "Does the task reduce the probability of a multiple failure to a tolerable level?" will be no, and the RCM decision process defaults to the secondary default actions discussed later.

Calculating FFI Using Availability and Reliability Only. If we use FFI to describe the failure-finding interval and M to describe the MTBF of the protective device, the above unavailability equation can be rearranged to give the following formula:

$$FFI = 2 \times unavailability \times M_{TIVE} \quad\quad (2)$$

where TIVE means protec*tive*.

This tells us that in order to determine the failure-finding interval for a single protective device, we need to know its mean time between failures and the desired availability of the device (from which we can determine the unavailability to be used in the formula).

For instance, assume that the riders of our motorbikes decide they are not satisfied with an availability of 95% and would prefer to see it increased to 99%. The associated unavailability is 1%. If the MTBF of the brake lights stays unchanged at 4 years, the checking interval needs to be changed from once a year to:

$$FFI = 2 \times 0.01 \times 4 \text{ years} = 2\% \text{ of } 48 \text{ months} = 1 \text{ month}$$

In other words, based on their availability expectations and the existing failure data, the bikers need to check whether their brake lights are working once a month. If they want an availability of 99.9%, they need to check about twice a week.

(Strictly speaking, the above calculations are only valid if the brake lights on all the bikes are used about the same number of times each week. If there is a wide variation, both the MTBF and the failure-finding interval should be calculated in terms of distance traveled, or even more precisely, in terms of the number of times the brakes—and hence the brake lights—are used. However, the key point to note at this stage is the connection between the checking interval, the desired availability, and the MTBF.)

For people who are uncomfortable with mathematical formulas, Equation 2 above can be used to develop a simple table, shown in Figure 8.33.

Desired Availability	99.99%	99.9%	99.8%	99.5%	99%	98%	95%	93%
Failure-finding Interval (% of MTBF)	0.02%	0.2%	0.4%	1%	2%	4%	10%	15%

FIGURE 8.33 Failure-finding intervals, availability, and reliability

Required Availability. Having established the relationship between availability, reliability, and failure-finding intervals, the next issue is to consider how we determine or decide what the desired availability should be. Chapter 7 explains how this can be achieved in three steps:

1. Establish the probability that the organization is prepared to tolerate for the *multiple failure* that could occur when the protective system (hidden function) is not working when called upon.
2. Determine how often the protective device will be called upon; in other words, determine the probability that the *protected function* will fail in the period under consideration.
3. Determine what availability the *hidden function* (protective device) must have in order to reduce the probability of the multiple failure to the desired level.

In addition to carrying out these steps, we need to determine or find out the mean time between failures of the hidden function. Once this has been done, we are in a position to establish the task frequency that corresponds to the level of availability that is required. Figure 8.34 summarizes the duty/standby pump example in Chapter 7, where:

- In step 1 above, the users decided that they wanted the probability of the multiple failure to be less than 1 in 1,000 in any one year.
- In step 2, they established that the rate of unanticipated failures of the duty pump could be reduced to an average of 1 in 10 years.
- This meant that the unavailability of the standby pump must not exceed 1%, so the availability of this pump has to be 99% or better (step 3).

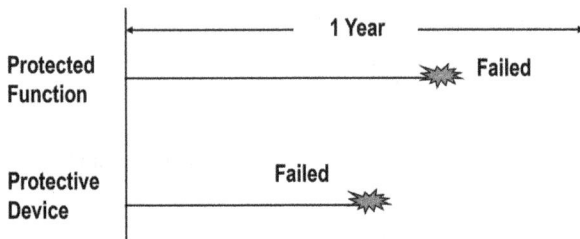

FIGURE 8.34   Desired availability of the protective device

Figure 8.34 suggests that to achieve an availability of 99% for the standby pump, someone would need to carry out a failure-finding task (in other words, check that it is fully functional) at an interval of 2% of its mean time between failures. Records might show that the standby pump has a mean time between failures of 8 years (or about 400 weeks), so the failure-finding task frequency should be:

$$2\% \text{ of } 400 \text{ weeks} = 8 \text{ weeks} = 2 \text{ months}$$

8.20 Rigorous Methods for Calculating FFI

A single formula for determining failure-finding intervals that incorporates all the variables considered so far can be developed by combining Equations (1) and (2) above, as explained in the following paragraphs. Let us begin by defining a few key terms:

- A probability of a multiple failure of 1 in 1,000,000 in any one year implies a *mean time between multiple failures* of 1,000,000 years. Let us call this M_{MF}. If this is so, then the probability of a multiple failure occurring in any one year is $1/M_{MF}$.
- We have seen that if the demand rate of the protected function is (say) once in 200 years, this corresponds to a probability of failure for the protected function of 1 in 200 in any one year, or a mean time between failures of the protected function of 200 years. Let us call this M_{TED} (where TED stands for protec*ted*), so the probability of failure of the protected function in any one year will be $1/M_{TED}$. This is also known as the demand rate.
- As before, M_{TIVE} is the *mean time between failures of the protective device*, and FFI is the *failure-finding task interval*.
- U_{TIVE} is the allowed unavailability of the protective device.

If we substitute the above expressions, Equation (1) becomes

$$\frac{1}{M_{MF}} = \left(\frac{1}{M_{TED}} \right) \times U_{TIVE} \qquad (3)$$

This can be rearranged as follows:

$$U_{TIVE} = \frac{M_{TED}}{M_{MF}} \qquad (4)$$

Equation (2) above states that

$$FFI = 2 \times U_{TIVE} \times M_{TIVE} \qquad (2)$$

So substituting U_{TIVE} from Equation (4) into Equation (2) gives

$$FFI = \frac{2 \times M_{TIVE} \times M_{TED}}{M_{MF}} \qquad (5)$$

This formula allows a failure-finding interval to be determined in a single step as follows:

If we apply this formula to the figures used in the duty/standby pump system mentioned above, M_{MF} is 1,000 years, M_{TIVE} is 8 years, and M_{TED} is 10 years, and so

$$FFI = \frac{2 \times 8 \times 10}{1,000} \approx 2 \text{ months}$$

Multiple Failure Modes in a Single Protective Device. Throughout this chapter, all the failure possibilities that could cause each protective device to fail have been grouped together as one single failure mode ("standby pump fails"). The vast majority of protective devices can be treated in this way, because all the failure modes that could cause a protective device to cease to function are checked when the function of the device as a whole is checked.

However, it is sometimes appropriate to carry out a detailed FMEA of the device in order to identify individual failure modes that might on their own cause the device to be unable to provide the required protection. This is usually done in two sets of circumstances:

- When some of the failure modes are known to be susceptible to proactive maintenance, but others are neither predictable nor preventable. In these cases, the appropriate on-condition or scheduled restoration/discard task should be applied to the failure modes that qualify, and failure-finding tasks should be applied to the *remainder* of the failure modes.
- When the protective device is new and the only failure data that are available (from data banks, component suppliers, or wherever) apply to parts of the device but not to the device as a whole.

In these cases, Equation (5) above can be modified to accommodate the MTBF of each component of the device.

When the Failure-Finding Task Can Cause the Failure. A major practical problem that affects the whole question of failure-finding is that the task itself can cause the very failure that it is supposed to detect. This usually happens in one of two ways:

- The task stresses the system in such a way that it eventually causes it to fail (as might be the case when a switch is tested, where the mere act of switching imposes stresses on the mechanism of the switch).
- If the system needs to be disturbed to do the task, there is always a chance that the person doing it will leave the system in a failed state.

In both cases, the device will be in a failed state from the moment the test is completed. If p is the probability that it will be left in such a state after a test, then p (as a decimal) will be its unavailability caused by the testing process. If M_{OTHER} is the mean time between *failures caused by phenomena other than the test,* it can be shown (for a single system) that

$$FFI = \frac{2 \times M_{OTHER}}{(1-p)} \times \left(\frac{M_{TED}}{M_{MF}} - p \right) \qquad (6)$$

In this equation, the expression $(1 - p)$ can be ignored if p is less than 0.05.

If the act of a breaker opening is the *only* cause of failure (in other words, there is no M_{OTHER}) *and* the failure conforms to an exponential survival distribution, the probability of the multiple failure is the demand rate (in years) multiplied by the number of cycles between failures of the protective device.

For example, if the demand rate is 40 years and the switch lasts an average of 600,000 cycles, then the probability of a multiple failure is

1 in (40 × 600,000) = 1 in 24,000,000 years

This is so, because if the only failure is caused only by opening the circuit, then the act of operating the breaker to check whether it has failed will simultaneously:

- Enable you to find out if the last operation of the breaker caused it to fail (functional check)

- Stress the breaker and so create the possibility that it will fail as a result of the check

So under this unique set of circumstances (random failure caused solely by operating the system), a failure-finding task that involves operating the item to check whether it has failed will have no effect on the probability of the multiple failure, regardless of how often the task is done. In other words, when considering technical feasibility for this task, the answer to the question "Is it practical to do the task at the required interval?" is no, because there is no suitable interval.

So in this case, if the organization wants the probability of a multiple failure of the switch described above to be less than 1 in 100,000,000 years, the only way the organization can achieve this is by reducing the demand rate on the switch and/or by installing either more switches or a more reliable switch.

All of this indicates that failure rates that are given as a number of operations should be treated with great caution, for the following reasons:

- They seldom indicate whether the failure under consideration is hidden or evident.
- They do not indicate whether the underlying failure pattern is age-related, in which case some form of scheduled restoration or scheduled discard might be appropriate, or whether it is random.
- Despite the previous comment, a failure mode caused solely by the operation of a switch is likely to be age-related. If this is so, then it is equally likely that a preventive task could be identified that reduces the probability of a multiple failure to the required level.

This suggests that as a rule, important switches—especially big circuit breakers—should not be treated as single failure modes. Rather, they should be subjected to a detailed FMEA, and the most appropriate maintenance policy should be developed for each failure mode.

Sources of Data for FFI Calculations

Most modern industrial undertakings possess several thousand protected systems, most of which incorporate hidden functions. The multiple failures associated with many of these systems will be serious enough to necessitate using one of the rigorous approaches to failure-finding.

If accurate data about the probability of failure of the protected function and the mean time between failures of the hidden function are available, the calculations can be performed quite quickly. If this information is not available—and very often it is not—then it is necessary to estimate what these variables are likely to be in the context under consideration. In rare cases, it might be possible to obtain data from one of the following:

- The manufacturers of the equipment
- Commercial data banks
- Other users of similar equipment

More often, however, the estimates have to be based on the knowledge and experience of the people who know the most about the equipment. In many cases these are operators and maintenance craftspeople. (When using data from external sources, take special note of the operating context of the items for which the data were gathered compared with the context in which your equipment is operating.)

Once a failure-finding task frequency has been established and the tasks are being done on a regular basis, it becomes possible to verify the assumptions used to determine the frequency quite rapidly. However, this does require the keeping of absolutely meticulous records, not only about when each failure-finding task is done, but also about:

- Whether or not the hidden function is found to be functional each time the task is done
- How often the protected function fails (this can often be inferred from the number of times the protected function makes use of the protective device—for instance, from the number of times a pressure relief valve actually has to relieve the pressure in the system)

On the basis of this information, the actual mean time between failures can be calculated and, if necessary, the task frequency revised accordingly.

Failure modes where the MTBF and/or the associated failure patterns are completely unknown—and a satisfactory guess cannot be made—should be put into an age-exploration program right away to establish the true picture. If the situation is such that the uncertainty cannot be tolerated while the data are being gathered—in other words, if the consequences of guessing wrong are simply too serious for the organization (or in some cases, society as a whole) to accept—then every effort should be made to change the consequences. This, in turn, will nearly always necessitate some form of redesign.

An Informal Approach to Setting Failure-Finding Intervals

Not every hidden function is important enough to warrant the time and effort needed to do a full, rigorous analysis. This applies mainly to multiple failures that do not affect safety or the environment. It could also apply to multiple failures that could affect safety but where the protected function is inherently very reliable and the threat to safety is marginal.

In these cases, it may be sufficient to take a general view of the entire protected system in its operating context and to go straight to a decision on a desired level of availability for the hidden function. This decision is then used in conjunction with the MTBF of the hidden failure to set a task interval, using the table in Figure 8.33. (Some organizations even go so far as to use an availability of 95% for all hidden functions where the associated multiple failure cannot affect safety or the environment. However, general policies of this nature can be dangerous, so they should only be used by people who have extensive experience with this type of analysis.)

Once again, if adequate records about hidden failures are not available—and they seldom will be—it will be necessary to guess at the MTBFs to begin with. But again, these records should be compiled as quickly as possible to validate the initial estimates.

Other Methods of Calculating Failure-Finding Intervals

The range of techniques for setting failure-finding intervals described so far in this chapter is by no means exhaustive. Many additional variants have been developed by the Aladon network of RCM specialists. These include formulas for:

- Voting systems
- Multiple, independent, fully redundant systems
- Deriving cost-optimized intervals for systems where the multiple failures do not affect safety or the environment

As this book is only intended to provide an introduction to this subject, these formulas are not included in this chapter.

The Practicality of Task Intervals

The methods described so far for calculating failure-finding intervals sometimes produce very short or very long intervals. In some cases, these intervals might be too long or too short.

A very short failure-finding task interval has two main implications:

- Sometimes the interval is simply far too short to be practical. Examples would be failure-finding tasks that call for major items of a plant to be shut down every few days.
- The task could cause habituation (which might happen if a fire alarm is tested too often).

In these cases, the proposed task is rejected, and we move on to the next stage of the RCM decision-making process, as discussed later.

We also encounter very long intervals—sometimes as long as a hundred years or more. Here the process is clearly suggesting that we really need not worry about doing the task at all. In these cases the proposed "task" should be stated as follows: "The risk-reliability profile is such that failure-finding is felt to be unnecessary."

In rare cases, task intervals emerge that are significantly longer than the demand rate (M_{TED}). It makes little sense to carry out a failure-finding task at intervals (FFI) that are longer than the system is effectively testing itself (M_{TED}), so in these cases, the answer to the question "Is it practical to do the task at the required interval?" will be no. However, bear in mind that if a failure-finding task is not done on a protected system (and if M_{TIVE} is more than four or five times greater than M_{TED}, which is usually the case), it can be shown that

$$M_{MF} = M_{TED} + M_{TIVE}$$

If this value of M_{MF} is too low to be acceptable, then the protection is inadequate, and the system will almost certainly have to be redesigned, as discussed in the Section 8.20.

Failure-Finding Tasks for Hidden Functions with Economic Consequences. From the previous formulas, it can be shown that the optimum value of FFI (combined cost is a minimum) for a single protective device is equal to

$$FFI = \left[\frac{2M_{TIVE}M_{TED}C_{FF}}{C_{MF}} \right]^{\frac{1}{2}} \tag{7}$$

where:

$$
\begin{aligned}
M_{TIVE} &= \text{MTBF of protective device} \\
M_{TED} &= \text{MTBF of protected function} \\
C_{FF} &= \text{cost of one failure-finding task} \\
C_{MF} &= \text{cost of one multiple failure}
\end{aligned}
$$

8.21 The Technical Feasibility of Failure-Finding

The issues discussed in Sections 8.18 and 8.19 of this chapter mean that for a failure-finding task to be technically feasible, it should be possible to do the task, it should be possible to do it without increasing the risk of the multiple failure, and it should be practical to do the task at the required interval.

Failure-finding is technically feasible if:

- It is possible to do the task.
- The task does not increase the risk of a multiple failure.
- It is practical to do the task at the required interval.

The objective of a failure-finding task is to reduce the probability of the multiple failure associated with the hidden function to an acceptable level. It is only worth doing if it achieves this objective.

Failure-finding is worth doing if it reduces the probability of the associated multiple failure to an acceptable level.

Failure-Finding Is a Proactive Risk Management Strategy!

Bear in mind that successful proactive maintenance prevents things from failing, whereas failure-finding accepts that they will spend some time—albeit not very much—in a failed state. This means that proactive maintenance is inherently more conservative (in other words, safer) than failure-finding, so the latter should only be specified if a more effective proactive task cannot be found. For this reason, it is wise to avoid RCM decision diagrams that put failure-finding ahead of proactive mainte-nance in the task selection process.

The RCM3 decision logic treats protective functions and protected functions together in order to reduce the risk associated with the fail-ure of the protected function. It must be true that only failures with catastrophic consequences (impacting safety or the environment and/or having very expensive economic consequences) are protected. The reliability of the protected function will be considered first through applying some proactive maintenance task. If a proactive task cannot be found that would reduce the risk associated with the failure of the protected function to a tolerable level, the optimization of existing protective devices (hidden functions) will be considered in order to do so. In RCM3, the protected function and protective device are no longer treated separately but are combined to proactively address the risk associated with the failure of the protected function.

We have also seen that scheduled failure-finding tasks only apply to hidden functions. This is because, by definition, the failure of an evident function inevitably becomes apparent to the operating crew, so there is no need to carry out regular checks to find out whether such a failure has occurred. So failure-finding tasks should only be considered if a functional failure will not become evident to the operating crew under normal circumstances and the failure is one for which a suitable proactive task cannot be found.

What If Failure-Finding Is Not Suitable?

If it transpires that a failure-finding task is not technically feasible or worth doing, we have exhausted all the possibilities that might enable us to extract the required performance from the existing asset. Where this leaves us is once again governed by the consequences of the multiple failure, as follows:

- If a suitable failure-finding task cannot be found and the multiple failure poses an intolerable risk to safety or the environment, a combination of tasks would then be considered in order to reduce the risk associated with the multiple failure to a tolerable level. Selecting a combination of tasks is uncommon for hidden functions but could in rare instances be required.

  An example of a combination of tasks is one where a protective device such as a check valve or damper would need periodic greasing (lubrication) while the system also needs to be tested through applying a failure-finding task (functional check).

  If a combination of tasks cannot be found that would reduce the risk of the multiple failure to a tolerable level, something must be changed in order to make the situation safe. In other words, a one-time change (redesign) is compulsory.

- If a failure-finding task cannot be found and the multiple failure does not affect safety or the environment, then it is acceptable to take no action, but a one-time change may be justified if the multiple failure has very expensive consequences (intolerable economic risk).

This decision process is summarized in Figure 8.35. (This diagram is a fuller description of this aspect of the process than the two boxes of the left-hand column in Figure 8.1.)

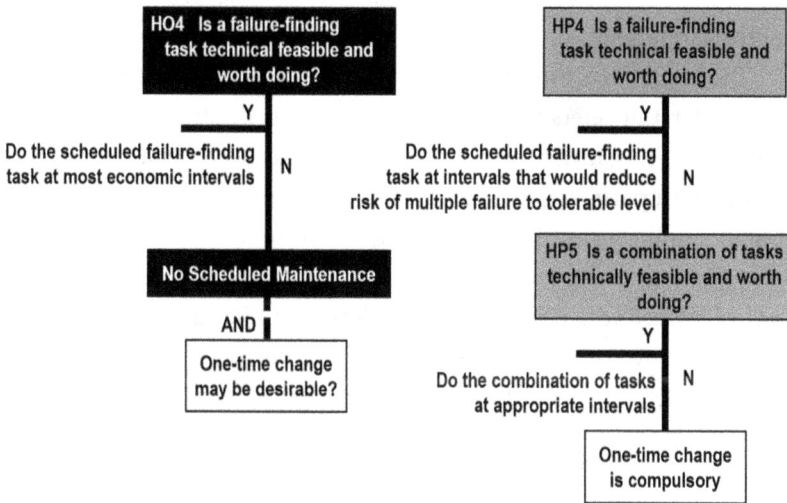

FIGURE 8.35   Failure-finding decision process

8.22 One-Time Change—Engineering Solution Applied Proactively

In RCM3, the engineering solution of a one-time change is applied proactively and no longer as a default action as previously required (RCM2 and the SAE standard). One-time changes still include physical redesigns and modifications (e.g., changing the physical properties of the item, changing the design, adding a protective device or redundancy), train-ing, and operating and maintenance procedures. We will explore each of these in the following sections.

Redesign and Modification

The question of equipment design has arisen again and again as we have traced the steps that must be followed to develop a successful maintenance program. In this part of the chapter, we consider two general issues that affect the relationship between design and main-tenance, and then we consider the part played by redesign in the task selection process. One-time changes don't only impact the con-sequences of the failure but may also impact the probability of the failure happening.

The term "one-time change" is used in its broadest sense in this chapter. First, it refers to any change in the specification of any item of equipment. This means any action that should result in a change in a drawing or a parts list. It includes changing the specification of a com-ponent, adding a new item, replacing an entire machine with one of a different make or type, or relocating a machine. It also means any other one-off change to a process or procedure that affects the operation of the plant. It even covers training as a method of dealing with a specific failure mode (which can be seen as "redesigning" the capability of the person being trained).

Defect Elimination. Defect elimination is the subject of many books and white papers and will not be discussed in detail in this section. It is the process of improving designs and processes to eliminate the reoccurrence of failures. These future failures are the design errors, the materials selection errors, the fabrication errors, the assembly errors, and any handling errors. When equipment or parts are installed, further causes of future failures arise from incorrect installation, incorrect assembly, incorrect mounting practices, inadequate protection against the elements (rain, dust, humidity), and deficient structural support (foundations and structures).

Some of these errors, along with commissioning errors and operating errors, cause failures early in the equipment's operating life and explain early-life or "start-up" failures. Those defects and errors that do not appear during the equipment's infant life will eventually surface and cause failures sometime later during its operating life.

The preferred terminology is to call the errors "defects," because that is what you get as a consequence of the mistake. But the truth is that a wrong action (or no action) was taken at some point, and as a consequence a defect was created. Every defect has a cause, and often more than one.

The process of defect elimination is to identify the error or defect cause (or mechanism) and apply the best and most effective corrective action—this will normally take the form of a one-time change.

Poka-Yoke. "Poka-yoke" is a Japanese term that means "mistake-proofing" or "inadvertent error prevention." The key word in the second translation, often omitted, is "inadvertent." There is no poka-yoke solution that protects against an operator's sabotage, but sabotage is a rare behavior among people. A poka-yoke is any mechanism that helps to avoid (*yokeru*) mistakes (*poka*). Its purpose is to eliminate product defects by preventing, correcting, or drawing attention to human errors as they occur. The concept was formalized, and the term adopted, by Shigeo Shingo as part of the Toyota Production System. It was originally described as "baka-yoke," but as this means "fool-proofing" (or "idiot-proofing"), the name was changed to the milder "poka-yoke."

A simple poka-yoke example is demonstrated when a driver of a car equipped with a manual gearbox must press on the clutch pedal (a process step, therefore a poka-yoke) prior to starting an automobile. The interlock serves to prevent unintended movement of the car. Another example of poka-yoke would be the car equipped with an automatic transmission, which has a switch that requires the car to be in park or neutral before the car can be started (some automatic transmissions require the brake pedal to be depressed as well). These serve as behavior-shaping constraints, as the action of "car in park (or neutral)" or "foot depressing the clutch/brake pedal" must be performed before the car is allowed to start. The requirement of a depressed brake pedal to shift most of the cars with an automatic transmission from park to any other gear is yet another example of a poka-yoke application. Over time, the driver's behavior is conformed with the requirements by repetition and habit.

Design and Maintenance

Changing anything is expensive. It involves the cost of developing the new idea (designing a new machine, drawing up a new operating procedure), the cost of turning the idea into reality (making a new part, buying a new machine, compiling a new training program), and the cost of implementing the change (installing the part, conducting the training program). Further indirect costs are incurred if equipment or people have to be taken out of service while the change is being implemented. There is also the risk that a change will fail to eliminate or even alleviate the problem it is meant to solve. In some cases, it may even create more problems.

As a result, the whole question of modifications should be approached with great caution. Two issues need particular attention:

- What do we consider first—design or maintenance?
- The relationship between inherent reliability and desired performance.

Which Comes First—Redesign or Maintenance?   Reliability, design, and maintenance are inextricably linked. This can lead to a temptation to start reviewing the design of existing equipment before considering its maintenance requirements. In fact, the RCM process considers maintenance first for two reasons:

- Most modifications take from 6 months to 3 years from conception to commissioning, depending on the cost and complexity of the new design. On the other hand, the maintenance person who is on duty today has to maintain the equipment as it exists today, not what should be there or what might be there some time in the future. So today's realities must be dealt with before tomorrow's design changes.

- Most organizations are faced with many more apparently desirable design improvement opportunities than are physically or economically feasible. By focusing on failure consequences, RCM does much to help us to develop a rational set of priorities for these projects, especially because it separates those that are essential from those that are merely desirable. Clearly, such priorities can only be established after the review has been carried out.

It is important to note, however, that one-time changes are no longer default actions but will be considered proactively in order to reduce the risks to tolerable levels.

Inherent Reliability Versus Desired Performance. Earlier chapters stressed that the initial capability of any asset is established by its design and by how it is made, and that maintenance cannot yield reliability beyond that inherent in the design. This led to two conclusions.

First, if the initial capability of an asset is greater than the desired performance, maintenance can help achieve the desired performance. Most equipment is adequately specified, designed, and built, so it is usually possible to develop a satisfactory maintenance program, as described in previous chapters. In other words, in most cases, RCM helps us to extract the desired performance from the asset as it is currently configured.

On the other hand, if desired performance exceeds inherent reliability, then no amount of maintenance can deliver the desired performance. In these cases, "better" maintenance cannot solve the problem, so we need to look beyond maintenance for the solutions. Options include:

- Modifying the equipment
- Changing operating procedures
- Lowering our expectations and deciding to live with the problem

This reminds us that maintenance is not always the answer to chronic reliability problems. It also reminds us that we must establish as soon and as precisely as possible what we want each piece of equipment to do in its operating context before we can start talking sensibly about the appropriateness of its design or its maintenance requirements.

One-Time Changes Applied Proactively

Figure 8.35 shows that a one-time change appears at the bottom of each risk category of the RCM3 decision diagram. In the case of failures posing intolerable risk to safety and/or the environment, one-time changes are compulsory, and in other cases, they may be desirable. Although one-time changes will be considered as a last strategy, they are considered proactively in RCM3 for failures where regular maintenance may not be sufficient to reduce intolerable risks to tolerable levels. One-time changes may also be the most cost-effective solution in the long run (depending on the failure rate). In this part of this chapter, we consider each case in more detail, starting with the safety case.

Safety or Environmental Consequences. If a failure could affect safety or the environment and no predictive or preventive task or combination of tasks can be found that reduces the risk of the failure to a tolerable level, and if the optimization of the existing protective device will not reduce the risk of the failure to a tolerable level, something must be changed, simply because we are dealing with a safety or environmental hazard that cannot be adequately prevented. In these cases, a one-time change is usually undertaken with one of two objectives:

- To reduce the probability of the failure mode occurring to a level that is tolerable. This is usually done by replacing the affected component with one that is stronger or more reliable.
- To change the item or the process in such a way that the failure no longer has safety or environmental consequences. This is most often done by installing a suitable protective device.

The protective devices were categorized earlier as follows:

- To alert operators of abnormal conditions
- To shut equipment down in the event of failure
- To eliminate or relieve abnormal conditions that are caused by failure and might otherwise cause more damage
- To take over from a function that has failed
- To prevent dangerous situations from arising

Remember that if such a device is added, its maintenance requirements must also be analyzed. Safety or environmental consequences can also be reduced by eliminating hazardous materials from a process, or even by abandoning a dangerous process altogether.

As mentioned in Chapter 7, when dealing with intolerable safety and/or environmental risks, RCM does not raise the question of economics. If the level of risk associated with any failure is regarded as intolerable, we are obliged either to prevent the failure or to make the process safe. The alternative is to accept conditions that are known to be unsafe or environmentally unsound. This is no longer acceptable in most industries.

Hidden Failures. In the case of hidden failures, the risk of a multiple failure can be reduced by modifying the equipment in one of four ways:

- **Make the hidden function evident by adding another device.** Certain hidden functions can be made evident by adding another device that draws the attention of the operator to the failure of the hidden function.

   For example, a battery used to power a smoke detector is a classical hidden function if no additional protection is provided. However, a warning light is fitted to most such detectors in such a way that the light goes out if the battery fails. In this way the additional protection makes the function of the battery evident. (Note that the light only tells us about the condition of the battery, not about the ability of the detector to detect smoke.)

Special care is needed in this area, because extra functions installed for this purpose also tend to be hidden. If too many layers of protection are added, it becomes increasingly difficult—if not impossible—to define sensible failure-finding tasks. A much more effective approach is to substitute an evident function for the hidden function, as explained next.

- **Substitute an evident function for the hidden function.** In most cases this means substituting a genuinely fail-safe protective device for one that is not fail-safe. This is surprisingly difficult to do in practice, but if it is done, the need for a failure-finding task falls away at once.

For example, one commonly used way to warn the driver of a vehicle that his brake lights have failed is to install a warning light that is switched on if the brake lights fail. (In many cases, this light is also switched on for a short while when the ignition is switched on. However, so are all the other lights on the dashboard. Under these circumstances one missing warning light is likely to be overlooked, so its function is effectively hidden.)

The system might also be configured in such a way that its full function can only be tested by disabling a brake light and seeing if the warning light comes on. This is a clumsy and invasive task that is likely to cause more problems than it solves, so it is likely to be dismissed on the grounds of impracticality. The multiple failures associated with this system could have serious safety consequences, so it is necessary to reconsider the design.

One way to eliminate this problem is to make the function of the brake lights and of the warning system evident. This can be done by substituting fiber-optic cables for the warning light and mounting the cables so that the driver looks through them at the brake lights every time he uses the brakes. (In fact, he sees a pinprick of light at the end of each cable.) In this situation, it is apparent to the driver if either a brake light or a cable fails. In other words, the function of this protective device is now evident, so failure-finding is no longer necessary.

• **Substitute a more reliable (but still hidden) device for the existing hidden function.** Figure 8.35 suggests that a more reliable hidden function (in other words, one that has a higher mean time between failures) will enable the organization to achieve one of three objectives:

  – To reduce the probability of the multiple failure without changing the failure-finding task intervals. This increases the level of protection.

  – To increase the interval between tasks without changing the probability of the multiple failure. This reduces resource requirements.

  – To reduce the probability of the multiple failure and increase the task intervals, giving increased protection with less effort.

• **Duplicate the hidden function.** If it is not possible to find a single protective device that has a high enough MTBF to give the desired level of protection, it is still possible to achieve any of the above three objectives by duplicating (or even triplicating) the hidden function. However, bear in mind that the function of all these devices would still need to be checked at an appropriate frequency.

Let us return to the example of a duty pump with a standby. It was explained on page 322 that if the users want the probability of a multiple failure to be less than 1 in 1,000, and the unanticipated failure rate of the duty pump is reduced to 1 in 10 years, then the availability of the standby pump has to be 99% or better. This led to the conclusion that a failure-finding task should be done on the standby pump every 2 months in order to achieve an availability of 99% (based on an MTBF for this pump of 8 years).

However, now let us assume that someone has decided that the probability of a multiple failure in this system should not exceed 1 in 100,000 (or 10^{-5}), rather than 1 in 1,000. If the mean time between unanticipated failures of the duty pump (M_{TED}) is unchanged at 10 years, applying Equation (4) shows that the unavailability (U_{TIVE}) of the standby pump should not exceed

$$U_{TIVE} = M_{TED}/M_{MF} = 10/100,000 = 10^{-4}$$

So, the unavailability of the standby pump must now not exceed 10^{-4} (0.01 %). If the MTBF of the standby pump is unchanged at 8 years, applying Equation (2) yields the following:

$$FFI = 2 \times 10^{-4} \times 8 \text{ years} = 14 \text{ hours}$$

Activating a standby pump this often is plainly impractical, so more thought has to be given to the design of this system.

In fact, Figure 8.36 shows that if we were to add a second standby pump, and ensure that the availability of each standby pump on its own exceeds 99% (corresponding to an unavailability of 1%, or 10^{-2}), the probability of the multiple failure would be

$$10^{-1} \times 10^{-2} \times 10^{-2} = 10^{-5}$$

or 1 in 100,000. Figure 8.34 suggests that this can be achieved by doing a failure-finding task on each standby pump at the original frequency of once every 8 weeks. In other words, a much higher level of protection is achieved without changing the task interval.

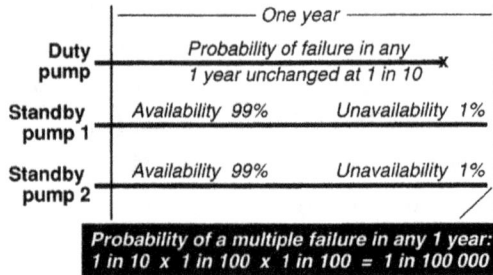

FIGURE 8.36   The effect of duplicating a hidden function

Operational and Non-operational Consequences.   If a technically feasible preventive task cannot be found that is worth doing for failures with operational or non-operational consequences, the immediate default decision is no scheduled maintenance. However, it may still be desirable to modify the equipment to reduce total costs. To achieve this, the plant could be modified to:

- Reduce the number of times the failure occurs, or possibly eliminate it altogether, again by making the component stronger or more reliable.
- Reduce or eliminate the consequences of the failure (for example, by providing a standby capability).
- Make a preventive task cost-effective (for instance, by making a component more accessible).

Note that in this case the failure consequences are purely economic, so modifications must be cost-justified, whereas they were the compulsory default action if the failure had safety or environmental consequences.

There is no one way to determine whether a modification will be cost-effective. Each case is governed by a different set of variables, which include a before-and-after assessment of maintenance and operating costs, the remaining technologically useful life of the asset, the likelihood that the modification will work, the number of other projects competing for the capital resources of the company, and so on.

A detailed cost-benefit study that takes all these factors into account can be very time-consuming, so it is helpful to know beforehand whether this effort is likely to be worthwhile. To help make a quick preliminary assessment, Nowlan and Heap (1978) developed the decision diagram shown in Figure 8.37.

No matter how reliable, all assets eventually are superseded by new technology or simply become obsolete. So, the first question to ask is whether the asset under consideration is rendered to be obsolete in the near future. If it is, then it is clearly not worth modifying it. On the other hand, if it is going to be around and useful for a while longer, the modification may be worthwhile and pay for itself. This is why the first question in Figure 8.37 asks:

Is the remaining technologically useful life of the equipment high?

Most organizations expect that modifications or new equipment will pay for itself within a certain period of time, say 2–5 years. This effectively sets the operational horizon for the equipment at, say, 5 years. This type of policy reduces the number of projects initiated

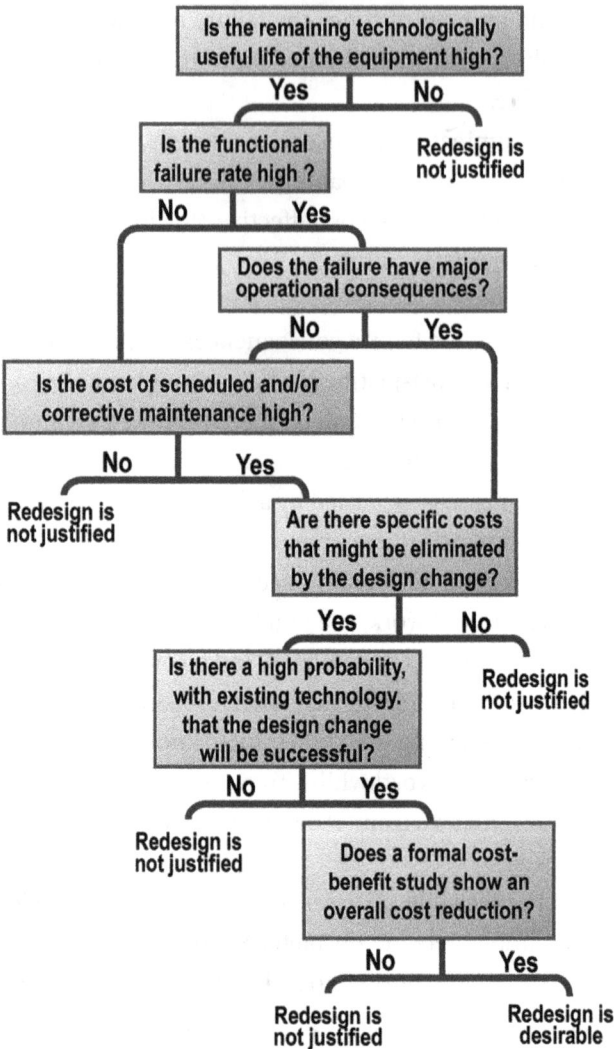

FIGURE 8.37   Decision diagram for a preliminary assessment of a proposed modification

on the basis of projected cost-benefit and ensures that only projects that will pay for themselves quickly are considered and submitted for approval. So if the answer to the first question in Figure 8.37 is no, redesign is probably not justified.

For example, Figure 8.38 shows a stainless-steel hopper that is periodically blocked by lumps. So far, the RCM process has revealed that this failure mode costs $400 in lost production every time it occurs and that it cannot be prevented by maintenance. It has been suggested that one way to eliminate the failure mode might be to install a stainless-steel grid above the hopper outlet at a cost of $6,000.

If the hopper were due to be superseded within 2 years, it is highly unlikely that this modification would be worth doing, especially in view of the fact that several months would elapse before it could be commissioned. On the other hand, if the hopper were to remain in service for several more years, the modification would be worth further consideration.

If the answer to the first question is yes, the next question to consider is whether the failure is happening enough to be considered a problem:

Is the functional failure rate high?

The answer to this question will eliminate items that fail so seldom that the cost of the redesign would probably be greater than the benefits derived from it (unless the preventive task is the reason for the low failure rate). A no answer to this question does not immediately abort the modification—the maintenance task itself may be so expensive that the modification is still justified.

For example, if the blockage in the hopper occurred once every 2 or 3 years, no one would pay much attention to it. If it occurred once a month, it would be worth investigating further.

If the failure rate is high, attention will be focused on the economic implications of the failure:

Does the failure have major operational consequences?

If the answer is yes, then the question of redesign should be taken further. A no answer to this question means that the failure only has a minor impact on operating cost, but we must still consider the maintenance cost associated with the failure by asking:

Is the cost of scheduled and/or corrective maintenance high?

Note that this question is approached from two directions. As we have seen, we may get to a no answer to the failure rate question only because a very costly preventive task is preventing functional failures. A no answer to the question of operational consequences means that failures might not be affecting operational capability, but they result in expensive repair costs. So a yes answer to either of these two questions brings us to the design change itself:

Are there specific costs that might be eliminated by the design change?

This question refers to the operational consequences and the direct costs of proactive and/or corrective maintenance. However, if these costs are not related to a specific design feature, it is unlikely that the problem will be solved by a design change. So an answer of no to this question means that it may be necessary to live with the economic consequences of the failure. On the other hand, if the problem can be pinned down to a specific cost element, then the economic potential of redesign is high.

In the case of the hopper, it is hoped that the grid would prevent the lumps from reaching the hopper outlet and thus eliminate the cost of $400 per blockage.

But will the new design work? In other words:

Is there a high probability, with existing technology, that the design change will be successful?

Although a specific redesign may be very desirable economically, it may be possible that the modification will not have the desired effect. A change directed at one failure mode may reveal other failure modes, requiring several attempts to solve the problem. Any design change that entails adding hardware also adds more failure possibilities—maybe too many and too complex. So if a proper assessment of the proposed change is performed and it indicates a low probability of success, the change is unlikely to be economically viable.

For instance, in the case of the hopper, we would need to be sure that lumps would not simply accumulate on the grid and coagulate into a possibly much more costly problem in the long term.

Any proposed design change that makes it this far is one that deserves a detailed cost-benefit study:

Does a formal cost-benefit study show an overall cost reduction?

Such a study compares the expected reduction in cost over the remaining useful life of the equipment with the cost of carrying out the modification. To be on the safe side, the expected benefit should be regarded as the projected savings if the first attempt at improvement is successful, multiplied by the probability of success of the first try. Alternatively, it might be considered that the design change will always be successful, but only some of the savings will be achieved.

If we are certain that the modification to the hopper will work, a discounted cash flow analysis of the figures provided for the hopper (at a discount rate of 10%) shows that the modification will pay for itself:

- In 5 years if the blockage occurs four times per year
- In 7 years if it occurs three times per year
- In more than 10 years if it occurs twice per year

This type of justification is not necessary, of course, if the reliability characteristics of an item are the subject of contractual warranties or if the changes are needed for reasons other than cost (such as safety).

Reliability Centered Design

A more formal approach to redesign and defect elimination is required for complex systems. Redesigns may be expensive and take time, as we discussed earlier, and while the redesign is being performed (planned, designed, built, installed, and commissioned), the equipment needs to be maintained in its current configuration. The change management must also be performed with precision to avoid additional failures and dangerous situations from arising following the design change. The Reliability Centered Design (RCD) process will only be introduced in this book, but the author encourages reliability engineers to follow a rigorous approach when redesigns are considered. The RCD process follows these steps:

1. Specify the business objectives (availability, reliability, safety, environmental integrity).
2. Identify the operating conditions.
3. Determine the functional requirements (what does the user want?).
4. Perform the baseline design.
5. Determine the critical systems (for system redesigns).
6. Identify the risks associated with the design (FMECA).
7. Develop a reliability block diagram (RBD).
8. Simulate the baseline design using real-life scenarios (operating conditions).
9. Answer the question, "Does the design meet the business objectives (cost-benefit study, reliability targets)?"
10. If yes, determine if the design exceeds expectations and whether the design can be optimized (consider alternatives).
11. If no, review the design and/or objectives (lower targets and objectives or redesign system to meet objectives).
12. Simulate, validate, and implement.
13. Review and ensure continuous improvement.

Software is required to perform proper analysis, simulation, and optimization.

CHAPTER 9

Managing Tolerable Risks—Why Care?

The final question of the RCM3 process deals with how to manage tolerable risks and ask the following question:

> What *can* be done to reduce or manage tolerable risks in a cost-effective way?

We have seen earlier in Chapter 7 that not all risks are intolerable. We have seen further that depending on the industry and organization's risk management policy, certain risks may be tolerable in one organization, where in another organization the same risk may not be. The individuals who decide what level of risk to tolerate may have different experiences, and therefore their tolerability may be vastly different. Regardless of what the criteria are for setting the threshold for risk tolerability, it will almost always vary among individuals and from one organization to the next.

The risk threshold determines which risks are tolerable and which are not. The previous chapter discussed in detail the strategies for proactive risk management for *intolerable risks*. These are summarized as follows:

- Proactive maintenance
  - Predictive maintenance
  - Preventive maintenance

- Combination of tasks (physical risks)
- Optimization of protective devices
- Failure-finding tasks and functional checks
- One-time changes:
  - Redesigns and modifications
  - Operating and maintenance procedures
  - Training

All of these are considered proactively to reduce intolerable risks to a tolerable level. The strategies recommended for managing tolerable risks may also include proactive maintenance and one-time changes, provided they can be done cost-effectively.

9.1 Strategies Dealing with Tolerable Risks

Economic Risks—Operational

When considering failures with economic impact only, many are tolerable, and many existing maintenance programs are taking care of failures that can actually be tolerated according to the organization's risk management policy. When analyzing existing maintenance tasks for failures with economic impact, many will not meet the worth-doing criteria—where maintenance simply costs more than the failure it is meant to prevent, or the task is not technically feasible. The author refers to these as "feel-good maintenance" or "recreational maintenance."

Physical Risk—Safety and the Environment

Some failures with safety and/or environmental impact may even be tolerable—despite how strange it may sound. At the time of writing the book, the author received a lot of resistance and criticism for saying this.

Let's consider an example:

A backup generator is installed in a soundproof and weatherproof enclosure. The maintenance staff is expected to test the generator from time to time to ensure the desired availability and to collect essential asset health information (condition of the oil through oil sampling and the ratio of glycol in the cooling system). In order to do this, the maintenance staff must open the panel on the side of the enclosure. The side panel is hinged at the top and is kept open through a mechanism that locks the panel in the open position. The mechanism is secured to the side panel and the enclosure through four screws on each side. Over time the screws corrode and eventually fail.

This has happened once in 7 years while maintenance work was being performed. The screws sheared, and the side panel slammed closed. The electrician got his hand caught between the falling side panel and the enclosure and bruised his hand with minor lacerations. The first reaction to this is that the failure will have safety consequences. Although true, the risk was tolerable due to the rare occurrence. This does not mean that nothing should be done to keep this from happening in the first place (or again). A minor modification (one-time change) applied proactively could avoid this failure. The screws were replaced with stainless-steel screws, thus changing the likelihood of the failure.

It is impossible to avoid all risks; organizations simply cannot afford it. Organizations are preoccupied with safety, but they operate equipment to be profitable, and while doing so, risks exist in almost every action taken. Risk management is the art of understanding the risks and managing the risks to tolerable levels while balancing cost and consequences.

9.2 No Scheduled Maintenance

We have seen that, for evident failures that do not affect safety or the environment, or hidden failures with economic impact where the multiple failure does not affect safety or the environment (economic risks), the decision to do no scheduled maintenance is acceptable provided that a cost-effective, technically feasible routine task cannot be found. In these cases, the items are left in service until a functional failure occurs, at which point they are repaired or replaced. In other words, "no scheduled maintenance" is only valid if:

- A suitable scheduled task cannot be found for a hidden function, and the associated multiple failure does not have safety or environmental consequences.
- A cost-effective preventive task cannot be found for failures posing an economic risk (operational or non-operational consequences).

Note that if a suitable preventive task cannot be found for a failure under either of these circumstances, it simply means that we do not carry out *scheduled maintenance* on that component in its present form. It does not mean that we simply forget about it. As we saw in the previous chapter, there may be circumstances under which it is worth changing the design of the component to reduce overall costs. No scheduled maintenance or run-to-failure does not mean no planned maintenance.

9.3 Walk-Around Checks

Although walk-around checks will probably not be considered for managing intolerable risks, they serve two purposes. The first is to spot accidental damage. These checks may include a few specific on-condition tasks for the sake of convenience, but damage in general can occur at any time and is not related to any definable level of failure resistance.

As a result, there is no basis for defining an explicit potential failure condition or a predictable P-F interval. Similarly, the checks are not based on the failure characteristics of any particular item but are intended to spot unforeseen exceptions in failure behavior.

Walk-around checks are also meant to spot problems due to ignorance or negligence, such as hazardous materials or foreign objects left lying around, spillage, and other items of a housekeeping nature. They also give managers an opportunity to ensure that general standards of maintenance are satisfactory and to check whether maintenance routines are being done correctly. Again, there are rarely any explicit potential failure conditions and no predictable P-F interval.

Some organizations distinguish between formal scheduled tasks and walk-around checks on the pretext that one is mainly technical and the other predominantly managerial, so they are sometimes done by different people. In fact, it does not matter who does them, as long as both are done frequently and thoroughly enough to ensure a reasonable degree of protection from the consequences of the failures concerned.

9.4 Spare Parts Optimization and Logistics

For tolerable risks, spare part optimization may be considered as a strategy as well as improvement of wrench time, planning and scheduling, and workforce efficiency. Extended downtime may cause failures that pose a tolerable risk to become intolerable due to excessive downtime and repair costs. Further consideration needs to be given to where to keep the spare part (logistics), how many spare parts to keep, and whether or not to repair or replace. These decisions may not directly impact the inherent reliability of the item but will surely reduce the downtime and therefore the business risk associated with the failure. Reliability Centered Spares (RCS) is a helpful methodology for determining the most effective stocking policy. The RCS process considers the following five basic questions:

- What are the maintenance requirements of the equipment?
- What happens if no spare part is available?
- Can the spares requirement be anticipated?
- What stockholding of the spare is needed?
- What if the maintenance requirements cannot be met?

Again, the RCS process requires proper analysis and software for detailed analysis of the spare part requirements and will not be covered in more detail in this book.

CHAPTER 10

The RCM3 Decision Process

10.1 Developing Risk Mitigation Strategies

Chapters 7 to 9 have provided a detailed explanation of the criteria used to answer the the last three of the eight questions that make up the RCM3 process. These questions are:

What are the risks associated with each failure (inherent risk quantified)?

What *must* be done to reduce intolerable risks to a tolerable level (using proactive risk management strategies)?

What *can* be done to reduce or manage tolerable risks in a cost-effective way?

This chapter summarizes the most important of these criteria. It also describes the RCM3 decision diagram, which integrates all the decision processes into a single strategic framework. This framework is shown in Figure 10.1 and is applied to each of the failure modes listed on the RCM3 Information Worksheet.

Finally, this chapter describes the RCM3 Decision Worksheet, which is the second of the two key working documents used in the application of RCM3 (the Information Worksheet shown in Figure 6.23 being the first).

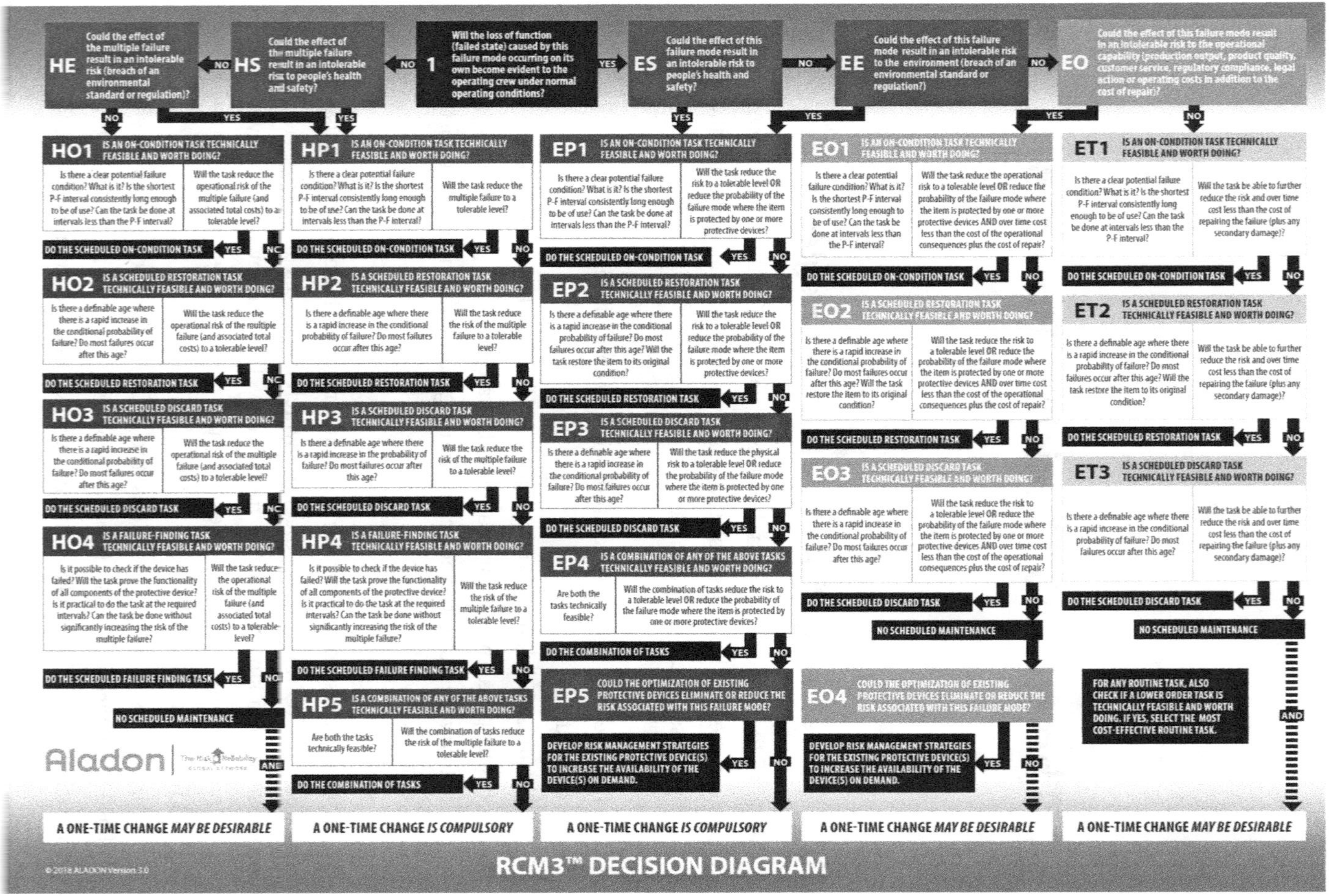

HE Could the effect of the multiple failure result in an intolerable risk to people's health and safety or breach an environmental standard or regulation?

HS Could the effect of the multiple failure result in an intolerable risk to people's health and safety?

1 Will the loss of function (failed state) caused by this failure mode on its own become evident to the operating crew under normal operating conditions?

ES Could the effect of this failure mode result in an intolerable risk to people's health and safety?

EE Could the effect of this failure mode result in an intolerable risk to the environment (breach of an environmental standard or regulation)?

EO Could the effect of this failure mode result in an intolerable risk to the environment, or is it likely to affect the organization's ability to meet a product quality, customer service, regulatory compliance, legal action or operating costs in addition to the cost of repair?

HO1 IS AN ON-CONDITION TASK TECHNICALLY FEASIBLE AND WORTH DOING?
Is there a clear potential failure condition? What is it? Is the shortest P-F interval consistently long enough to be of use? Can the task be done at intervals less than the P-F interval? — Will the task reduce the operational risk (total risks) to a tolerable level?
DO THE SCHEDULED ON-CONDITION TASK

HO2 IS A SCHEDULED RESTORATION TASK TECHNICALLY FEASIBLE AND WORTH DOING?
Is there a definable age where there is a rapid increase in the conditional probability of failure? Do most failures occur after this age? — Will the task reduce the operational risk of the multiple failure (and associated total costs) to a tolerable level?
DO THE SCHEDULED RESTORATION TASK

HO3 IS A SCHEDULED DISCARD TASK TECHNICALLY FEASIBLE AND WORTH DOING?
Is there a definable age where there is a rapid increase in the conditional probability of failure? Do most failures occur after this age? — Will the task reduce the operational risk of the multiple failure (and associated total costs) to a tolerable level?
DO THE SCHEDULED DISCARD TASK

HO4 IS A FAILURE FINDING TASK TECHNICALLY FEASIBLE AND WORTH DOING?
Is it possible to check if the device has failed? Will the task prove the functionality of all components of the protective device? Is it practical to do the task at the required intervals? Can the task be done without significantly increasing the risk of the multiple failure? — Will the task reduce the risk of the multiple failure to a tolerable level?
DO THE SCHEDULED FAILURE FINDING TASK

NO SCHEDULED MAINTENANCE

A ONE-TIME CHANGE MAY BE DESIRABLE

HP1 IS AN ON-CONDITION TASK TECHNICALLY FEASIBLE AND WORTH DOING?
Is there a clear potential failure condition? What is it? Is the shortest P-F interval consistently long enough to be of use? Can the task be done at intervals less than the P-F interval? — Will the task reduce the multiple failure to a tolerable level?
DO THE SCHEDULED ON-CONDITION TASK

HP2 IS A SCHEDULED RESTORATION TASK TECHNICALLY FEASIBLE AND WORTH DOING?
Is there a definable age where there is a rapid increase in the conditional probability of failure? Do most failures occur after this age? — Will the task reduce the risk of the multiple failure to a tolerable level?
DO THE SCHEDULED RESTORATION TASK

HP3 IS A SCHEDULED DISCARD TASK TECHNICALLY FEASIBLE AND WORTH DOING?
Is there a definable age where there is a rapid increase in the conditional probability of failure? Do most failures occur after this age? — Will the task reduce the risk of the multiple failure to a tolerable level?
DO THE SCHEDULED DISCARD TASK

HP4 IS A FAILURE FINDING TASK TECHNICALLY FEASIBLE AND WORTH DOING?
Is it possible to check if the device has failed? Will the task prove the functionality of all components of the protective device? Is it practical to do the task at the required intervals? Can the task be done without significantly increasing the risk of the multiple failure? — Will the task reduce the risk of the multiple failure to a tolerable level?
DO THE SCHEDULED FAILURE FINDING TASK

HP5 IS A COMBINATION OF ANY OF THE ABOVE TASKS TECHNICALLY FEASIBLE AND WORTH DOING?
Are both the tasks technically feasible? — Will the combination of tasks reduce the risk of the multiple failure to a tolerable level?
DO THE COMBINATION OF TASKS

A ONE-TIME CHANGE IS COMPULSORY

EP1 IS AN ON-CONDITION TASK TECHNICALLY FEASIBLE AND WORTH DOING?
Is there a clear potential failure condition? What is it? Is the shortest P-F interval consistently long enough to be of use? Can the task be done at intervals less than the P-F interval? — Will the task reduce the risk to a tolerable level OR reduce the probability of the failure mode where the item is protected by one or more protective devices?
DO THE SCHEDULED ON-CONDITION TASK

EP2 IS A SCHEDULED RESTORATION TASK TECHNICALLY FEASIBLE AND WORTH DOING?
Is there a definable age where there is a rapid increase in the conditional probability of failure? Do most failures occur after this age? Will the task restore the original condition? — Will the task reduce the probability of the failure mode where the item is protected by one or more protective devices?
DO THE SCHEDULED RESTORATION TASK

EP3 IS A SCHEDULED DISCARD TASK TECHNICALLY FEASIBLE AND WORTH DOING?
Is there a definable age where there is a rapid increase in the conditional probability of failure? Do most failures occur after this age? — Will the task reduce the physical risk to a tolerable level OR reduce the probability of the failure mode where the item is protected by one or more protective devices?
DO THE SCHEDULED DISCARD TASK

EP4 IS A COMBINATION OF ANY OF THE ABOVE TASKS TECHNICALLY FEASIBLE AND WORTH DOING?
Are both the tasks technically feasible? — Will the combination of tasks reduce the risk to a tolerable level OR reduce the probability of the failure mode where the item is protected by one or more protective devices?
DO THE COMBINATION OF TASKS

EP5 COULD THE OPTIMIZATION OF EXISTING PROTECTIVE DEVICES ELIMINATE, OR REDUCE THE RISK ASSOCIATED WITH THIS FAILURE MODE?
DEVELOP RISK MANAGEMENT STRATEGIES FOR THE EXISTING PROTECTIVE DEVICE(S) TO INCREASE THE AVAILABILITY OF THE DEVICE(S).

A ONE-TIME CHANGE IS COMPULSORY

ET1 IS AN ON-CONDITION TASK TECHNICALLY FEASIBLE AND WORTH DOING?
Is there a clear potential failure condition? What is it? Is the shortest P-F interval consistently long enough to be of use? Can the task be done at intervals less than the P-F interval? — Will the task be able to further reduce the risk and over time cost less than the cost of repairing the failure (plus any secondary damage)?
DO THE SCHEDULED ON-CONDITION TASK

ET2 IS A SCHEDULED RESTORATION TASK TECHNICALLY FEASIBLE AND WORTH DOING?
Is there a definable age where there is a rapid increase in the conditional probability of failure? Do most failures occur after this age? Will the task restore the item to its original condition? — Will the task be able to further reduce the risk and over time cost less than the cost of repairing the failure (plus any secondary damage)?
DO THE SCHEDULED RESTORATION TASK

ET3 IS A SCHEDULED DISCARD TASK TECHNICALLY FEASIBLE AND WORTH DOING?
Is there a definable age where there is a rapid increase in the conditional probability of failure? Do most failures occur after this age? — Will the task be able to further reduce the risk and over time cost less than the cost of repairing the failure (plus any secondary damage)?
DO THE SCHEDULED DISCARD TASK

FOR ANY ROUTINE TASK, ALSO CHECK IF A LOWER ORDER TASK IS TECHNICALLY FEASIBLE AND WORTH DOING. IF YES, SELECT THE MOST COST-EFFECTIVE ROUTINE TASK.

NO SCHEDULED MAINTENANCE

A ONE-TIME CHANGE MAY BE DESIRABLE

EO1 IS AN ON-CONDITION TASK TECHNICALLY FEASIBLE AND WORTH DOING?
Is there a clear potential failure condition? What is it? Is the shortest P-F interval consistently long enough to be of use? Can the task be done at intervals less than the P-F interval? — Will the task reduce the operational risk to a tolerable level OR reduce the probability of the failure mode where the item is protected by one or more protective devices AND over time cost less than the cost of the operational consequences plus the cost of repair?
DO THE SCHEDULED ON-CONDITION TASK

EO2 IS A SCHEDULED RESTORATION TASK TECHNICALLY FEASIBLE AND WORTH DOING?
Is there a definable age where there is a rapid increase in the conditional probability of failure? Do most failures occur after this age? Will the task restore the item to its original condition? — Will the task reduce the operational risk to a tolerable level OR reduce the probability of the failure mode where the item is protected by one or more protective devices AND over time cost less than the cost of the operational consequences plus the cost of repair?
DO THE SCHEDULED RESTORATION TASK

EO3 IS A SCHEDULED DISCARD TASK TECHNICALLY FEASIBLE AND WORTH DOING?
Is there a definable age where there is a rapid increase in the conditional probability of failure? Do most failures occur after this age? — Will the task reduce the operational risk to a tolerable level OR reduce the probability of the failure mode where the item is protected by one or more protective devices AND over time cost less than the cost of the operational consequences plus the cost of repair?
DO THE SCHEDULED DISCARD TASK

EO4 COULD THE OPTIMIZATION OF EXISTING PROTECTIVE DEVICES ELIMINATE, OR REDUCE THE RISK ASSOCIATED WITH THIS FAILURE MODE?
DEVELOP RISK MANAGEMENT STRATEGIES FOR THE EXISTING PROTECTIVE DEVICE(S) TO INCREASE THE AVAILABILITY OF THE DEVICE(S) ON DEMAND.

NO SCHEDULED MAINTENANCE

A ONE-TIME CHANGE MAY BE DESIRABLE

RCM3™ DECISION DIAGRAM

Aladon | The Asset Reliability

© 2018 ALADON Version 3.0

FIGURE 10.1 ... RCM3™ Decision Diagram

10.2 The RCM3 Decision Process

The RCM3 Decision Worksheet is illustrated in Figure 10.2. The rest of this chapter demonstrates how this worksheet is used to record the answers to the questions in the RCM3 decision diagram and, in the light of these answers, to record:

- What routine maintenance (if any) is to be done, how often it is to be done, and by whom
- For which failures a one-time change should be considered to reduce the risk to a tolerable level (normally associated with failures where no effective form of maintenance exists or for failures that are serious enough to warrant redesign)
- Failures that are protected, where the optimization of the protective systems will reduce the risk of the multiple failure to a tolerable level (when optimization of the protected function on its own will not reduce the risk to a tolerable level)
- Cases where a deliberate decision has been made to let failures happen

The Decision Worksheet is divided into 27 columns. The columns headed "F," "FS," and "FM" identify the failure mode under consideration. They are used to cross-reference the Information and Decision Worksheets, as shown in Figure 10.3.

The headings on the next 13 columns refer to the questions on the RCM3 decision diagram in Figure 10.1, as follows:

- The column headed "1" is used to determine whether the failure is evident or hidden.
- The columns headed "ES/HS," "EE/HE," and "EO" are used to record the answers to the questions concerning the risk associated with each failure mode (consequence category).
- The next three columns (headed "HP1" etc., "HP2" etc., and "HP3" etc.) record whether a proactive task has been selected, and if so, what type of task.

FIGURE 10.2 RCM3 Decision Worksheet

RCM3™ Information Worksheet © 2018 Aladon V2.0	Location ID **K-03-001-00BZ**		No	Compiled by
	Location Description **Benzene Storage & Supply System**		Ref.	Reviewed by

Function	Failed State	Failure Mode		Failure Effect
		Cause	Mechanism	
1 To supply benzene on demand to the process at a minimum rate of 300 gallons/min.	A Does not supply benzene at all	1 Pump impeller stuck.	Jammed by foreign object.	**Local Failure Effect** A foreign object is present in the system f object enters the pump, blocks the impelle
				Next-Higher-Level Effect The pump motor trips on overload and the in the control room. With the pump down benzene to the process.
				End Effect Repair time up to 4 hours to replace the ir would be lost at a cost of $5,000. The imp during the 5 hours between delivery cycle
				Potential Worst-Case Effect If the existing protection fails, the motor w greatly increased. Although it is unlikely th damage and downtime will be high.
1	A	2 Pump line shaft bearing collapse.	Normal wear.	**Local Failure Effect** Through normal use the bearing starts to wears. Eventually the bearing seizes.

RCM3™ Decision Worksheet © 2018 Aladon V2.0	Location ID **K-03-001-00BZ**		No	Compiled by
	Location Description **Benzene Storage & Supply System**		Ref.	Reviewed by

Reference				Risk Category Evaluation				Risk Management Strategy Evaluation									Risk Management Strategy Description
F	FS	FM	1	ES HS	EE HE	EO		HP1 HO1 EP1 EO1 ET1	HP2 HO2 EP2 EO2 ET2	HP3 HO3 EP3 EO3 ET3	HP4	HP5	HO4	EP4	EP5	EO4	
1	A	1	Y	N	N	Y		N	N	N						Y	No scheduled maintenance. Optimize existing protective device to reduce the risk of the multiple failure. Develop FFI for motor overload protection.
1	A	2	Y	N	N	Y		Y									Use ultrasound to detect pump line shaft bearing noise. Plan replacement when excessive bearing noise is detected.
1	A	3	Y	N	N	Y		N	Y								Lubricate the pump line shaft bearing with three shots of EP2 grease

FIGURE 10.3 Cross-referencing the Information and Decision Worksheets

- If no proactive task can be found that would reduce intolerable risks to a tolerable level or when tolerable risks cannot be reduced in a cost-effective way, it becomes necessary to answer the questions related to optimization of existing protective devices, combination of tasks, and one-time changes. The columns headed "HP4," "HP5," "HO4," "EP4," "EP5," and "EO4" are used to record the answers to these questions.

The last 11 columns record the proactive task or risk management strategy that has been selected (if any), the frequency with which it is to be done and who has been selected to do it, and the revised risk. It also records the unit of measure (UOM), the trade best suited and qualified to do the task, the number of tradespeople required to do the task, and the duration of the task (how long it will take to do the task). The "Risk Management Strategy Description" column is also used to record the cases where one-time change is required or where it has been decided that the failure mode does not need scheduled mainte-nance. The "Mod Type" (modification type) column is used to describe whether the one-time change recommendation is a redesign or modi-fication, a procedural change, or a training requirement.

The last four columns are used to record the revised risk. "M_{ted}," the first of the four columns under "Revised Risk Ranking," is used to record the mean time between failures of a protected function. The second col-umn, "P," is used for recording the probability of failure (with consider-ation of the new risk management strategy), and the third column, "C," is used for recording the consequence of failure. The last column, "R," is used to record the revised risk.

In the following paragraphs, each of these sections of the Decision Worksheet is reviewed in the context of the associated questions on the decision diagram.

Failure Consequences

The precise meanings of questions 1, ES, HS, EE, HE, and EO in Figure 10.1 are discussed at length in Chapter 6. These questions are asked for each failure mode, and the answers are recorded on the Decision Worksheet on the basis shown in Figure 10.4.

Figure 10.5 shows how the answers to these questions are recorded on the Decision Worksheet. Note that:

- Each failure mode is dealt with in terms of one risk category only. So if an asset poses a risk to the environment (breach of an envi-ronmental standard or regulation), we do not also evaluate its risk

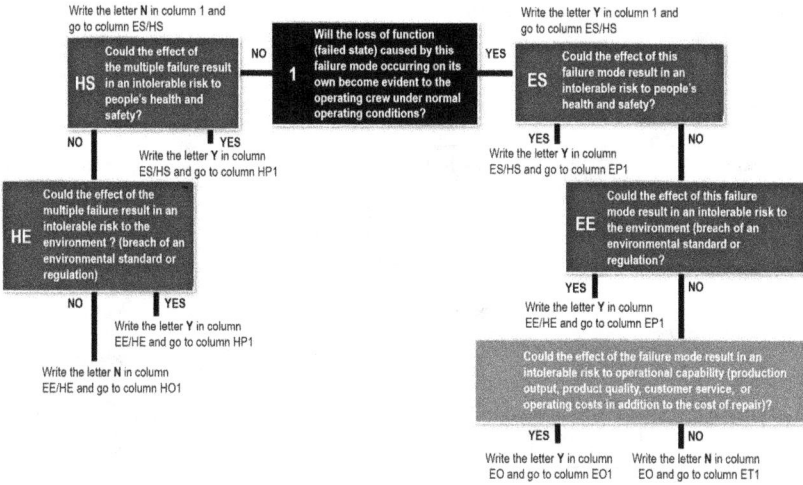

FIGURE 10.4 Using the Decision Worksheet to record risk categories

to operational capability (at least when performing the first analysis of any asset). This means that if, for instance, a "Y" is recorded in Column "EE," nothing is recorded in Column "EO."

- Once the risk category of the failure mode has been categorized, the next step is to seek a suitable proactive risk management strategy. Figure 10.5 summarizes the criteria used to determine whether failures are hidden or evident and how to select the risk category for each failure mode as described in the failure effect description.

Proactive Risk Management Strategies

The eighth question in the RCM3 process—"What *must* be done to reduce intolerable risks to a tolerable level?"—requires all intolerable risks to be addressed and mitigated. This is a drastic departure from traditional RCM processes, which required every failure mode to be treated (regardless of the consequence severity). The approach of treating intolerable risks saves both time and resources. Additionally, traditional RCM processes consider proactive maintenance as the only proactive failure management strategies and consider one-time changes and functional

RCM3™ Decision Worksheet

© 2018 Aladon V2.0

Location ID

Location Description

Reference			Risk Category Evaluation				Risk Managen			
F	FS	FM	1	ES HS	EE HE	EO	HP1 HO1 EP1 EO1 ET1	HP2 HO2 EP2 EO2 ET2	HP3 HO3 EP3 EO3 ET3	:
1	A	1	N							— A hidden failure—mostly a protective device
1	A	2	N	Y						— A hidden failure—multiple failure poses an intolerable risk to people's *health and safety*
1	A	3	N	N	Y					• A hidden failure—multiple failure poses an intolerable risk to the *environment*
1	A	4	N	N	N					▪ A hidden failure—multiple failure poses an intolerable risk to *operational capability*
1	A	5	Y							An evident failure—failures that become evident to the operating crew under normal operating conditions
1	A	6	Y	Y						An evident failure—failure poses an intolerable risk to people's *health and safety* (physical risk)
1	A	7	Y	N	Y					▪ An evident failure—failure poses an intolerable risk to the *environment* (physical risk)
1	A	8	Y	N	N	Y				An evident failure—failure poses an intolerable risk to *operational capability* (economic risk)
1	A	9	Y	N	N	N				An evident failure—failure with a risk that is tolerable (nonoperational consequence)

Note:

These questions are answered using the information that was recorded in the "Failure Effect" columns.

FIGURE 10.5   Failure risk categories

checks as default actions (where no proactive task can be found that satisfies both the *technically feasible* and *worth-doing* criteria). In order to proactively reduce the risk to tolerable levels, maintenance alone may not be enough. One-time changes (redesign, training, and operating and maintenance procedures) and the optimization of protective systems may also be required. These strategies will also be considered and applied proactively and not as default actions. This allows the review groups to make more decisions and find real solutions during the analysis process while saving time—the cost-effective optimization of tolerable risks can be done outside the RCM meetings through meeting the subject-matter experts

Proactive Maintenance—Predictive and Preventive Maintenance Tasks

The eighth to tenth columns on the Decision Worksheet are used to record whether a proactive task has been selected, as follows:

- The column headed "HP1, HO1, EP1, EO1, ET1" is used to record whether a suitable on-condition task could be found to anticipate the failure mode in time to mitigate the risk.
- The column headed "HP2, HO2, EP2, EO2, ET2" is used to record whether a suitable scheduled restoration task could be found to prevent the failures and mitigate the risk.
- The column headed "HP3, HO3, EP3, EO3, ET3" is used to record whether a suitable scheduled discard task could be found to prevent the failures and mitigate the risk.

In each case, a task is only suitable if it is worth doing and technically feasible. Chapters 8 and 9 explained in detail how to establish whether a task is technically feasible; these criteria are summarized in Figure 10.6.

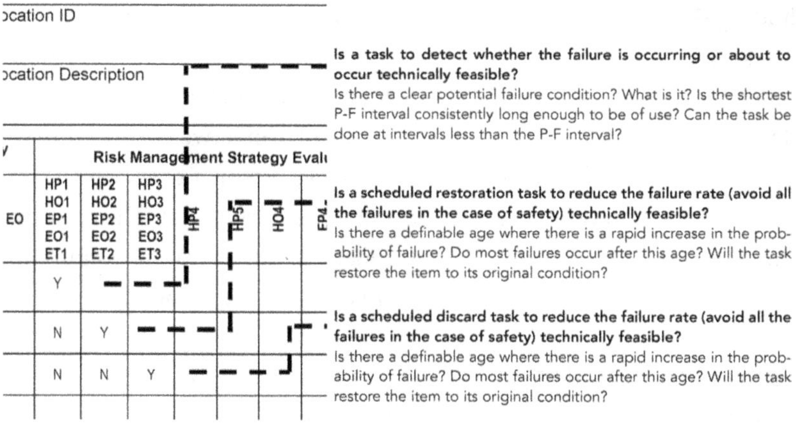

Note:

The "worth-doing" criteria must be satisfied also, before proactive tasks will be selected.

FIGURE 10.6 Technical feasibility criteria for "proactive tasks"

In essence, for a task to be technically feasible and worth doing, it must be possible to provide a positive answer to all the questions shown in Figure 10.6 that apply to that category of tasks, and the task must fulfill the worth-doing criteria in Figure 10.5. If the answer to any of these questions is no or unknown, then that task as a whole is rejected. If all the questions can be answered positively, then a "Y" is recorded in the appropriate column.

If a task is selected, a description of the task, the frequency with which it must be done, and who must do the task are recorded as explained later in this chapter, and the analysts move on to the next failure mode. However, as mentioned in Chapter 8, bear in mind that if it seems that a lower-order task may be more cost-effective than a higher-order task, then the lower-order task should also be considered and the more effective of the two chosen.

Proactive Detection—Failure-Finding Tasks and Functional Checks

The RCM3 process differentiates between two types of detective maintenance actions: failure-finding tasks and functional checks. Failure-finding refers to the task of determining whether the hidden function is working or not, and of repairing it if it has failed. As discussed before, hidden failures are normally associated with some sort of protective device. Functional checks entail the testing of protective systems that fail evident, e.g., heat tracing on a pipe, backup horn on a truck, self draining condensation trap, and demisters in an electric motor. The failure of these devices may become evident under certain operating conditions. Both of these will be discussed in more detail in the following sections.

Failure-Finding Tasks. The RCM3 process treats protected functions differently from how unprotected functions are treated. In order to reduce risks to tolerable levels, it may require that both the protected system and the protective devices are optimized. The RCM3 process focuses primarily on increasing the reliability of the protected function before the optimization of the protective device is considered. The columns headed "HP4" and "HO4" on the Decision Worksheet are used to record the answers to the question of whether it is possible to reduce the risk of multiple failure through optimizing protective devices. The basis on which of these questions are answered is shown in Figure 10.7. (Note that these questions are only asked if the answers to the previous three questions are all no.)

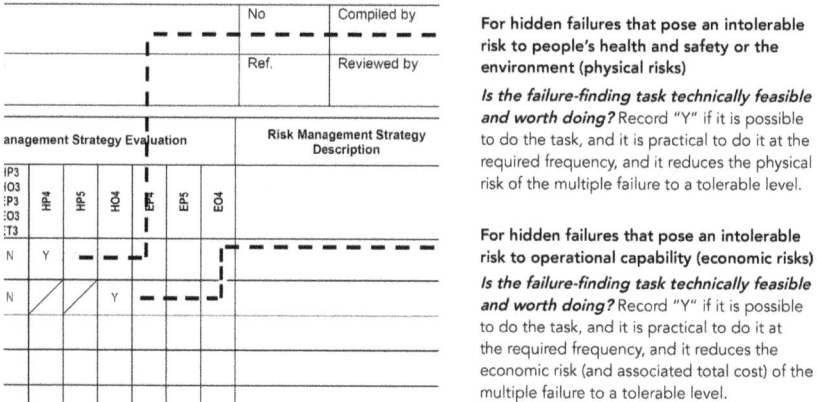

FIGURE 10.7 Technical feasibility criteria for failure-finding tasks

Functional Checks. Although functional checks are also performed on some type of protective device, we find that these devices fail evident under certain operating conditions. These may not be fail-safe and therefore should be treated differently. This may be best described using an example.

Example: A potable water supply pipe is wrapped with heat tracing to prevent freezing of the water in the pipe when the ambient temperature drops below 32°F. The heater element is thermostatically controlled and turns on when the outside temperature drops below 32°F. In the event that the heater element fails (or fails to turn on), the water will freeze and damage the pipe or valves. This will become evident when the temperature drops below freezing and the heater element circuit is also in a failed state and therefore is not a hidden function by the true definition. In order to reduce the risk of the pipes freezing, this heating device must be checked periodically and will typically be tested before the winter season starts. In order to secure the required availability of this device, the functional check is best managed through a task that is initiated by specific conditions rather than a routine task done at specific intervals.

Figure 10.8 illustrates how the failure will become evident during normal operation even though the device is not fail-safe. Ensuring the device is available prior to the period where it will be in use may reduce the risk of the failure to a tolerable level. Depending on the consequence severity, the task may have to be performed more often.

RCM3™ Decision Worksheet © 2018 Aladon V2.0	Location ID		No	Compiled by
	Location Description		Ref.	Reviewed by

Reference			Risk Category Evaluation				Risk Management Strategy Evaluation									Risk Management Strategy Description
F	FS	FM	1	ES HS	EE HE	EO	HP1 HO1 EP1 EO1 ET1	HP2 HO2 EP2 EO2 ET2	HP3 HO3 EP3 EO3 ET3	HP4	HP5	HO4	EP4	EP5	EO4	
1	A	1	Y	N	N	Y	N	N	N						Y	Develop a risk management strategy for existing protective device – Perform testing to ensure the device is working properly. Investigate whether a one-time change could make the failure of the device evident.

FIGURE 10.8 Technical feasibility criteria for functional checks

It is important to note that the operating context will again determine whether the failure of the device will become evident. If this installation is in a location where the temperature never drops below freezing, the failure of the heater will never become evident (unless we specifically test for it), but it would probably not be installed to start with. If it is installed, it will be a superfluous function in this operating context, and the recommendation would be to remove it. In the example in Figure 10.8, we also assume that the failure will not impact safety or the environment and will only affect the operational capability.

Optimize Existing Protective Devices

In Chapters 7 and 8 we described how the RCM3 process favors the optimization of the protected function before we rely on the optimization of the protective device. However, making the protected function more reliable may not be enough to reduce the risk of the failure to a tolerable level (or reduce the probability of the failure mode from happening), and optimization of the protective device may be required; in other cases, no proactive task may be technically feasible and worth doing (see Figure 10.9). This may also be a regulatory requirement that requires protective systems (please refer to Chapter 7).

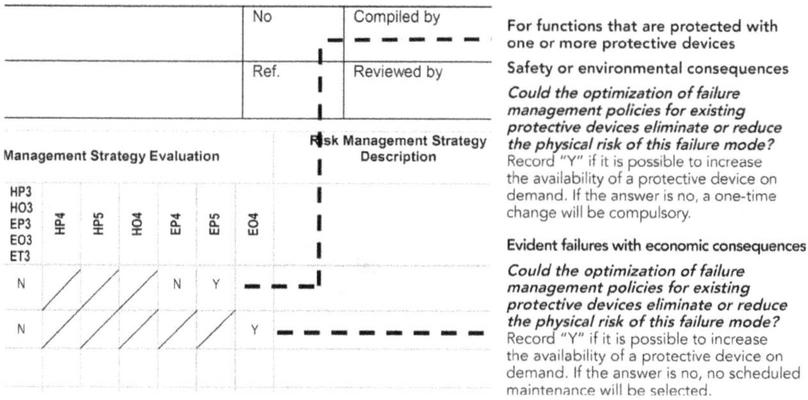

FIGURE 10.9   Criteria for optimizing existing protective devices

Combination of Tasks

In Chapter 7 the consideration for combination of tasks was discussed in detail. A combination of tasks will only be considered where the associated failure impacts safety or the environment. Both the tasks must be technically feasible in order to be considered (see Figure 10.10).

One-Time Change

A one-time change will be considered to reduce the risk in the event that no proactive task can be found that would do so on its own or when a combination of tasks will also not accomplish this. The one-time change is always the last consideration. This has been discussed in detail in Section 8.20. For failures that impact safety and the environment, the one-time change will be compulsory, and for the rest it may be desirable (see Figure 10.11).

| RCM3™ Decision Worksheet © 2018 Aladon V2.0 | Location ID | | | | | | | | | | | | | | | No | Compiled by |
| Location Description | | | | | | | | | | | | | | | | Ref. | Reviewed by |

Reference			Risk Category Evaluation					Risk Management Strategy Evaluation										Risk Management Strategy Description
F	FS	FM	1	ES HS	EE HE	EO	HP1 HO1 EP1 EO1 ET1	HP2 HO2 EP2 EO2 ET2	HP3 HO3 EP3 EO3 ET3	HP4	HP5	HO4	EP4	EP5	EO4		
3	A	1	N	Y			N	N	N	N	Y						Do the combination of tasks – Clean the ultrasonic sensor tube every 3 months and functionally test the device annually
6	A	2	Y	Y			N	N	N			Y					Do the combination of tasks - Inspect the tank for signs of corrosion annually and wire brush and paint the vessel every 10 years

For hidden failures with safety or environmental consequences

For evident failures with safety or environmental consequences

Is a combination of any of the above tasks technically feasible and worth doing? For hidden failures with physical risks, a combination of tasks will be considered if a predictive or preventive task, together with a failure-finding task, will reduce the risk of the failure to a tolerable level (this is very rare). If the answer is no, a one-time change is compulsory.

Is a combination of any of the above tasks technically feasible and worth doing? For evident failures with physical risks, the answer is yes if any *two or more* proactive tasks will reduce the risk of the failure to a tolerable level (this is very rare). Both tasks must be technically feasible on their own. If the answer is no, a one-time change is compulsory.

FIGURE 10.10 Criteria for combination of task

		Risk Management Strategy Evaluation								No	Compiled by	Risk Management Strategy Description	
iption										Ref.	Reviewed by		
	EO	HP1 HO1 EP1 EO1 ET1	HP2 HO2 EP2 EO2 ET2	HP3 HO3 EP3 EO3 ET3	HP4	HP5	HO4	EP4	EP5	EO4			
		N	N	N	N	N					One-time change compulsory		
		N	N	N			N	N			One-time change compulsory		

For hidden failures with physical risk (safety or environmental consequences)

If the answers to questions HR4 (failure-finding task) and HR5 (combination of tasks) are no, a one-time change (redesign) is compulsory.

For evident failures with physical risk (safety or environmental consequences)

If the answers to questions ER4 (combination of tasks) and ER5 (optimizing existing protective devices) are no, a one-time change (redesign) is compulsory.

FIGURE 10.11 Criteria for compulsory one-time changes

No Scheduled Maintenance

No scheduled maintenance does not imply no planned maintenance. No scheduled maintenance will only be considered for failures having tolerable risks. The criteria for no scheduled maintenance are given in Figure 10.12.

RCM3™ Decision Worksheet	Location ID																No		Compiled by
© 2018 Aladon V2.0	Location Description																Ref.		Reviewed by

Reference			Risk Category Evaluation					Risk Management Strategy Evaluation									Risk Management Strategy Description
F	FS	FM	1	ES HS	EE HE	EO		HP1 HO1 EP1 EO1 ET1	HP2 HO2 EP2 EO2 ET2	HP3 HO3 EP3 EO3 ET3	HP4	HP5	HO4	EP4	EP5	EO4	
3	A	1	N	N	N		N	N	N			N					No scheduled maintenance
6	A	2	Y	N	N	Y	N	N	N							N	No scheduled maintenance
9	A	1	Y	N	N	N	N	N	N								No scheduled maintenance

FIGURE 10.12 Criteria for no scheduled maintenance

For the above three cases, the consequences of failure are purely economic, and no suitable proactive tasks have been found. As a result, the decision is no scheduled maintenance, and a one-time change may be desirable.

Completing the last part of the Decision Worksheet (see Figure 10.13) will be discussed in the following paragraphs.

No	Compiled by	Date	Sheet	
Ref.	Reviewed by	Date	of	

Aladon
The Risk Reliability
GLOBAL NETWORK

	Risk Management Strategy Description	Initial Task Int	UOM	Trade Code	Qty	Dur (Hrs)	Mod Type	Revised Risk Ranking			
EO4								Pted	P	C	R

FIGURE 10.13 Completing the Decision Worksheet task description and revised risk

Proposed Risk Management Strategy

If a proactive task, combination of tasks, optimization of existing protective devices, failure-finding task, or functional check has been selected during the decision-making process, a description of the task should be recorded in the column headed "Risk Management Strategy Description." Ideally, the task should be described as precisely on the Decision Worksheet as it will be on the document that reaches the person doing the task. If this is not possible, then the task should at least be described in enough detail to make the intent absolutely clear to whoever writes up the detailed task description. This issue is discussed in more detail in the next chapter.

If the decision process calls for a one-time change (design change, procedural change, or proposed training), then the proposed task should provide a brief description of the design change. The actual form of the new design should be left to the designers. This issue is also discussed further in the next chapter.

Note: When the revised risk is determined (see later section), the assumption is that the proposed risk management strategy has been implemented. In the case of one-time changes, the strategy implementation may take some time, especially when redesigns and modifications are recommended. It is very important to review the assumptions made during the analysis to ensure that the revised risk levels are met.

Finally, if a decision has been taken to allow the failure to occur, in most cases the words "no scheduled maintenance" should be recorded in the "Risk Management Strategy Description" column. When no scheduled maintenance is selected, it may also be desirable to consider a one-time change in order to reduce the risk to a tolerable level. One-time changes to spare part policies, procedures, or even physical redesign may be considered to reduce or eliminate the economic consequences of failures.

Initial Task Interval

Task intervals are recorded on the Decision Worksheet in the "Initial Task Int" column. When completing the Decision Worksheet, record each task interval on its own merits—in other words, without reference to any other tasks. This is because the reason for doing a task at a particular frequency can change over time; indeed, the reason for doing the task at all could disappear. So, if the frequency of task X is based on the frequency of task Y and task Y is later eliminated, the frequency of task X becomes meaningless.

As explained in the next chapter, if we are confronted with a number of tasks that need to be done at a wide range of different frequencies, the time to consider consolidating them into a smaller number of work packages is when compiling maintenance schedules. However, the initial task frequencies should always remain on the Decision Worksheet to remind us how the schedule frequencies were derived (in other words, to preserve the audit trail).

Note also that task intervals can be based on any appropriate measure of exposure to stress. This includes calendar time, running time, distance traveled, stop-start cycles, output or throughput, or any other readily measurable variable that bears a direct relationship to the failure mechanism. However, calendar time tends to be used where possible because it is the simplest and cheapest to administer.

Finally, some task frequencies may be governed by regulatory requirements. It is important to remember that regulations are reactive in nature and that the task frequency may not be enough to reduce the risk to a tolerable level. It may be necessary to record when regulatory task frequencies are used.

Unit of Measure

The unit of measure is recorded in the column marked "UOM." When detailing the task interval, some routine tasks may take years between two tasks, while others may be hours or days. If the unit of measure is number of cycles or stop-start counts, these should also be recorded.

Who Must Do The PM?

The trade or discipline responsible for doing the task should be best qualified to provide the person for the task. The trade is recorded in the "Trade Code" column on the Decision Worksheet. It is used to list who should do each task. Note that the RCM3 process considers this issue one failure mode at a time. In other words, it does not approach the subject with any preconceived ideas about who should (or should not) do maintenance work. It simply asks who has the competence and confidence to do this task correctly.

The answer could be anyone at all. Tasks might be allocated to maintainers, operators, insurance inspectors, the quality function, specialist technicians, vendors, structural inspectors, or laboratory technicians. Trade codes are assigned to each trade, e.g., "M" for mechanical trades, "E" for electrical trades, "I&C" for instrumentation and controls, and so on.

Quantity of Tradespeople Required to Do the Task

Some tasks may require more than one tradesperson to actually complete the task safely and effectively. The number of tradespeople required to do the task is recorded in the column marked "Qty," and if more than one trade is required, the other trade(s) and the number of people required should also be listed.

It is not unusual that company policy requires two or more tradespeople to go to remote locations when maintenance is needed. This can be for safety or security reasons. However, if only one tradesperson will be doing the work, the quantity will be one. The policy of not allowing one person to go to these locations unaccompanied should be recorded in the operating context.

Duration of the Task

The duration of the PM task is recorded in the column marked "Dur (Hrs)." If two or more trades or two or more tradespeople are required, the total number of hours must be recorded. The duration of the task could be used to calculate the worth of doing of maintenance with economic consequences. The total hours required to do maintenance is also used to determine if sufficient resources are available to perform the minimum amount of safe maintenance as required to reduce the physical and economic risk to a tolerable level as determined by the organization's asset management strategy.

One-Time Change Recommendations

As discussed in earlier chapters, a one-time change is considered where no proactive task can be found to reduce or eliminate the risks associated with failures and where none of the other actions (i.e., failure-finding for hidden failures) were technically feasible. For evident or hidden failures that affect safety or the environment and where no proactive tasks, failure-finding tasks (for hidden functions), or combination of tasks have been selected, a *one-time change is compulsory*. In other cases, where the failures only have economic consequences, a *one-time change may be desirable*. A one-time change may be made to one or more of the following:

- Redesign or modification to physical asset (i.e., adding protection) [R]
- Training and capability [T]
- Procedural [P]
- Spare parts [S]

The code letter (bracketed above) that is associated with each recommendation type is recorded in the column marked "Mod Type."

Revised Risk Ranking

The "Revised Risk Ranking" head in the worksheet spans four columns. The first of these columns, marked "P_{ted}," is used to record the MTBF of the protected function. In order to reduce the risk of a multiple failure to a tolerable level, the availability of the protective device is calculated based on the demand rate.

The next column marked "P," is used to record the value that corresponds to the probability of failure on the risk matrix introduced in Chapter 7 (Figure 7.7). The probability of failure is now the value obtained when considering that a proactive risk mitigation strategy (proactive maintenance task, combination of tasks, or proposed modification) is being done to reduce the unanticipated failure (if applicable).

The column marked "C" is used to record the consequence associated with each failure mode. This value is also obtained from the risk matrix in Figure 7.7. Unless the consequence severity is changed through some modification or redesign, this value will normally remain the same if proactive maintenance only has been considered.

The last column, marked "R," is used to determine the revised or residual risk after the risk management strategy has been implemented. This is also taken from the risk matrix in Figure 7.7 and is a combination of the probability and consequence severity from above. All revised risks should be tolerable (some extreme exception may apply).

10.3 Completing the Decision Worksheet

To illustrate how the Decision Worksheet is completed, we consider three failure modes that have been discussed at length in previous chapters. These are:

- The bearing that seizes on a pump with no standby
- The bearing that seizes on an identical pump that does have a standby
- The failure of the standby pump set as a whole

The associated decisions are recorded on the Decision Worksheet shown in Figure 10.14. Please note three important points about this example:

- The first two pumps could suffer from many more failure modes than the failure under consideration. Each of these other failures would also be listed and analyzed on its own merits.

RCM3™ Decision Worksheet
© 2018 Aladon V2.0

Location ID		No	Compiled by	Date	Sheet
Location Description		Ref.	Reviewed by	Date	of

Aladon — The Risk Reliability GLOBAL NETWORK

Information Reference			Risk Category Evaluation				Risk Management Strategy Evaluation									Risk Management Strategy Description	Initial Task Int	UOM	Trade Code	Qty	Dur (Hrs)	Mod Type	Revised Risk Ranking			
F	FS	FM	1	ES HS	EE HE	EO	HP1 HO1 EP1 EO1 ET1	HP1 HO1 EP1 EO1 ET1	HP3 HO3 EP3 EO3 ET3	HP4	HP5	HO4	EP4	EP5	EO4								Pted	P	C	R
Standalone Pump – Pump A																										
1	A	1	Y	N	N	Y	Y	Y								Perform vibration analysis	1	M	M	1	0.5	NA		1	3	6M
1	A	2				*etc......*																				
Duty Pump with Standby – Pump B																										
1	A	1	Y	N	N	Y	N	N	N							No scheduled maintenance							2		3	9M
1	A	2				*etc......*																				
Standby Pump - Pump C																										
2	A	1	N	N	N	N	N	N	N			Y				Switch to standby pump and ensure standby is capable of filling the tank. Switch back to duty pump after the check.	4	W	O	1	2	NA			NA	

FIGURE 10.14 An RCM3 Decision Worksheet with sample entries

- A number of other preventive tasks could have been chosen to anticipate the failure of the bearing. The decisions in the example are for the purpose of illustration only.
- The standby pump is treated as a "black box." In practice, if such a pump were known to suffer from one or more dominant failure modes, these failures would be analyzed individually.

In essence, not only do the RCM worksheets show what course of action has been selected to deal with each failure mode, but they also show why it was selected. This information is invaluable if the need to do any maintenance task is challenged at any time.

The ability to trace each task right back to the functions and desired performance of the asset also makes it a simple matter to keep the maintenance program up to date. This is because users can readily identify and reassess tasks that are affected by a change in the operating context of the asset (such as a change in shift arrangements or a change in safety regulations) and avoid wasting time reassessing tasks that are unlikely to be affected by the change.

10.4 Software and RCM

The information contained in the RCM3 Information Worksheet and Decision Worksheet lends itself readily to being stored in a computerized database. In fact, if a large number of assets are to be analyzed, it is almost essential to use recognized software for this purpose. Software is essential for rapidly implementing the recommendations and for ensuring the RCM analyses are kept evergreen. Software can also be used to sort the proposed tasks by interval and skill set, and to generate a variety of other reports (failure modes by consequence category, tasks by task category, and so on). Finally, storing the analyses in a database makes it infinitely easier to revise and refine the analyses as more is learned and as the operating context changes (as it surely will—see Section 13.5 of Chapter 13).

However, note that the software should only ever be used to capture and sort RCM information and to assist with integration into work management systems. For reasons discussed in Chapter 13, computers should never be used to drive the RCM process.

Software Functionality

As mentioned above, software should not be used to drive the RCM process, but it is advisable to use software to ensure efficient and effective facilitation. The RCM3 software should have the following capability:

- Import facility to import asset structure and hierarchy (or to create from scratch)
- Asset identification (asset group and asset type)
- Positive identification of location, equipment, and part numbers
- Functionality to establish asset criticality and priority (used as a focusing tool for determining asset strategy development and work order prioritization)
- To capture operating context and link attachments
- Risk matrix (configurable) with definable risk categories
- Capability to quantify inherent risk (risk framework)
- RCM worksheets (Information Worksheet [FMEA] and Decision Worksheet)
- RCM decision logic with audit facility
- Strategy development and task development, task frequency, and resource assignment
- Capability to quantify the revised risk following strategy development
- Risk and cost comparison (to compare the economic impact of the mitigation strategy with revised risk profile)
- Calculating whether strategy is worth doing for economic risks
- Calculating failure-finding intervals

- Reporting capability (worksheets, risk/cost profile, strategy details, task frequencies, resource requirements, standard jobs)
- Templating and equipment templates
- Export capabilities (Word, PDF, and configurable in Excel)
- Integration to EAM systems

CHAPTER 11

Implementing RCM Recommendations

11.1 Implementation—the Key Steps

The formal application of the RCM process ends with completed Decision Worksheets. These specify a number of routine tasks that need to be done at regular intervals to ensure that the asset continues to do whatever its users want it to do, together with the default actions that must be taken if an appropriate routine task cannot be found.

The people who participate in this process learn a great deal about how the asset works and about how it fails. This, on its own, frequently causes the participants to change their behavior in ways that often lead directly to remarkable improvements in asset performance. However, in order to derive the maximum long-term benefit from RCM, steps must be taken to implement the recommendations on a formal basis. These steps should ensure that:

- All the recommendations are approved formally by the managers with overall responsibility for the assets.
- All routine tasks are described clearly and concisely.
- All actions that call for one-off changes are identified and implemented correctly.

- Routine tasks and operating procedure changes are incorporated into appropriate work packages.
- The work packages and one-off changes are implemented. Specifically, this in turn entails:
  - Incorporating the work packages into systems that ensure that they will be performed by the right people at the right time and that they will be done correctly
  - Ensuring that any faults found are dealt with speedily

These steps are summarized in Figure 11.1. The most important of them are discussed in more detail in the rest of this chapter.

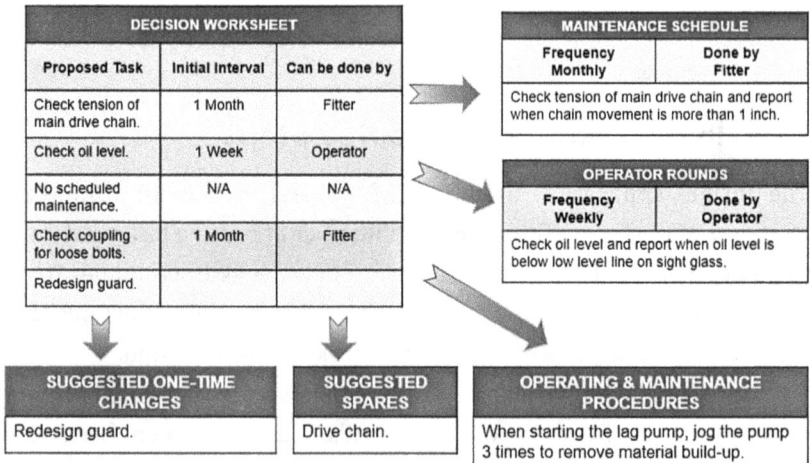

DECISION WORKSHEET		
Proposed Task	Initial Interval	Can be done by
Check tension of main drive chain.	1 Month	Fitter
Check oil level.	1 Week	Operator
No scheduled maintenance.	N/A	N/A
Check coupling for loose bolts.	1 Month	Fitter
Redesign guard.		

MAINTENANCE SCHEDULE	
Frequency Monthly	Done by Fitter
Check tension of main drive chain and report when chain movement is more than 1 inch.	

OPERATOR ROUNDS	
Frequency Weekly	Done by Operator
Check oil level and report when oil level is below low level line on sight glass.	

SUGGESTED ONE-TIME CHANGES
Redesign guard.

SUGGESTED SPARES
Drive chain.

OPERATING & MAINTENANCE PROCEDURES
When starting the lag pump, jog the pump 3 times to remove material build-up.

FIGURE 11.1   Implementation steps after RCM analysis

11.2 The RCM Audit

If it is correctly applied, the RCM process provides the most robust framework currently available for formulating asset management strategies. These strategies profoundly affect the safety, environmental integrity, and economic well-being of the organization using the assets.

However, if something does go badly wrong in spite of the best efforts of the people applying the process, every decision will be subjected to a thorough and sometimes intensely hostile review by organizations ranging from regulatory authorities through insurers and shareholders to representatives of victims (or their survivors). As a result, any organization that uses RCM should take great care to ensure that the people who apply it know what they are doing, and also to satisfy itself that the decisions are *sensible* and *defensible*. The latter step is known as the RCM audit.

RCM audits entail a formal review of the contents of the RCM Information and Decision Worksheets. This section of the chapter looks at who should do the audit, when it should be done, and what it entails.

Who Should Do the Audit

Senior managers bear the overall responsibility for the asset if something goes badly wrong, so it is in their own interests and that of their employers to satisfy themselves that reasonable steps are being taken to prevent such occurrences. Senior managers do not necessarily have to do the audits themselves but may delegate them to anyone in whose judgment they have enough confidence. However, if this is done, it should always be understood that the auditors are acting on behalf of senior management, so the latter still bear the ultimate responsibility for the decisions. (Whoever carries out the audits should also be thoroughly trained in RCM.)

If the auditors disagree with any findings or conclusions, they should discuss the matter with the people who performed the analysis. In so doing, the auditors should be prepared to accept that they themselves may be wrong.

When the Audit Should Be Done

Audits should be carried out as soon as possible after each review has been completed (preferably within 2 weeks), for three reasons:

- The people who did the analysis are keen to see the results of their efforts put into practice. (If this happens too slowly, they start to lose interest, and more importantly, they begin to question whether management was serious about involving them in the first place.)
- People can still recall easily why they made specific decisions.
- The sooner the decisions are implemented, the sooner the organization derives the full benefits of the exercise.

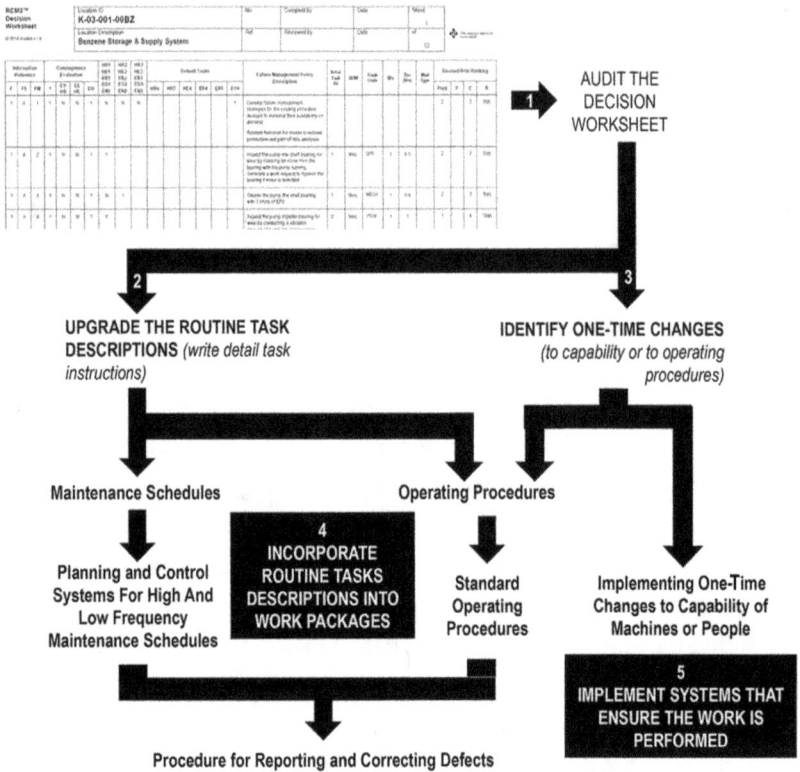

FIGURE 11.2 Implementation after the RCM audit

When overall agreement is reached about each analysis, the decisions are implemented as described in the rest of this chapter.

What the Audit Entails

An RCM analysis needs to be audited from the perspective of *method* and *content*. When reviewing the *method*, the auditor seeks to ensure that the RCM process has been correctly applied. When reviewing the *content*, the auditor seeks to ensure that the correct information has been gathered and conclusions drawn both about the asset itself and about the process that it is part of. Issues that most often need attention are as follows.

Levels of Analysis. The analysis should be carried out at the right level. The most common fault is to analyze assets at too low a level, and the usual symptom is large numbers of items with only one or two functions defined per item.

Functions. All the functions of the asset, together with the standards of performance desired by the user, should be clearly and correctly described:

- By and large, each function statement should define only one function, although it may incorporate more than one performance standard. As a rule, each function statement should contain only one verb (unless it is a protective device).
- Performance standards should be quantified, and they should indicate what the asset must be able to do in its present operating context rather than its rated capacity (what it can do).
- All protective devices should be listed, and their functions should be correctly described ("to do X if Y occurs").
- The functions of all gauges and indicators should be listed, together with desired levels of accuracy.

Failed States. All the functional failures associated with each function should be listed (usually complete failure plus the negative of each performance standard in the function statement).

Failure Modes. Ensure that failure modes that have happened, or that are reasonably likely, have not been omitted. Failure mode descriptions should also be specific. In particular:

- They should include a verb, not just be specified as a component.
- The verb should be a word other than "fails" or "malfunctions" unless it is appropriate to treat the failure of a subassembly as single failure mode (option 3 on page 174).
- Switch and valve failures should indicate whether the item fails in the open or closed position.

Failure modes should relate directly to the functional failure under consideration, and failure modes and effects should not be transposed, as in:

Failure Mode (Cause and Mechanism)	Failure Effect
Motor trips on overload	Pump impeller jammed by foreign object

Another common mistake is to combine two substantially different failure modes in one description, as follows:

Wrong	Correct
Screen damaged or worn	Screen damaged by foreign object
	Screen worn due to normal wear

Failure Effects. Failure effect descriptions should make it possible to decide:

- Whether (and how) the failure will be evident to the operating crew
- How often the failure would happen (if nothing is done to prevent it)
- Whether (and how) the failure poses a threat to safety or damage to the environment
- What effect (if any) the failure has on production or operations (output, product quality, customer service)
- Whether or not the function is protected and what would happen if the protection fails to operate

Failure effects should not incorporate actual "consequence" statements like "This failure affects safety," or "This failure is evident." However, they should list likely total downtime as opposed to repair time, and they should indicate what must be done to rectify the failure (replace, repair, reset, etc.).

Finally, auditors should satisfy themselves that anything that is said to be "analyzed separately" actually is analyzed separately.

Consequence Severity Evaluation. Special care should be taken to ensure that the hidden function question has been answered correctly. In particular, the correct meanings should have been attached to the terms "on its own" and "under normal circumstances" in this question, as explained on pages 000 and 000. Special attention should also be paid to the evaluation of the safety and environmental consequences of evident failures, as well as to the effectiveness of any tasks that might have been selected to manage risk in these two categories.

Risk Evaluation. The review group should be able to evaluate whether the risk associated with each failure is tolerable or not. The risk definition and risk matrix should be the official interpretation as adopted by the organization where all stakeholders were involved (engineering, operations, maintenance, and safety) and not just any arbitrary risk matrix. Interpretation of the risk matrix must be verified. Intolerable risk *must* be addressed through applying the RCM3 decision logic.

Strategy Selection. Any proactive strategy that has been selected should not only satisfy the criteria for technical feasibility, as explained in Chapter 8, but also address the risk associated with the failure. Look out for these three key points:

- If the answer to question 1 is no and the answers to questions HR3 and HR4 are no, then question HR5 must be answered. If the answer to HR5 is yes, the proposed task should not be "no scheduled maintenance."
- If the answer to question 1 is yes and the answer to question ES or EE is yes, the proposed task should not be "no scheduled maintenance."
- If the failure has operational or non-operational consequences, the task must be cost-effective.

If functions are protected, the protection directly related to the failure mode under consideration should be optimized (answers to ER5 and EO4 are yes) if the risk associated with failure of the protected function cannot be reduced using a proactive maintenance task.

Proposed strategies should be described in enough detail to leave the auditor in no doubt about what is intended. In particular, routine task descriptions should not simply list the type of task ("scheduled on-condition tasks," "scheduled failure-finding," etc.).

The task description should also relate directly and solely to the failure mode in question. It should not incorporate a combination of tasks because this usually signifies two different failure modes (unless the answer to question HR4 or ER4 is yes). For example:

Wrong	Correct
Inspect chain for wear and adjust tension	Adjust tension of chain or Inspect chain for wear

Initial Interval.   Task intervals should clearly have been set according to the criteria provided in Chapter 8. In particular, look out for a tendency to confuse P-F intervals with useful life in on-condition task intervals.

Revised Risk.   Ensure that after recommendations have been made, the risk has been revised to reflect the impact of the strategy. The revised risk should be within tolerable levels as defined by the organization's risk policy. As mentioned before, the risk evaluation should be done using the organization's approved risk matrix. Only in extreme cases should the revised risk exceed tolerable levels as defined by the organization's risk strategy. It is also worth noting that until the mitigation strategies are actually implemented, the inherent risk defined prior to the risk strategy development is still present. Implementation is key for reducing the risk to tolerable levels.

11.3 Risk Management Strategy (Task) Descriptions

Before any task reaches the person who has to do it, it must be described in enough detail to leave no doubt about what is to be done. Clearly, the degree of detail required will be influenced by the overall level of skill and experience of the workers involved. However, bear in mind that the more that is left out of a task description, the greater the chance that someone will miss a key step or choose to do the wrong task altogether. In this context, special care needs to be taken with the description of any failure-finding task that calls for a hazardous situation to be simulated in order to test the function of a protective device.

Task descriptions should also explain what action must be taken if a defect is encountered. (For instance, should the defect be reported to a supervisor or to the maintenance department—or should it be rectified immediately?) Instructions like "Check component A for condition B and replace if necessary" should be used with caution, because the "check" part of the task might only take a few seconds, while the "replace" part could take several hours. This can play havoc with the duration of planned downtime. Instructions of this sort should, in fact, be written as "Check component A for condition B and report defects to supervisor." Only use "if necessary" for quick servicing routines, such as "Check gearbox oil level using dipstick and top up with approved oil if necessary."

Examples of the right and the wrong way to specify tasks are shown below:

Wrong	Correct
Check coupling	Check feed screw coupling for loose bolts and replace if necessary *or* Visually check agitator coupling flange for cracks and report defects to the maintenance supervisor, etc.
Calibrate gauge	Fit 0–20-bar test gauge to test point and check if reading on pressure gauge Pl 1204 is within 5 psi of the reading on the test gauge when the test gauge reads 25 psi. Arrange to replace out-of-spec gauges when plant is shut down for cleaning *or* Remove pressure gauge Pl 1204 to workshop and calibrate following procedure in manual 27A

Each task should be defined as clearly as possible on the Decision Worksheet. This saves the duplication of effort that occurs if detailed procedures have to be written up later by someone else. It also reduces the possibility of transcription errors. However, if time does not permit the procedures to be specified during the RCM analysis, then they must be specified later. This can often be done as part of an ISO 9000–type initiative.

Note that if detailed task descriptions are to be prepared later, this should ideally be done by someone who participated in the original RCM analysis. If this is not possible, the third party should understand clearly that he or she is being asked to define the tasks on the Decision Worksheet in more detail, and *not* to re-audit the analysis.

Basic Information

In addition to a clear description of the task itself, the document on which the task is listed should also clearly state the following:

- *A description of the asset* to which it applies together with an equipment number where relevant
- *Who should do the task* (operator, electrician, fitter, technician, etc.)
- *The frequency* with which the task is to be done
- *Whether (and if necessary, how) the equipment should be stopped* and/or isolated while the task is being done, together with any other safety precautions that must be taken
- *Special tools and prescribed spares* to have on hand since these items can save much unproductive walking to and from after the job has started

ISO Standards and RCM

ISO 9000. A major objective of RCM is to identify what work people should be doing. (In other words, to ensure that "they do the right job.") On the other hand, a major thrust of quality systems like ISO 9000 is to define what people should be doing as clearly as possible in order to minimize the chance of errors. (In other words, to ensure that "they do the job right.")

This suggests that the process of transferring tasks from RCM Decision Worksheets to end-user documents can be seen as the point where the output of an RCM analysis becomes the input to an ISO 9000 procedure writing exercise. It also suggests that if both initiatives are to be undertaken, it makes sense to apply RCM first.

ISO 31000 and ISO 55000. The international standards for risk (ISO 31000) and asset management (ISO 55000) provide guidance on what organizations must do to ensure compliance and proper management of their physical assets. The standards are descriptive and comprehensive in describing what organizations should be doing (what to do), but the standards do not specify how to do it. RCM3 is the process that is describing not just what to do but also how to do it. If applied correctly, RCM3 will assist organizations to implement the standards. RCM3 places reliability and risk management mainstream with an organization's management systems.

11.4 Implementing One-Time Changes

At the end of a typical RCM analysis, it is not unusual to find that between 2% and 10% of the failure modes default to redesign. Chapter 8 mentioned that in the context of RCM, a one-time change could be any of the following three areas:

- A change to the physical configuration of an asset or system
- A change to a process or operating procedure
- A change to the capability of a person, usually by training

Once they have been accepted by the auditors, these changes need to be implemented as thoroughly and as quickly as possible. As mentioned earlier in this book, the risk will not be mitigated unless *all* the recommendations are implemented. Modifications and redesigns may take a long time before they are approved and completed. The equipment may actually be required to operate while the designs are being modified. Special care should be taken during these periods. Key issues in each of these three areas are discussed below.

Changes to the Physical Asset

All modifications should be:

- **Properly justified.** Chapter 8 explained that modifications should be justified in terms of their consequences. Modifications intended to deal with single or multiple failures that have safety or environmental consequences should reduce the risk (frequency and/or severity) of the consequences to a level that is acceptable. Figure 8.36 showed an algorithm that can be used to justify modifications intended to deal with failures that only have economic consequences.
- **Correctly designed by suitably qualified engineers.** As a rule, attempts should not be made to redesign assets during the RCM process, but the designer should consult afterward with the people who did the review in order to develop a correctly focused specification.
- **Properly implemented.** Steps must be taken to ensure that modifications are carried out as intended in terms of time, cost, and quality and that all drawings, manuals, and parts lists are updated correctly.
- **Properly managed.** Modifications should not interfere with essential routine maintenance activities in other parts of the plant, and the maintenance requirements of every modified item of equipment should be correctly assessed and implemented.

Implementing these changes requires the support and cooperation of the engineering/projects function, the training department, and operations management. This means that care must be taken to ensure that the relevant people in these departments understand why the changes are needed and are prepared to play their part in helping to implement them.

Changes to the Way in Which the Plant Is Operated

One-off changes to the way in which a plant must be operated are handled in the same way as routine tasks that are incorporated into operating procedures, as explained in the next part of this chapter.

Changes to the Capability of People

As explained in Chapter 6, the RCM process frequently reveals failure modes caused by slips or lapses on the part of operators or maintainers (skill-based human errors). These immediately become apparent to any operators or maintainers who participate directly in the process, and they usually modify their behavior appropriately as soon as they learn what they are doing wrong.

However, we also need to ensure that people who have not participated directly in the process acquire the relevant skills. In most cases, the most efficient way to do this is to revise or extend existing training programs or to develop new programs. In most organizations, this will be done in consultation with the training department.

11.5 Work Packages

Once the maintenance procedures have been fully specified, they need to be packaged in a form that can be planned and organized without too much difficulty, and they need to be presented in a neat and compact form to the people who will be doing the tasks. This can be done in two ways:

- High-frequency maintenance procedures to be done by operators can be incorporated into the operating procedures of the equipment.
- The balance of the maintenance routines is packaged into separate schedules and checklists.

Standard Operating Procedures

The previous part of this book mentioned that any changes that must be made to the way in which an asset is operated should be documented in standard operating procedures, or SOPs. (In situations where SOPs do not exist already, it will almost certainly be necessary to develop them in order to ensure that the changes are implemented.) In many cases, SOPs are also the simplest and cheapest way to manage high-frequency tasks that need to be done by operators.

As a rule, tasks should only be incorporated into operating procedures if they need to be done at intervals of 1 week or less. Tasks that need to be done by operators at longer intervals should be packaged into separate schedules and planned, organized, and controlled in the same way as maintenance schedules.

Maintenance Schedules

A maintenance schedule is a document listing a number of maintenance tasks to be done by a person with a specified level of skill on a specified asset at a specified frequency.

Compiling a maintenance schedule from RCM Decision Worksheets is a fairly simple process. However, a few additional factors need to be taken into account, as explained in the following paragraphs.

Consolidating Frequencies. In Chapter 10 it was mentioned that if a wide range of different task intervals appear on a Decision Worksheet, they should be consolidated into a smaller number of work packages when compiling the schedules based on the worksheets.

The most expensive tasks, in terms of the direct cost of doing them and the amount of downtime needed to do them, tend to dictate basic schedule intervals. However, planning is simplified if schedule intervals are multiples of one another.

Note also that if a task frequency is changed in this fashion, it should always be incorporated into a schedule of a higher frequency. Task intervals should never be arbitrarily increased, because doing so could move an on-condition task frequency outside the P-F interval for that failure, or it could move a scheduled discard task past the end of the life of the component.

Contradictions. When a low-frequency schedule incorporates a higher-frequency schedule, should the latter be incorporated as a global instruction, or should it be rewritten in full? In other words, should, for example, an annual schedule include an instruction like "Do the 3-month schedule," or should all the tasks in the 3-month schedule be written out in the annual schedule?

In fact, it is wise to rewrite the schedules in order to avoid the problem of contradictions.

For instance, consider what could happen in a situation where a 3-month schedule includes the instruction "Check gearbox oil and top up if necessary," and the annual schedule for the same machine starts with the instruction "Do the 3-month schedule" and later says "Drain, flush, and refill gearbox."

Too many anomalies and contradictions of this nature rapidly erode the credibility of the system in the eyes of the people doing the work, so it is worth taking a little extra time to ensure that they don't occur.

Adding Tasks. When compiling schedules on the basis described above, there is often a great temptation to start adding tasks to the completed schedule. This is most often done on the basis that "when we do A and B, we might as well do X, Y, and Z." This should be avoided for the following reasons:

- Extra tasks increase the routine workload. If too many tasks are added, the workload is increased to the point where either there is insufficient labor to do all the tasks, or the equipment cannot be released for the amount of time required to do them, or both.
- The people doing the schedules soon realize that X, Y, and Z are not strictly necessary, and they judge the schedule as a whole accordingly. As a result, they start looking for reasons why they cannot do the schedule as a whole. When they find them, tasks A and B are also not done, and the whole maintenance program begins to fall apart.

This problem is common in shutdowns. Many shutdown tasks are done, not because they are really needed, but because the plant is stopped, and it is possible to get at the equipment. This adds greatly to the cost and sometimes to the duration of the shutdown. Unnecessary work also leads to an increase in start-up failures when the plant starts up again.

(This does not mean that people who do routine tasks should concentrate only on the specified tasks and ignore any other potential and functional failures that they may encounter. Of course, they should keep their eyes and ears open. The point is that the schedule itself should only specify what really needs to be done at that frequency.)

11.6 Maintenance Planning and Control Systems

High- and Low-Frequency Maintenance Schedules

Once the tasks have been grouped into sensible work packages, the next step is to set up planning and control systems that ensure that they are done by the right person at the right time. A key factor that influences the design of such systems is the frequency of the schedules.

In particular, high- and low-frequency schedules are handled differently because both the work content and the planning horizons differ.

High-frequency schedules are defined as schedules performed at intervals of up to 1 week. These schedules usually consist of simple on-condition and failure-finding tasks. They have a low work content and hence can be done quickly. Most of them can also be done while the plant is running, so they can be done at more or less any time. These two factors mean that the associated planning systems can be kept very simple.

However, high-frequency schedules also exist in large numbers, so if careful thought is not applied to their administration, they can easily get out of hand. For example, daily schedules that have to be done for 350 days of the year on 1,000 items of equipment could generate 350,000 instructions annually if each schedule is issued separately (either electronically or on paper) every time it has to be done. This is clearly nonsense, and the problems it creates are a common reason why high-frequency schedules are often administered badly or not at all.

But high-frequency tasks are the backbone of successful routine maintenance, so some way must be found to ensure that they are done without creating an excessive administrative burden.

Low-frequency schedules are those done at intervals of a month or longer. Their longer planning horizon makes them less amenable to simple planning systems of the type used for high-frequency schedules. They usually have a higher work content, so more time is needed to do them, and the plant usually has to be stopped while they are done. As a result, they need more complex planning and control systems.

Some of the options that can be used to manage both types of schedules are:

- Schedules done by operators
- Schedules done by the quality function
- High-frequency schedules done by maintenance people
- Low-frequency schedules done by maintenance people

These options are explored in the next sections of this chapter.

Schedules Done by Operators

From the maintenance viewpoint, the most valuable attribute of operators is that they are near the equipment for much of the time. This puts them in an ideal position to do many on-condition, functional checks, and failure-finding tasks. These are often very high-frequency tasks—some will be daily or even once or twice per shift—so special care must be taken to keep the associated administrative systems as simple as possible.

Simple reminder systems that can be used for operator tasks instead of formal check sheets include:

- Incorporating the maintenance checks into standard operating procedures, as discussed earlier
- Mounting the schedule permanently onto a wall or on a control cabinet where the operators can see it easily or provide the schedules electronically on mobile devices.
- Training the operators in such a way that the inspections become second nature (a high-risk approach that is not usually recommended) or enforce routines through electronic means.

Formal written checklists should only be used for operator checks when the failure consequences are likely to be particularly severe and when there is reason to doubt whether the tasks will be done without a formal reminder. The checklists can be the same as those described later for high-frequency tasks done by maintenance people. The author does not endorse operator inspection rounds to simply keep the operators busy. The author refers to these activities as feel good maintenance or recreational maintenance. The tasks must still be technically feasible and worth doing.

Schedules and Quality Checks

We have seen how more and more performance standards incorporate product quality standards. This means that more and more potential and functional failures can be revealed by product quality checks. These checks are often being done already (for example, using SPC as discussed in Chapter 8). Key points to note are as follows:

- Quality checks must be recognized as a valid and valuable source of maintenance information.
- Steps must be taken to ensure that quality-related potential failures are attended to as soon as they are noticed. This issue is discussed later.

High-Frequency Schedules Done by Maintenance

Despite all the earlier comments about the merits of using operators to do high-frequency maintenance work, many of these tasks still need to be done by maintenance people. These usually need to be more formally planned than operators' checks, because maintenance people cover more machines spread over a wider area than operators have to cover, and they usually do a wider variety of tasks. One approach is to divide the plant into sections and prepare a checklist of the type shown in Figure 11.3 for each section.

MAINTENANCE CHECKLIST	PLANT SECTION	Boiler House								TO BE DONE BY Operator	WEEK ENDING DATE
ITEM NO:	DESCRIPTION	SCHEDULE	M	T	W	T	F	S	S		
03030401	Coal Handling System	M-265									
03030402	Boiler No 1	M-388									
03030402	Boiler No 1	M-389									
03030403	Boiler No 2	M-388									
03030403	Boiler No 2	M-389									
03030404	Ash Handling System	M-539									
03030405	Feed Water System	M-462									
03030406	Flue Gas System	M-391									
ALLOCATED TO		TIME								COMPLETED BY	SUPERVISOR

FIGURE 11.3

Note the following points about this type of checklist:

- The checklist only lists the schedules to be done, not individual tasks. The schedules are issued separately, often in book form or on handheld devices. In this way, only one checklist is issued per section per week, rather than dozens of schedules every day.
- Roughly the same amount of work should be planned for each day, and it should not exceed between half an hour and an hour per day.
- The checklist shown can be used to plan at intervals between daily and weekly. Jobs can be planned for alternate days and twice per week, so the checklist encompasses a wider range of the shorter P-F intervals.
- The checklist can start and finish on any 5- or 7-day cycle—it is not essential to stick to the Monday–Sunday cycle shown in the example.
- The checklists embody the schedule plans, and they are issued automatically every week, so there is no need for any sort of planning system.
- The checklists are not used for any tasks that are to be done at intervals of longer than a week.
- Each checklist involves one or two documents per week per section. This amounts to no more than 50 documents per week for a facility containing 1,000 items subject to these checks.

Some high-frequency tasks require readings to be taken, either manually (logging a meter reading) or electronically (vibration analysis). Readings of this nature are tasks, while the checklist described above is designed for complete schedules. This can cause problems, especially if we start issuing a separate document for each of these records alongside the checklists. This should be avoided, because the numbers of documents simply start climbing again. Possible alternatives are as follows:

- Develop a special document for all the readings in each section, and attach this one document to the checklist for that section each week.

- Use one person to take all such readings in the entire plant.
- Ask the people taking the readings to record only those readings that are outside acceptable limits in the "Remarks" column of the checklist (unless the readings are recorded automatically, as in the case of certain condition monitoring devices).
- Automate the recording process.

Issuing High-Frequency Schedules.   The checklists are issued to the relevant maintainer the week before they are to be done. Preferably, they should be the first activity done by that person each day. If the maintainer cannot complete the planned tasks on any day, the tasks are done the following day. Note the following additional features of a well-run checklist system:

- If the maintainer cannot complete the planned tasks on any day, the tasks are done the following day. If the maintainer is continually unable to complete the prescribed checks, something is fundamentally wrong, and the situation should be investigated.
- The maintainer notes any potential or functional failures in the "Remarks" column of the checklist—not on the schedules themselves.
- The maintainer initiates corrective action at the end of each daily round. In some facilities, this may be the responsibility of the maintainer; in others, he or she may have to work through a supervisor. The action will vary from arranging for the plant to stop at once to arranging for the fault to be corrected at the next shutdown. This decision is based on the possible consequences of the failure and the net P-F interval. (Note that these issues should have been considered as part of the RCM process when the routine task was originally specified.)
- As in the case of operators, it is important either that action is taken or that the maintainer is told why action is unnecessary or being deferred; otherwise, the maintainers also lose interest in the system.

- At the end of each cycle, the completed checklists can be stored as a record that the tasks were done, so it is not necessary to reenter them into a history recording system. However, problems that are encountered, and the actions taken to deal with them, should be documented, as discussed in Chapter 14.

Controlling High-Frequency Schedules. A problem associated with most checklist systems is the "tearoom tick syndrome." This means that people indicate that the checklist has been done when, in fact, it has not. To avoid this problem, supervisors should conduct random overinspections. These entail doing the schedules on the checklist in the company of the maintainer who normally does it. If the checklist is not being done correctly, unreported failures soon become apparent, and the supervisor takes appropriate action.

Low-Frequency Schedules Done by Maintenance

We have seen that high-frequency schedules can be planned, organized, and controlled using one carefully structured checklist. In contrast, the long planning horizon associated with low-frequency schedules means that the steps needed to plan, organize, and control them are carried out separately. What is more, the procedures used to plan schedules based on elapsed time differ markedly from those used for schedules based on running time, but similar procedures can be used to organize and control the two types of schedule. As a result, we consider the planning process separately in the following paragraphs but consider the subsequent steps together.

Elapsed-Time Planning. The principles of elapsed-time planning are well known and are used for many purposes in addition to maintenance planning. For low-frequency schedules, elapsed-time planning is usually based on a planning board or its computerized equivalent.

Most of these systems use an overall planning horizon of 1 year, divided into 52 weeks. However, bear in mind when setting up such systems that some failure-finding tasks in particular can have cycle times of up to 10 years, and the planning horizon of any associated planning system must accommodate such tasks. When setting up these systems, note also that low-frequency schedules nearly always involve equipment stoppages, and these can have operational consequences in exactly the same way as the stoppages that they are supposed to prevent, so special care must be taken to minimize these consequences. Points to watch for include:

- Peaks and troughs in the production cycle. The most time-consuming schedules should be planned for periods of lowest activity, in order to minimize their effect on operations.
- Two machines that require the same special resource at the same time (such as a crane).
- Cases where it is only possible to do a schedule if other machines are stopped at the same time. This applies especially to services like a steam raising plant and air compressors.

On the other hand, wherever such constraints permit, try to spread the routine maintenance workload as evenly as possible over the year in order to stabilize labor requirements.

A final point about elapsed-time-based low-frequency schedule planning is that it looks deceptively simple to use computers for this purpose. However, bear in mind that the issues discussed above introduce a wide range of completely unrelated constraints into the calendar-time planning process. For this reason, take great care when designing or acquiring calendar-time-based systems that plan schedules on the basis of predetermined parameters, or that automatically re-plan schedules that have not been done. The author has encountered a number of such systems that simply move schedules from week to week, regardless of policy constraints. This becomes chaotic, especially when schedules that should only be done in the low season are gradually moved into the middle of the high season and so on.

Running-Time Planning. Running-time planning involves the following steps:

- The number of cycles each machine has completed in each period is recorded (the cycles can be measured in terms of time, distance traveled, units of output, etc.).

- This record is fed into the planning system.
- The cumulative total of hours run is updated to reflect the time run since the last schedule was done.

Running-time planning systems lend themselves readily to the use of computers because they entail processing and storing large quantities of data. If computers are to be used for running-time planning, data capture should also be automated if possible. The system should also be designed to provide a continuously updated forecast of the scheduled workload on each workshop as far as possible into the future. This gives managers time to smooth any peaks and troughs that appear in the forecast.

Organizing Low-Frequency Schedules. Most planning systems start organizing low-frequency schedules the week before the schedules are due (except for shutdown schedules). The organizing process usually contains the following elements:

- A list is prepared that shows the schedules due the following week. They are usually separated by craft and plant section.
- Meetings are held with the operations department to agree on which day and at what time the schedules will be done (especially those that require equipment downtime).
- The schedules themselves are issued to the relevant supervisors, who plan who will do them and arrange any other resources that may be needed as they would for any other incoming maintenance job.

Controlling Low-Frequency Schedules. Low-frequency schedules are subject to the same performance controls as any other type of maintenance work. This applies to the time taken to do the schedules, standards of workmanship, and so on.

Two additional factors need to be considered. First, the planning system should indicate when any schedule is overdue. As mentioned earlier, such schedules should not be reprogrammed automatically, but should be managed on an exception basis.

Second, maintenance schedules should be reviewed continuously in the light of changing circumstances (especially circumstances that affect the consequences of failure) and new information. In this context, bear in mind that the more everyone associated with the equipment is involved in determining its maintenance requirements to begin with, the more everyone is likely to offer thoughtful and constructive feedback about these requirements in the future. This issue is discussed in more detail in the next chapter.

11.7 Reporting Defects

In addition to ensuring that the tasks are done, we need to ensure that any potential failures that are found are rectified before they become functional failures, and we also need to ensure that hidden functional failures are rectified before the multiple failure has a chance to occur. This means that anyone who might discover a potential or functional failure must have unrestricted access to a simple, reliable, and direct procedure for reporting it immediately to whoever is going to repair it.

This communication takes place instantaneously if the person who operates the machine is also the person who maintains it. The speed and accuracy of the response to defects that can be achieved under these circumstances are a major reason why people who operate machines should also be trained to maintain them (or vice versa). A second benefit of this approach is that formal defect reporting systems are only needed for failures that the operator/maintainer is unable to deal with on his or her own.

If this organizational structure is not possible or not practical, the next best way to ensure that defects are attended to quickly is to allocate maintenance people permanently to a specific asset or group of assets. Not only do such people get to know the machines better, which improves their diagnostic skills, but the speed of response also tends to be quicker than it would be if they work in a central workshop. It is also still possible to keep the defect reporting systems simple and informal.

If it is not possible to organize close maintenance support of either sort, then it becomes necessary to implement more formal defect reporting systems. In general, the further away the maintenance function is from the assets it is to maintain—in other words, the more heavily centralized it is—the more formal the defect reporting process becomes. This is also true of defects that can only be dealt with during major shutdowns.

Basically, formal defect reporting systems enable anyone to inform the maintenance department in writing (electronically or manually) about the existence of a potential or functional failure. The main criteria of such systems should always be simplicity, accessibility, and speed.

Manual defect reporting systems are usually based on simple job cards. (These job cards can also be used by the maintenance department to plan and record work, but this aspect of their use is beyond the scope of this book.) If a computerized defect reporting system is used, the screen is formatted in much the same way.

The final point about systems of this sort is that people must be properly motivated to use them. This means that defects that are reported must be acted upon, or the user must be told why if no action is taken. Nothing will kill such a system more quickly than if defects are reported and nothing apparently happens.

11.8 Eliminating Defects

Although defect elimination is not the subject of this book, it is equally important to apply a formal process of determining the cause of failure and how to best correct and eliminate repeat failures. A formal process of root cause failure analysis and defect elimination must be adopted by the organization and preferably integrated with the RCM3 process to ensure continuous improvement and perpetuity of the reliability improvement process.

CHAPTER 12

Actuarial Analysis and Failure Data

12.1 The Six Failure Patterns

Throughout this book, numerous references have been made to the six patterns of failure, shown again in Figure 12.1. Frequent use has also been made of terms like "age," "life," and "MTBF." This chapter explores these concepts and the relationship between them in more detail. It also considers what role (if any) technical history records and other failure data play in formulating maintenance policies. We start with a detailed look at failure patterns B and E, because they represent the most widely held views of age-related and random failure. Next, we review patterns C and D, and finally we look at patterns F and A. Section 12.2 of this chapter summarizes the uses and limitations of failure data.

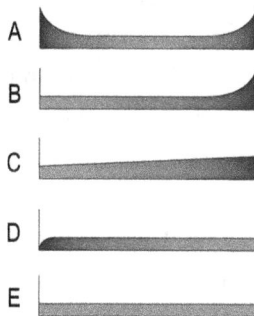

FIGURE 12.1   Six patterns of failure

Failure Pattern B

Chapters 2 and 6 mentioned that failure pattern B depicts age-related failures. Chapter 6 explained that although these failures are the result of a more-or-less linear process of deterioration, there will still be considerable differences in the behavior of any two components that are subject to the same nominal stresses. Figure 12.2 shows how this behavior translates into failure pattern B.

An example of a component that might behave as shown in Figure 12.2 is the impeller of a pump that is used to pump a moderately abrasive liquid. Part 1 of Figure 12.2 shows the wear-out characteristics of a dozen such impellers. Ten of them deteriorate at roughly the same rate, and last between 11 and 16 periods before failing. However, two of the impellers fail much sooner than expected; impeller A fails much sooner perhaps because it was not properly case-hardened and impeller B because the properties of the liquid changed for a while, causing it to wear more quickly than usual. Note that this failure distribution only applies to impellers that fail due to wear. It does not apply to impellers that fail for other reasons.

In Part 2 of Figure 12.2, the distribution of failure frequencies is plotted against operating age for a large sample of components. It shows that apart from a few premature failures, the majority of the components are likely to conform to a normal distribution about one point.

For example, assume that we have accumulated actual failure data for a sample of 110 impellers, all of which have failed due to wear. Ten of these impellers failed prematurely, one in each of the first 10 periods. The other 100 impellers all failed between periods 11 and 16, and the frequency of these failures conforms to a normal distribution. (For a normal distribution, the failure frequencies in the last six periods would be roughly as shown if they are rounded to the nearest whole number.) On the basis of these figures, the mean time between failures of the impellers due to wear is 12.3 periods.

Part 3 of Figure 12.2 shows the survival distribution of the impellers based on this frequency distribution. For example, 98 impellers

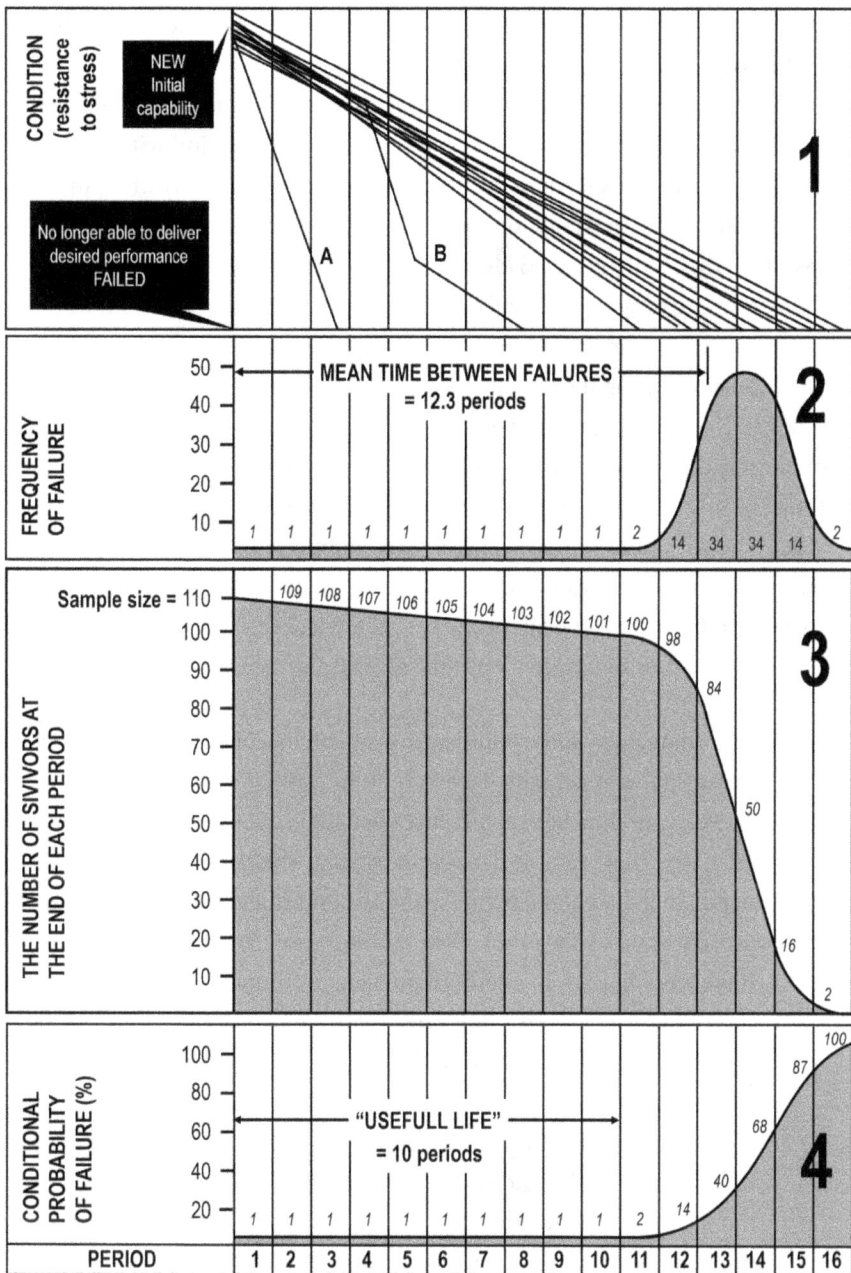

FIGURE 12.2   Failure pattern B

lasted for more than 11 periods, and 16 impellers lasted for more than 14 periods.

Part 4 of Figure 12.2 is failure pattern B. It shows the probability that any impeller that has survived to the beginning of a period will fail during that period. This is known as the conditional probability of failure.

Allowing for a small degree of rounding error, this shows, for instance, that there is a 14% chance that an impeller that has survived to the beginning of period 12 will fail in that period. Similarly, 14 out of the 16 impellers that make it to the beginning of period 15 will fail in that period—a conditional probability of failure of 87%.

The frequency curve in Part 2 and the probability curve in Part 4 are depicting the same phenomenon, but they differ markedly in the way they show it. In fact, the conditional probability of failure curve provides a better illustration of what is really happening than the frequency curve, because the latter could deceive us into thinking that things are getting better after the peak of the frequency curve.

These curves illustrate a number of additional points:

- The frequency and conditional probability curves show that the word "life" can actually have two quite distinct meanings. The first is the mean time between failures (which is the same as the average life if the whole sample has run to failure). The second is the point at which there is a rapid increase in the conditional probability of failure. For want of any other term, this has been named the "useful life."
- If we were to plan to overhaul or replace components at the mean time between failures, half would fail before they reached it. In other words, we would only be preventing half of the failures, which is likely to have unacceptable operating consequences. Clearly, if we wish to prevent most of the failures, we would need to intervene at the end of the "useful life." Figure 12.2 shows that the useful life is shorter than the mean time between failures—if the bell curve is wide, it can be very much shorter.

  As a result, it can only be concluded that *the mean time between failures is of little or no use in establishing the frequency of scheduled restoration and scheduled discard tasks for items that conform to* failure pattern B. The key variable is the point at which there is a rapid increase in the probability of failure.

- If we do replace the component at the end of its useful life as defined above, the average service life of each component would be shorter than if we let it run to failure. As discussed on page 000, this would increase the cost of maintenance (provided that there is no secondary damage associated with the failures). For instance, if we were to replace all the surviving impellers in Figure 12.2 at the end of period 10, the average service life of the impellers would be about 9.5 periods, instead of 12.3 periods if they were allowed to run to failure.

- The fact that there are two "lives" associated with pattern B–type failures means that we must take care to specify which one we mean whenever we use the term "life."

  For example, we might phone the manufacturer of a certain component to ask what its life is. We may have in mind the useful life, but if we don't spell out exactly what we mean, the manufacturer might in all good faith give us the mean time between failures. If this is then used to establish a replacement frequency, all kinds of problems arise, often resulting in wholly unnecessary unpleasantness.

These issues apart, perhaps the biggest problem associated with pattern B is that very few failure modes actually behave in this fashion. As mentioned in Chapters 2 and 6, it is much more common to find failure modes that show little or no long-term relationship between age and failure.

Failure Pattern E

Figure 8.24 illustrated three components that failed on a random basis. A number of reasons why failures can occur on this basis were discussed in Chapter 8. This part of this chapter explores some of the quantitative aspects of random failure in more detail and then goes on to review some of the implications of failure pattern E. To start with, Figure 12.3 shows the relationship between the frequency and conditional probability of random failures.

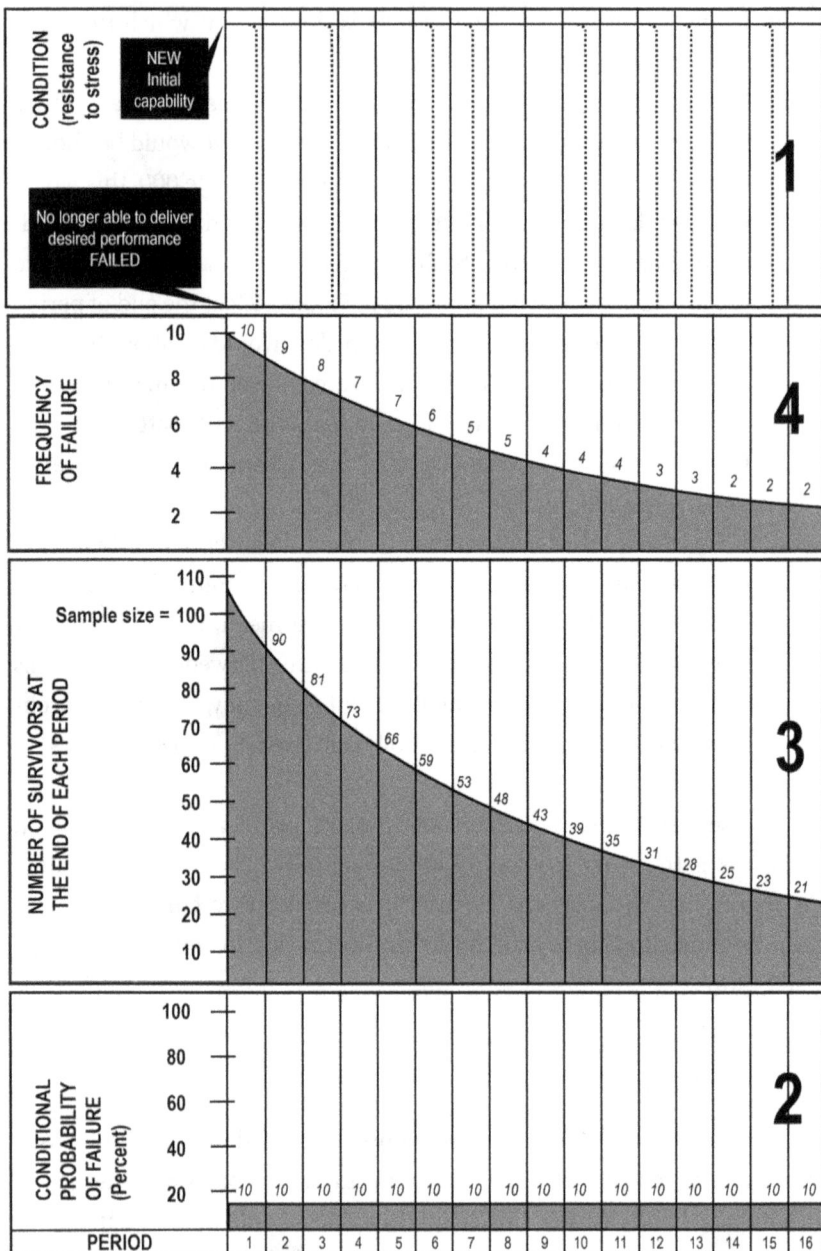

FIGURE 12.3   Failure pattern E

In Part 1 of Figure 12.3, the dotted lines represent a number of components—in this case, ball bearings—which fail at random. As in Figure 8.24, each failure is preceded by a (somewhat elongated) P-F curve.

Random failure means that the probability that an item will fail in any one period is the same as it is in any other. In other words, the conditional probability of failure is constant, as shown in Part 2 of Figure 12.3.

For example, if we accept the empirical evidence that rolling element bearings usually conform to a random failure pattern—a phenomenon first observed by Davis (1952)—the conditional probability of failure is constant, as shown in Figure 12.3, Part 2. Specifically, this shows that there is a 10% probability that a bearing that has made it to the beginning of any period will fail during that period.

Part 3 of Figure 12.3 shows how a conditional probability of failure that is constant translates into a survival distribution that is exponential.

For example, if we started with a sample of 100 bearings and the probability of failure in the first period is 10%, then 10 bearings would fail in period 1 and 90 bearings would survive for more than one period.

Similarly, if there is a 10% probability that the bearings that survive beyond the end of period 1 will fail in period 2, then 9 bearings would fail in period 2, and 81 bearings would make it to the beginning of period 3. Part 3 of Figure 12.3 shows how many bearings would survive to the beginning of each subsequent period for the first 16 periods.

Theoretically, this process of decay would continue until infinity. In practice, however, we usually stop at unity—in other words, when the survival curve drops below 1.

In the example shown in Figure 12.3, a rate of decay of 10% per period means that unity is reached after about 43 periods. This suggests that one lone bearing might last for 43 periods, but the vast majority will have failed long before then.

Finally, Part 4 of Figure 12.3 shows the frequency curve derived from the survival curve in Part 3. This curve is also exponential. (The shape of this frequency curve often causes it to be confused with failure pattern F, which is a conditional probability curve based on a different frequency distribution.)

The fact that the frequency and survival curves both carry on declining indefinitely means that the conditional probability curve also remains flat indefinitely. In other words, at no stage does pattern E show a significant increase in the conditional probability of failure, so at no stage can an age be found at which we should contemplate scheduled rework or scheduled discard. Further points about pattern E are as follows.

MTBF and Random Failures. Despite the fact that it is impossible to predict how long any one item that conforms to failure pattern E will last (hence the use of the term "random" failure), it is still possible to compute a mean time between failures for such items. It is given by the point at which 63% of the items have failed.

For example, Part 3 of Figure 12.3 indicates that 63% of the items have failed about halfway through period 10. In other words, the MTBF of the bearings in this example is 9.5 periods. The fact that these items have a mean time between failures but do not have a "useful life" as defined earlier means that we must be doubly careful when talking about the life of an item.

Comparing Reliability. The MTBF provides a basis for comparing the reliability of two different components that both conform to failure pattern E, even though the failure is random in both cases. This is because the item with the higher MTBF will have a lower probability of failure in any given period.

For example, assume that brand X bearings conform to the failure distribution shown in Figure 12.3. If the conditional probability of failure of brand Y is only 5% in each period, brand Y bearings would only be half as likely to fail and so would be considered much more reliable.

In the case of items that conform to failure pattern B, a more reliable component has a longer useful life than one that is less reliable. So in simple language it could be said of the pattern B components that one type lasts longer than the other, while in the case of the pattern E components, one type fails less often than the other.

(In practice, the reliability of bearings is measured by the "B10" life. This is the life below which a bearing supplier guarantees that no more than 10% of its bearings will fail under given conditions of load and speed. This corresponds to one period on Part 2 of Figure 12.3. It also suggests that if a bearing conforms to a truly exponential survival distribution, then the MTBF of bearings due to normal wear and tear should be about 9.5 times the B10 life. So if bearing brand Y is twice as reliable as bearing brand X, the B10 life—which is also known as the L10 life or the N10 life—of brand Y will be twice as long as that of brand X. This is useful when making procurement decisions about bearings, but it still does not tell us how long any one bearing will last in service.)

P-F Curves and Random Failures. Figure 8.24 and Part 1 of Figure 12.3 both show random failures preceded by P-F curves. This is not meant to suggest that all failures that happen on a random basis are preceded by such a curve. In fact, a great many failure modes that conform to pattern E are not preceded by any sort of warning, or if they are, the warning period is often much too short to be of any use. This is especially true of most of the failures that affect light current electrical and electronic items.

This does not detract in any way from the validity of the analysis. It simply means that no form of preventive maintenance—on-condition, scheduled restoration, or scheduled discard—is technically feasible for these components, and they have to be managed on an appropriate default basis as discussed in Chapters 8 and 9.

A Note on Weibull Distributions. At this stage, it is worth commenting on the Weibull distribution. This distribution is widely used because it has a great variety of shapes that enable it to fit many kinds of data, especially data relating to product life. The Weibull frequency distribution (or more correctly, probability density function) is

$$f(t) = (\beta/a^{\beta})t^{\beta-1}\exp[-(t/a)^{\beta}]$$

where β is called the shape parameter because it defines the shape of the distribution, and a is the scale parameter. The equation defines the spread of the distribution and corresponds to the 63rd percentile $[100\,(1 - e^{-1})]$ of the cumulative distribution.

The Weibull probability density function and corresponding conditional probability curves are shown in Figure 12.4. (This shows why the conditional probability of failure is also known as the "hazard rate.") When $\beta = 1$, the Weibull distribution is the exponential distribution. When β is

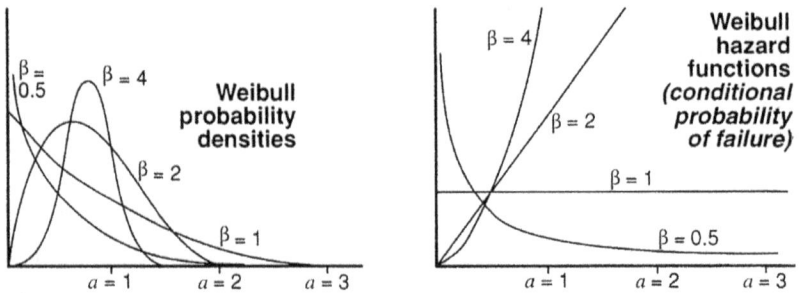

FIGURE 12.4 Weibull distributions

between 3 and 4, it closely approximates the normal distribution. Later in this chapter we see how it describes other failure patterns.

Failure Pattern C

Failure pattern C shows a steadily increasing probability of failure, but no one point at which we can say, "That's where it wears out." This part of the chapter looks at a possible reason why pattern C occurs and then shows how it is derived.

The possible cause of pattern C that we consider is fatigue. Classical engineering theory suggests that fatigue failure is caused by cyclic stress, and that the relationship between cyclic stress and failure is governed by the S-N curve, as shown in Figure 12.5.

Figure 12.5 suggests that if the S-N curve is known, then we should be able to predict the life of the component with great accuracy for a given amplitude of cyclic stress. However, this is not so in practice because the average amplitude of the cyclic stress is not constant, and the ability of the component to withstand the stress—in other words, the location

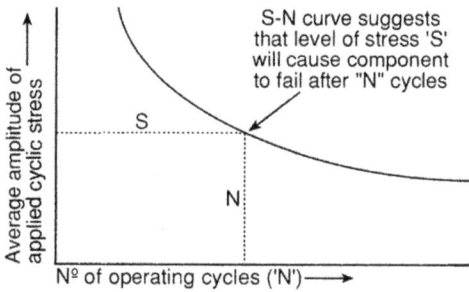

FIGURE 12.5 The S-N curve

of the S-N curve—will not be exactly the same for every component. The top illustration of Figure 12.6 suggests that the average amplitude of the applied stress might conform to a normal distribution about some mean, which is designated by S in Figure 12.6. This distribution is shown by curve P. Similarly, the distribution of the S-N curves might be designated by normal curve Q. The combination of these two curves will be such that the ages at which failure occurs will conform to a distribution skewed to the left. How much it is skewed depends on the shape of the S-N curve itself. For the sake of argument, the second part of Figure 12.6 suggests that it will conform to a Weibull distribution with shape parameter $\beta = 2$. (Strictly speaking, this should be called a "shifted" Weibull distribution because it does not start at time zero. The shifted Weibull distribution is discussed further at the end of this section.)

On the basis of this distribution, Part 2 of Figure 12.6 goes on to suggest how many failures might occur in each period if we were to test a sample of 1,000 components to failure. (The fact that the numbers marked with an asterisk are not integers explains why this curve should be called a probability density rather than a frequency distribution.)

Part 3 of Figure 12.6 translates Part 2 into a survival curve, while Part 4 shows the conditional probability of failure based on the preceding two curves. Both of the latter curves are derived in the same way as the corresponding curves in Figure 12.2.

Note that in these examples, the slope of pattern C appears to be quite steep. However, bear in mind that the actual slope is governed by the Weibull scale parameter a, which can be measured in anything

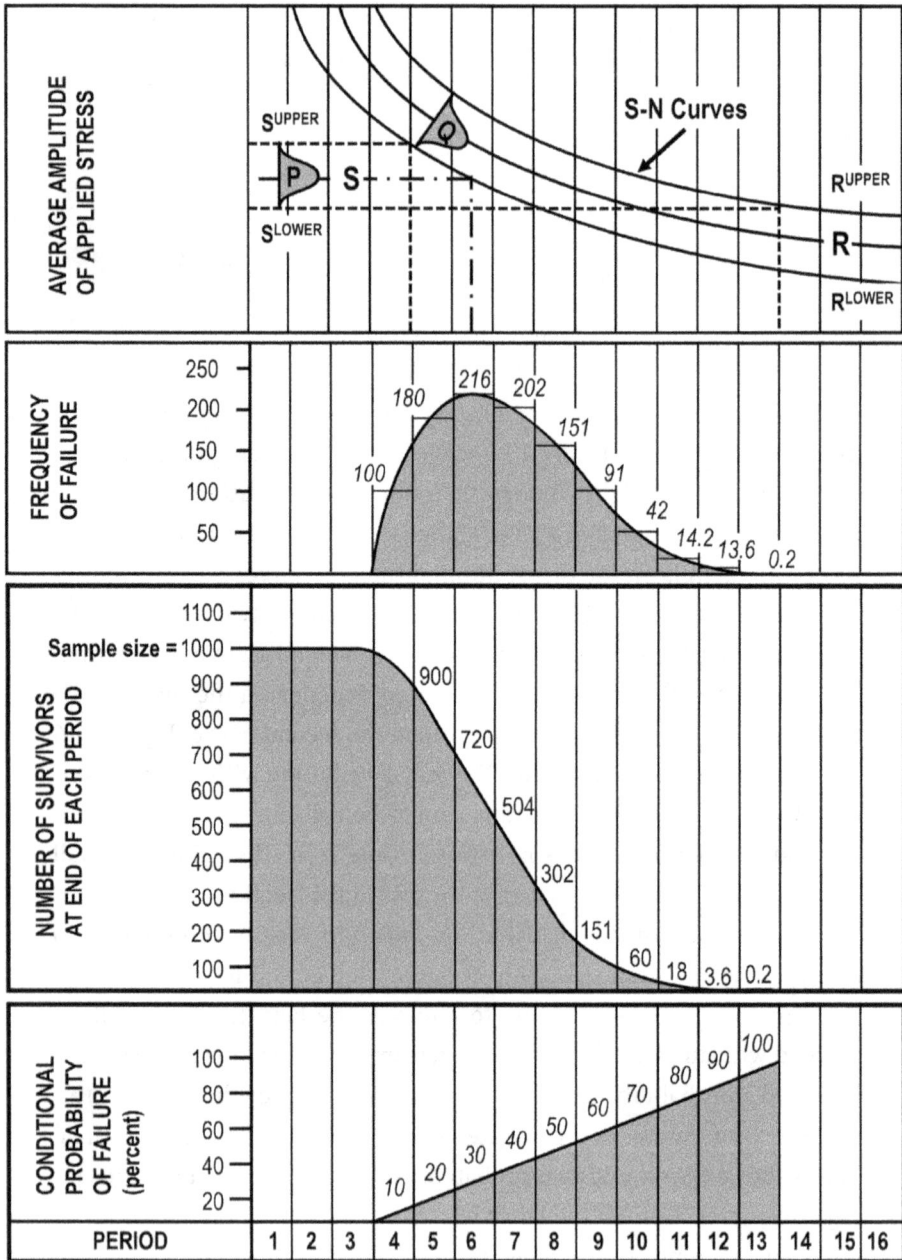

FIGURE 12.6   Failure patten C

ranging from weeks to decades (or even centuries), so the slope of pattern C can vary from quite steep to almost flat.

Also note that pattern C is not only associated with fatigue. For instance, it has been found to fit the failure of the insulation in the windings of certain types of generators.

Conversely, not all fatigue-related failures necessarily conform to failure pattern C.

For instance, if curve P in Figure 12.6 were skewed toward the S^{lower} limit and curve Q were skewed toward the R^{upper} limit, the failure frequency curve would be biased further toward the right. This would give a Weibull shape parameter greater than 2, which tends toward a normal distribution and so gives a conditional probability of failure curve that resembles pattern B.

On the other hand, if the slower limit is below the point at which R^{lower} becomes asymptotic, then the frequency distribution will develop a long tail on the right. This corresponds to a Weibull distribution where β is between 1 and 2, which, in turn, generates failure pattern D.

Finally, the discussion on page 000 mentioned that a large number of factors influence the rate at which fatigue failures develop in ball bearings. This would make the spread of any distribution very wide, which would, in turn, lead to an almost flat conditional probability curve. Add to this the variety of additional bearing failure modes listed on page 000, which have the same symptoms as fatigue, and the overall probability density effectively becomes fully exponential, which leads to failure pattern E as we have seen.

So fatigue could manifest itself as failure pattern B, C, D, or even E.

A Word about the Shifted Weibull Distribution. The shifted Weibull distribution means that the conditional probability curve starts at a point to the right of time $t = 0$. Figure 12.6 shows that this is the point where there is a rapid increase in the conditional probability of failure, which is, of course, the useful life as defined earlier. In Figure 12.6 this is three periods. However, earlier depictions of pattern C show a conditional probability of

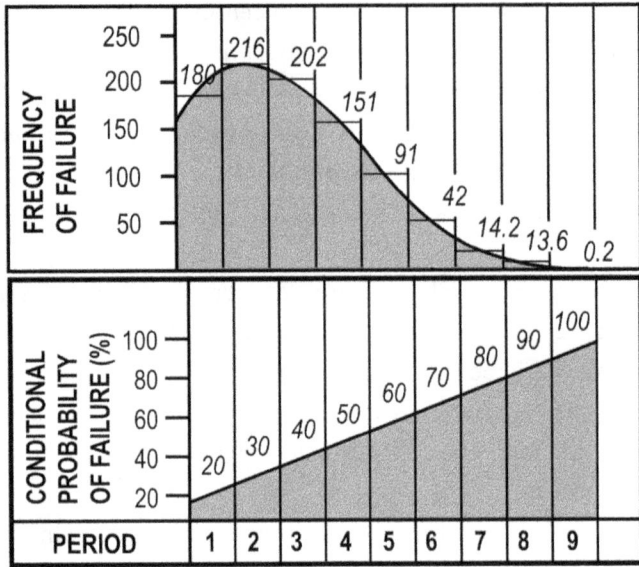

FIGURE 12.7 Truncated Weibull distribution and failure pattern C

failure starting above zero. This might occur in practice if a failure mode led to a truncated Weibull distribution (one that hypothetically starts to the left of time $t = 0$) with a shape parameter of $\beta = 2$, as shown in Figure 12.7.

Failure Pattern D

As mentioned above, failure pattern D is the conditional probability curve associated with a Weibull distribution whose shape parameter β is greater than 1 and less than 2.

Failure Pattern F

Pattern F is perhaps the most interesting, for two reasons:

- It is the only pattern where the probability of failure actually declines with age (apart from A, which is a special case).
- It is the most common of the six patterns, as mentioned on page 00. For these reasons, it is worth exploring in more detail the factors that give rise to this pattern.

The shape of failure pattern F indicates that the highest probability of failure occurs when the equipment is new or just after it has been over-hauled. This phenomenon is known as a start-up failure (also known as infant mortality), and it has a wide variety of causes. These are summa-rized in Figure 12.8 and discussed in the subsequent paragraphs.

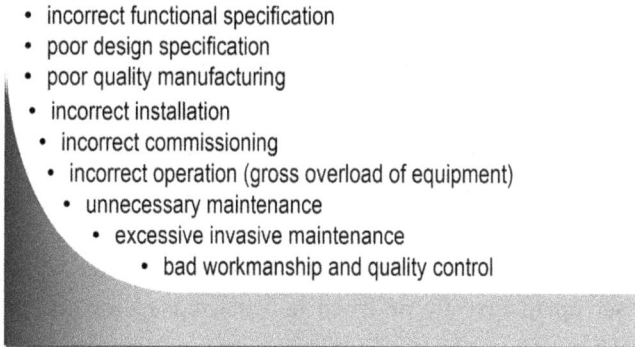

- incorrect functional specification
- poor design specification
- poor quality manufacturing
- incorrect installation
- incorrect commissioning
- incorrect operation (gross overload of equipment)
- unnecessary maintenance
- excessive invasive maintenance
- bad workmanship and quality control

FIGURE 12.8   Failure pattern F

Design. Start-up failure problems attributable to design occur when part of an item is simply incapable of delivering the desired performance and hence tends to fail soon after being put into service. When they affect an existing asset, these problems can only really be solved by redesign, as discussed in Chapter 8. They can be forestalled to some extent by:

- **Using proven technology.** The author encountered one company that professed to be "in a headlong rush to be second" in adopting new technology, because it found that being first usually means a huge investment in debugging new equipment—an involuntary investment made in the form of equipment downtime. On the other hand, being second can be competitively disadvantageous in the long term.
- **Using the simplest possible equipment to fulfill the required function.** This is done on the premise that bits that aren't there can't fail.

Manufacture and Installation. Start-up failures attributable to equip-ment manufacture occur either because the manufacturer's quality standards are too loose or because the parts concerned have been badly installed. These problems can only be solved by rebuilding the affected assemblies or replacing the affected parts. Two ways to forestall these problems are:

- To implement suitable SQA (supplier quality assurance) and PQA (project quality assurance) schemes. Such schemes usually work best when they are run by someone other than the prime contractor.
- To request extended warranties, perhaps with the full-time on-site support of the vendor's technicians until the equipment has been working as intended for a specified period.

Commissioning. Commissioning problems occur either when equipment is set up incorrectly or when it is started up incorrectly. These problems are minimized if care is taken to ensure that everyone involved in commissioning knows exactly how the plant is supposed to work and is given enough time to ensure that it does so.

Routine Maintenance. Many start-up failures are caused by routine maintenance tasks that are either unnecessary or unnecessarily invasive. The latter are tasks that disrupt or disturb the equipment, and so they needlessly upset basically stable systems. The way to avoid these problems is to stop doing unnecessary tasks and, in cases where scheduled maintenance is necessary, to select tasks that disturb the equipment as little as possible.

Maintenance Workmanship. Clearly, if something is badly put together, it will fall apart quickly. This problem can only be avoided by ensuring that anyone who is called upon to do a preventive or corrective maintenance task is trained and motivated to do it correctly the first time.

Start-Up Failures and RCM. The above discussion suggests that start-up failure problems are usually solved by one-off actions rather than by scheduled maintenance (with the exception of a few cases where it may be feasible to use on-condition tasks to anticipate failures). However, despite the minimal role played by routine maintenance, using RCM to analyze a new asset before putting it into service still leads to substantial reductions in start-up failures for the following reasons:

- A detailed study of the functions of the asset usually reveals a surprising number of design flaws that, if not corrected, would make it impossible for the asset to function at all.
- Craftspeople and operators learn exactly how the asset is supposed to function and so are less inclined to make mistakes that cause failures.
- Many weaknesses that would otherwise lead to premature failures are identified and dealt with before the asset enters service.
- Routine maintenance is reduced to the essential minimum, which means fewer destabilizing interventions, but this essential minimum ensures that the early life of the asset is not plagued by failures that could have been anticipated or prevented.

Failure Pattern A

It is now generally accepted that failure pattern A—the bathtub curve—is really a combination of two or more different failure patterns, one of which embodies start-up failures and the other of which shows increasing probability of failure with age. Some commentators even suggest that the central (flat) portion of the bathtub constitutes a third period of (random) failure between the other two, as shown in Figure 12.9.

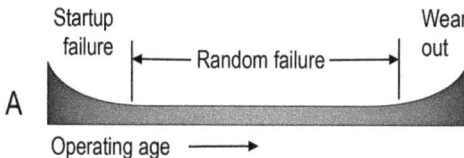

FIGURE 12.9 The classical view of the "bathtub"

This means that failure pattern A actually depicts the conditional probability of two or more different failure modes. From the failure management viewpoint, each of these must be identified and dealt with in the light of its own consequences and its own technical characteristics.

Similar conclusions can be drawn about failure pattern B as it is shown in Figure 12.2. This is because the failures that occur between periods 11 and 16 are caused by normal wear, while those occurring between periods 1 and 10 are caused by other random factors that still cause the impeller to wear out but cause it to do so faster than normal.

This starts to raise a number of questions about the meaning of these patterns, which are considered at length in the next part of this chapter.

12.2 Technical History Data

The Role of Actuarial Analysis in Establishing Maintenance Policies

A surprising number of people believe that effective maintenance policies can only be formulated on the basis of extensive historical information about failure. Thousands of manual and computerized technical history recording systems have been installed around the world on the basis of this belief. It has also led to great emphasis being placed on the failure patterns described in Section 12.1 of this chapter. (The fact that the bathtub curve still appears in nearly every significant text on maintenance management is testimony to the almost mystical faith that we place in the relationship between age and failure.)

Yet from the maintenance viewpoint, these patterns are fraught with practical difficulties, conundrums, and contradictions. They include:

- Complexity
- Sample size and evolution
- Reporting failure
- The ultimate contradiction

These are summarized below.

Complexity. Most industrial undertakings consist of hundreds, if not thousands, of different assets. These are made up of dozens of different components, which between them exhibit every extreme and intermediate aspect of reliability behavior. This combination of complexity and diversity means that it is simply not possible to develop a complete analytical description of the reliability characteristics of an entire undertaking—or even any major asset within the undertaking.

Even at the level of individual functional failures, a comprehensive analysis is not easy. This is because many functional failures are caused not by two or three but by two or three dozen failure modes. As a result, while it may be fairly easy to chart the incidence of the functional failures, it is a major statistical undertaking to isolate and describe the failure pattern that applies to each of the failure modes that fall within the envelope of each functional failure. What is more, many failure modes have virtually identical physical symptoms, which makes them easy to confuse with each other. This, in turn, makes sensible actuarial analysis almost impossible.

Sample Size and Evolution. Large industrial processes usually possess only one or two assets of any one type. They also tend to be brought into operation in series rather than simultaneously. This means that sample sizes tend to be too small for statistical procedures to carry much conviction. For new assets that embody high levels of leading-edge technology, they are always too small.

These assets are also usually in a continuous state of evolution and modification, partly in response to new operational requirements and partly in an attempt to eliminate failures that either have serious consequences or that cost too much to prevent. This means that the amount of time that any asset spends in any one configuration is relatively short.

So actuarial procedures are not much use in these situations because the database is both very small and constantly changing. (As discussed later, the main exception is undertakings that use large numbers of identical components in a more-or-less identical manner.)

Reporting Failure. The problem of analyzing failure data is further complicated by differences in reporting policy from one organization to another. One area of confusion is the distinction between potential and functional failures.

For instance, in the tire example discussed on pages 000 and 000, one organization might classify and record the tires as "failed" when they are removed for retreading after the tread depth drops below 3/8 inch. However, as long as the tread depth is not allowed to drop below 1/8 inch, this "failure" is actually a potential failure as defined in Chapter 8. So other organizations might choose to classify such removals as "precautionary," because the tires have not actually failed in service, or they might even choose to classify the removals as "scheduled," because the tires are scheduled for replacement at the earliest opportunity after the potential failure has been discovered. In both of the latter cases, it is likely that the removals will not even be reported as failures.

On the other hand, if for some reason the tread depth does drop below 1/8 inch, then there is no doubt that the tire has failed.

Similar differences might be caused by different performance expectations. Chapter 5 defined a functional failure as the inability of an item to meet a desired standard of performance, and these standards can, of course, differ for the same asset if the operating context is different.

For instance, page 105 gives the example of a pump that has failed if it is unable to deliver 300 gallons per minute in one context and 350 gallons per minute in another.

This shows that what is a failure in one organization—or even one part of an organization—might not be a failure in another. This can result in two quite different sets of failure data for two apparently identical items.

Further differences in the presentation and interpretation of failure data can be caused by the different perspectives of the manufacturers and users of an asset. The manufacturer usually considers this its responsibility to provide an asset capable of delivering a warranted level of performance (if there is one) under specific conditions of stress. In other words, the **manufacturer** warrants a certain basic design capability and often makes this conditional upon the performance of certain specified maintenance routines.

On the other hand, we have seen that many failures occur because users operate the equipment beyond its design capabilities (in other words, the "want" exceeds the "can," as discussed on pages 102–103). While users are naturally inclined to incorporate data about these failures in their own history records, manufacturers are naturally reluctant to accept responsibility for them. This leads many manufacturers to "censor" failures caused by operator error and not include them in failure data. As Nowlan and Heap (1978) put it, the result is that users talk about what they actually saw, while the manufacturer talks about what they should have seen.

The Ultimate Contradiction (the Resnikoff Conundrum). An issue that bedevils the whole question of technical history is the fact that if we are collecting data about failures, it must be because we are not preventing them. The implications of this are summed up most succinctly by Resnikoff (1978) in the following statement:

"The acquisition of the information thought to be most needed by maintenance policy designers is in principle unacceptable and is evidence of the failure of the maintenance program. This is because critical failures entail potential (in some cases, certain) loss of life, **but there is no rate of loss of life which is acceptable to (any) organization as the price of failure information to be used for designing a maintenance policy**. Thus, the maintenance policy designer is faced with the problem of creating a maintenance system for which the expected loss of life will be less than one over the planned operational lifetime of the asset. This means that, both in practice and in principle, the policy must be designed without using experiential data, which will arise from the failures which the policy is meant to avoid."

Despite the best efforts of the maintenance policy designer, if a critical failure should happen to occur, Nowlan and Heap (1978) go on to make the following comments about the role of actuarial analysis:

"The development of an age-reliability relationship, as expressed by a curve representing the conditional probability of failure, requires a considerable amount of data. When the failure is one which has serious consequences, this body of data will not exist, since preventive measures must of necessity be taken after the first failure. Thus, actuarial analysis cannot be used to establish the age limits of greatest concern—those necessary to protect operating safety."

In this context, note also the comments made on page 264 about safe-life limits and test data. These data are usually so scanty that the safe-life limit (if there is one) is established by dividing the test results by some conservatively large arbitrary factor rather than by the tools of actuarial analysis.

The same limitation applies to failures that have really serious operational consequences. The first time such a failure occurs, immediate decisions are usually made about preventive or corrective action without waiting for the data needed to carry out an actuarial analysis.

All of which brings us to the ultimate contradiction concerning the prevention of failures with serious consequences and historical information about such failures: *that successful preventive maintenance entails preventing the collection of the historical data that we think we need in order to decide what preventive maintenance we ought to be doing.*

This contradiction applies in reverse at the other end of the scale of consequences. Failures with minor consequences tend to be allowed to occur precisely because they don't matter very much. As a result, large quantities of historical data will be available concerning these failures, which means that there will be ample material for accurate actuarial analyses. These may even reveal some age limits. However, because the failures don't matter much, it is highly unlikely that the resulting scheduled restoration or scheduled discard tasks will be cost-effective. So while the actuarial analysis of this information may be precise, it is also likely to be a waste of time.

The chief use of actuarial analysis in maintenance is to study reliability problems on the middle ground, where there is an uncertain relationship between age and failures that have significant economic consequences but no safety consequences. These failures fall into two categories:

- Those associated with large numbers of identical items whose functions are, to all intents and purposes, identical, and whose failure might only have a minor impact when taken singly but whose cumulative effect can be an important cost consideration.

  Examples of items that fall into this category are street lights, vehicle components (especially from large fleets), and many of the components used by the armed forces and in the electricity, water, and gas distribution industries. Items of this type are used in sufficient numbers for precise actuarial analyses to be carried out, and detailed cost-benefit studies are justified (in many cases, if only to minimize the amount of traveling involved in maintaining the items).

- Those that are less common but are still thought to be age-related, and where both the cost of any preventive task and the cost of the failure are very high. As mentioned on page 000, this applies especially to gradually increasing failure probabilities typified by failure pattern C.

The Way Forward. The above paragraphs indicate that except for a limited number of fairly specialized situations, the actuarial analysis of the relationship between operating age and failure is of very little use from the maintenance management viewpoint. Perhaps the most serious shortcoming of historical information is that it is rooted in the past, whereas the concepts of anticipation and prevention are necessarily focused on the future.

So a fresh approach to this issue is needed—one that switches the focus from the past to the future. In fact, RCM is just such an approach. First, it deals with the specific problems identified above:

- **Defining failure.** By starting with the definition of the functions and the associated performance standards of each asset, RCM enables us to define with great precision what we mean by "failed."

By distinguishing clearly between built-in capability and desired performance, and between potential failures (the failing state) and functional failures (the failed state), it eliminates further confusion.

- **Complexity.** RCM breaks each asset down into its functions and each function into failed states, and only then it identifies the failure modes that cause each failed state. This provides an orderly framework within which to consider each failure mode. This, in turn, makes failure modes much easier to manage than if we were to start out at the failure mode level (which is the starting point of most classical FMEAs and FMECAs).

- **Evolution.** By providing a comprehensive record of all the performance standards, functional failures, and failure modes associated with each asset, RCM makes it possible to work out very quickly how any change in the design or in the operating context is likely to affect the asset, and to revise maintenance policies and procedures only in those areas where changes need to be made.

- **The ultimate contradiction.** RCM deals with the ultimate contradiction in several ways. First, by obliging us to complete the Information Worksheet described in Chapter 6, it focuses attention on what could happen. Contrast this with the actuarial emphasis on what has happened. Second, by asking how, and how much, each failure matters as set out in Chapter 7, it ensures that we focus on failures that poses intolerable risks and that we do not waste time on those that don't. Finally, by adopting the structured approach to the selection of proactive risk management strategies and default strategies described in Chapters 8 to 9, RCM ensures that we do what is necessary to prevent serious failures from happening, and as far as humanly possible, avoid having to analyze them historically at all.

Second, the RCM process focuses attention on the information needed to support specific decisions. It does not ask us to collect a whole lot of data in the hope that they will eventually tell us something. This point is discussed in more detail in the next section of this chapter.

Specific Uses of Data in Formulating Maintenance Policies

In spite of all the above comments, the successful application of RCM needs a great deal of information. As explained at length in Chapters 4 to 9, much of this information is descriptive or qualitative, particularly on the RCM Information Worksheet. However, in view of the emphasis that has been placed on quantitative issues in this chapter, Table 12.1 summarizes the principal types of quantitative data that are used to support different stages of the maintenance decision process. It does so under the following headings:

- **Datum.** The piece of information of interest.
- **Application.** A very brief summary of the use to which each datum is put. Note that some data are used in conjunction with others to reach a final decision, and that many are only used when qualitative data are not strong enough to make an intuitive decision possible.
- **Comments.** The place where each datum is most likely to be found. Note that in some cases, the datum is established by the user of the asset.
- **Pages.** The pages in this book where the use of each datum is discussed at greater length.

A number of final points concerning quantitative data include:

- Management information
- A note on the MTBF
- Technical history

These points are reviewed below.

Management Information. Table 12.1 only describes data that are used directly to formulate policies designed to deal with specific failure modes. It does not include data used to track the overall performance

Table 12.1 Summary of Key Maintenance Decision-Support Data

DATUM	APPLICATION	COMMENTS	PAGES
Desired standards of performance	These standards define the objectives of maintenance for each asset. They cover output, product quality/customer service, energy efficiency, safety, and environmental integrity	Set by the users of the assets (and by regulators for environmental and some safety standards)	31 79–86
ASSESSING OPERATIONAL AND NON-OPERATIONAL FAILURE CONSEQUENCES			
Downtime	Assessing whether each failure will affect production/operations, and if so, how much	Not the same thing as MTTR (mean time to repair)	154–156
Cost of lost production	Used together with downtime to evaluate total cost of each failure that affects operations	Only needed when the cost-benefit of scheduled maintenance is not intuitively obvious	215 217–220
Cost of repair	Used together with MTBF to evaluate cost-effectiveness of scheduled maintenance	Only needed when the cost-benefit of scheduled maintenance is not intuitively obvious (Operational and non-operational consequences only)	217–220
Mean time between failures	Used with downtime, cost of lost production (if any) & repair cost to compare cost of scheduled maintenance with the cost of a failure over a period of time	Only needed when the cost-benefit of scheduled maintenance is not intuitively obvious (Operational and non-operational consequences only)	51 217–220
ASSESSING SAFETY AND ENVIRONMENTAL FAILURE CONSEQUENCES			
Tolerable risk of a single failure	Used to assess whether scheduled maintenance is worth doing for failures that could have a direct adverse effect on safety or the environment	Almost always assessed by the users of the assets/likely victims on an intuitive basis	210–214
ESTABLISHING ON-CONDITION TASK FREQUENCIES			
Potential failure	Point at which imminent failure becomes detectable	Based on the nature of the P-F curve and the monitoring technique: usually quantified for performance monitoring, condition monitoring, and SPC	276–289
P-F Interval	Used to establish the frequency of on-condition tasks	"How quickly it fails": very seldom formally recorded	276–289 290–300

Table 12.1 Summary of Key Maintenance Decision-Support Data (*Continued*)

DATUM	APPLICATION	COMMENTS	PAGES
SCHEDULED RESTORATION AND SCHEDULED DISCARD TASK FREQUENCIES			
Age at which there is a rapid increase in the conditional probability of failure	Used to establish the frequency of most scheduled restoration and scheduled discard tasks	"Useful life": Based on formal records if these are available: more often based on consensus of people who have the most knowledge of the asset	256–269
Actuarial analysis of relationship between age and failure	Optimizing restoration/ discard intervals for large numbers of identical parts whose failure is known to be age-related, or for expensive pattern C–type failures	Worth doing for no more than 1–2% of failure modes in most industries: needs extensive and reliable historical data: used for failure modes that have operational and non-operational consequences only	405
HIDDEN FAILURE CONSEQUENCES AND FAILURE-FINDING TASK FREQUENCIES			
Tolerable probability of a multiple failure	Used to establish maintenance policies for protected systems	Set by the users of the asset: only used when a rigorous analysis is to be done	322–326
Mean time between failures of a protected function (M_{TED})	Used together with "tolerable probability of a multiple failure" to determine the desired availability of a protective device	Based on past and anticipated future performance of the protected function: only used to support a rigorous analysis—not needed for the intuitive approach (*see below*)	322–326
Desired availability of a protective device	Used together with the MTBF of the protective device to establish a failure-finding task interval	Derived from the two above variables if the task frequency is to be derived on a rigorous basis: otherwise set directly by the users of the asset on the basis of an intuitive assessment of the risks of the multiple failure	315 322–326
Mean time between failures of a protective device (M_{TIVE})	Used with desired availability to establish a failure-finding task interval	Based on records of *failures found* if these are available: if not, any suitable data source should be used to begin with (including educated guesses), but a suitable database should be started immediately	321

of the maintenance function and usually classified as "management information." Examples of such information are plant availability statistics, safety statistics, and information about expenditure on maintenance against budgets.

Monitoring the overall performance of the maintenance function is, of course, an essential aspect of maintenance management. This topic is discussed in more detail in Chapter 14.

A Note on the MTBF. In recent times, the concept of the "mean time between failures" seems to have acquired a stature that is quite disproportionate to its real value in maintenance decision making. For instance, it has nothing to do with the frequency of on-condition tasks, and it has nothing to do with the frequency of scheduled restoration and scheduled discard tasks. However, it does have certain very specific uses. Table 12.1 mentions six of these:

- To *quantify risk* for statistically random failures (in a broad sense).
- To establish the frequency of failure-finding tasks.
- To help decide whether scheduled maintenance is worth doing in the case of failure modes that poses economic risks (operational or non-operational consequences only). In other words, it helps us to decide whether such tasks need to be done, but not how often they need to be done.
- To help establish the desired availability of a protective device.
- To determine budgets and life-cycle costs.
- To determine spare part strategies.

In the first and second cases, the MTBF is always needed to make the appropriate decision, but in the third and fourth cases, it is only used if the nature and risk associated with the failures are such that a rigorous analysis must be carried out.

The MTBF also has a number of uses outside the field of maintenance policy formulation:

- **In the field of design.** To carry out a detailed cost-justification of a proposed modification, as mentioned briefly on page 349.

- **In the field of procurement.** To evaluate the reliability of two different components that are candidates for the same application.
- **In the field of management information.** As discussed in Chapter 14, assessing the overall effectiveness of a maintenance program by tracking the mean time between unanticipated failures of any asset.

A detailed exploration of the first two of these issues is beyond the scope of this book. The third is dealt with in Chapter 14.

Technical History. Together with the above comments about the MTBF, Table 12.1 can be used to help decide what sorts of data really need to be recorded in a technical history recording system.

Perhaps the most important information that needs to be recorded on a formal basis is what is found each time a failure-finding task is done. Specifically, we need to record whether the item was found to be fully functional or whether it was in a failed state. Such records enable us to determine the mean time between failures of the protective device (M_{TIVE} on page 319) and hence to check the validity of the associated failure-finding task interval. This information should be recorded for all hidden functions—in other words, for all protective devices that are not fail-safe.

In addition to hidden failures, Table 12.1 identifies two further areas where historical failure data can be used to make (or to validate) decisions about maintenance policies.

The first is the occurrence of failure modes that have significant operational consequences. This information can be used to compute the mean time between the failures in order to assess the cost-effectiveness of scheduled maintenance. However, as mentioned in Table 12.1, this only needs to be done if the cost-benefit of proactive action is not intuitively obvious. If it is, such action—be it scheduled maintenance or redesign—would be taken, and so there should be no more failures to record (except perhaps as potential failures if the proactive action is an on-condition task).

Table 12.1 mentions that in rare cases, it may also be worth capturing these data in order to carry out full actuarial analyses with a view to optimizing scheduled restoration and scheduled discard frequencies.

The second area where historical failure data can be used to make (or validate) decisions about maintenance policies is the mean time between failures of a protected function (M_{TED} on page 325). This is needed if a failure-finding interval is to be determined on a rigorous basis. It can be determined by recording the number of times a protective device is called upon to function when the failure of the protected function occurs. For instance, a record can be made every time the overpressurization of a boiler causes a relief valve to start passing.

If any of these data are to be captured, the failure reporting systems should be designed to identify the datum that is required—usually the failure mode—as precisely as possible. This can be done by asking the person who does the task (or who discovers the failure in the case of failure-finding) to either:

- Complete a suitably designed form that is then used to enter the data into a manual or computerized history recording system.

or

- Enter the data directly if an online computer system is used to store it.

In most organizations, the records themselves can be stored in:

- A simple proprietary PC-based database

or

- A specialized computerized or manual maintenance history recording system

The design of such systems is also beyond the scope of this book. However, Table 12.1 suggests that if technical history recording systems are used to capture specific data for specific reasons, rather than to record everything in the hope that it will eventually tell us something, they become useful and powerful contributors to the practice of maintenance management rather than the expensive white elephants that so many of them tend to be.

CHAPTER 13

Applying the RCM Process

13.1 Who Knows?

The eight basic questions that make up the RCM3 process have been considered at length in Chapters 3 to 9. After looking more deeply at the information needed to answer the questions, Chapter 12 concluded that in most industries, historical records are seldom (if ever) comprehensive enough to be used for this purpose on their own. Yet the questions must still be answered, so the required information still has to be obtained from somewhere.

More often than not, "somewhere" actually turns out to be "someone"—someone who has intimate knowledge and experience of the asset under consideration. There are also occasions when the information-gathering process reveals widely differing viewpoints that have to be reconciled before decisions can be made.

Later sections of this chapter describe how small groups can be used to gather the information, reconcile differing views, and make the decisions. However, before considering these groups, this part of the chapter reviews the information needed to answer each question and considers who is most likely to possess it. It does so with reference to earlier sections of the book where the questions have been discussed in detail.

What are the functions and associated performance standards of the asset in its present operating context?

RCM is based on the premise that every asset is acquired to fulfill a specific function or functions, and that maintenance means doing whatever is necessary to ensure that the asset continues to perform each function to the satisfaction of its users. In most cases, the most important representatives of the users are operations and production managers. In order to ensure that RCM3 generates a maintenance program that delivers what these managers want, they need to participate actively in the entire process. (In areas such as safety, hygiene, or the environment, the advice of appropriate specialists may also be needed.)

However, we have also seen that the built-in capability of the asset—what it can do—is the most that maintenance can actually deliver. Maintenance and design people, often at supervisory levels, tend to be the custodians of this information, so they too are a key part of this process.

If this information is shared at a single forum, maintainers begin to appreciate much more clearly what operators are trying to achieve, while users gain a clearer understanding of what maintenance can—and cannot—deliver.

In what ways does it fail to fulfill its functions?

The example on page 112 showed why it is essential that the performance standards used to judge functional failures should be set by maintenance and operations people working together.

What causes each functional failure?

Chapter 6 explained how maintenance is really managed at the failure mode level. It went on to stress the importance of identifying the causes and mechanism of each functional failure. The example on page 161 showed how these causes are often most clearly understood by the shop-floor and supervisory people who work most closely with each machine (especially the craftspeople and technicians who have to diagnose and repair each failure). In the case of new equipment, a valuable source of information about what can fail is a field technician who is employed by the vendor and who has worked on the same or similar equipment.

What happens when each failure occurs?
Section 6.5 of Chapter 6 lists a wide variety of information that needs to be recorded as failure effects. These include:

- The evidence that the failure has occurred, which is most often obtained from the *operators* of the equipment.
- How often it occurs is also information that will be provided by *operators and maintainers. OEMs* are also a great source of data.
- The amount of time the machine is usually out of action each time the failure occurs, again obtained from *operators or first-line supervisors.*
- The hazards associated with each failure, which may need *specialist advice* (especially concerning such issues as the toxicity and flammability of chemicals, or the hazards associated with mechanical items such as pressure vessels, lifting equipment, and large rotating components).
- What must be done to repair the failure, which is usually obtained from the *craftspeople or technicians* who carry out the repairs.
- The financial impact (downtime cost, litigation, penalties, lost sales, etc.) can be obtained from the *production supervisor* or *operations manager.*

In what way does each failure matter?
The inherent risk associated with each failure is discussed at length in Chapter 7 and summarized in the five questions at the head of Figure 10.1 on page 356. The assessment of failure consequences can only be done in close consultation with *production/operations* people, for the following reasons:

- **Physical risk-based decisions—safety and environmental consequences.** If the effects of a failure mode are explained reasonably thoroughly, it is usually quite easy to assess whether it is likely to affect safety or the environment. The main difficulty in this area lies in deciding what level of risk is acceptable. The discussion about who should evaluate risk on page 194 suggests that this decision should be made by a group consisting of the likely victims of the failure, the people who would bear the responsibility if it were to occur, and, if necessary, an expert on the specific characteristics of the failure.

- **Hidden failures.** The analysis of hidden functions requires at least four items of information, especially if a rigorous approach is used to determine failure-finding task intervals (see Chapter 8). This information is summarized below:

  - **Evidence of failure.** The first question on the RCM decision diagram asks if the loss of function caused by this failure mode on its own will become evident to the operating crew under normal circumstances. This question can only be answered with assurance by consulting the operating crew concerned.

  - **Normal circumstances.** As explained on page 246, different people can attach quite different meanings to the term "normal" in the same situation, so it is wise to ask this question in the presence of the operators and their supervisors.

  - **Acceptable probability of a multiple failure.** This should also be established by the group discussed on page 236.

  - **The mean time between failures of a protected function.** This is needed if the desired availability of a protected device is to be determined on a rigorous basis. If this information has not been recorded in the past, it can sometimes be obtained by asking the operators of the equipment how often the protective device is called upon to operate as a result of the failure of the protected function.

- **Operational risk-based decisions—economic consequences.** A failure has operational consequences if it affects output, product quality, or customer service, or if it leads to an increase in costs other than the direct costs of repair. Clearly, the people who are in the best position to assess these consequences are operations managers and supervisors, perhaps with help from cost accountants.

- **Non-operational risk-based decisions—tolerable risks.** The people who are usually in the best position to assess direct repair costs are first- and second-line maintenance supervisors.

Once the consequence severity has been determined, the review group will be in the position to assess the risk associated with each failure. The next step, based on the inherent risk, will be to evaluate whether the risk is tolerable or not. The development of the risk definition (risk matrix) is not covered in this section. The development of the risk matrix is covered in Chapter 7.

What must be done to reduce intolerable risks?
The information needed to assess the technical feasibility of different types of proactive tasks was discussed in Chapters 6 to 9, and the key questions are summarized on page 363. If clear actuarial data are not available to provide answers, then the questions must again be answered on the basis of judgment and experience, as follows:

- **On-condition tasks.** Page 288 stressed how important it is to consider as many different potential failures as possible when seeking on-condition tasks. The monitoring possibilities range from sophisticated condition monitoring techniques through product quality and primary effects monitoring to the human senses, so we should consult operators, craftspeople, supervisors, and, if necessary, specialists in the different techniques.

  A similar group would need to consider the duration and consistency of the associated P-F intervals, as explained on page 289.

  The amount of time needed to avoid the consequences of the failure (in other words, the net P-F interval) is established jointly by *maintenance and operations supervisors.*
- **Scheduled restoration and scheduled discard.** In the absence of suitable historical data, the people who are usually most likely to know whether any failure mode is age-related, and if so whether

and when there is a point at which there is a rapid increase in the conditional probability of failure, are again the *operators, craftspeople, and supervisors* who are closest to the asset.

Whether it is possible to restore the original resistance to failure of the asset is usually decided by *maintenance supervisors* or, in doubtful cases, by *technical specialists*.

- **Failure-finding.** If the frequency of a failure-finding task is to be established without performing a rigorous analysis of the protected system, the desired availability of the protective device should be determined by a *group of the sort* described on page 231.

  In the absence of formal records, the MTBF of the protective device can be derived initially either by asking the manufacturer of the device for this information or by asking anyone who might have done any functional checks in the past what they found when they did the checks. As mentioned on page 233, this is usually an operator or maintainer.

  Maintenance craftspeople and supervisors are usually the people who are best qualified to assess whether it is possible to do a failure-finding task in accordance with the criteria set out on page 327.

- **One-time changes.** The question of one-time changes (redesign) is discussed at length in Chapter 8. Note that the formal RCM process is only meant to identify situations where redesign is either compulsory or desirable. RCM review groups should not attempt to develop new designs during RCM meetings for two reasons:
  - The design process requires skills that are usually not present at an RCM forum.
  - Done properly, developing even one new design takes a great deal of time. If this time is spent during RCM review meetings, it slows down and can even paralyze the rest of the program. (This is not to suggest that designers should not consult the users and maintainers of the assets—just that it should not be done as part of the RCM review process.)

What can be done to reduce tolerable risks in a cost-effective way?
The RCM3 process requires all intolerable risks to be addressed proactively using the robust decision logic. Tolerable risks should also be addressed provided they can be managed or lowered in a cost-effective way. In order to save time and expensive resources, the facilitator may address tolerable risks outside the review meeting through one-on-one discussions with subject-matter experts—the *operators, maintainers, engineers, and manufacturers.*

The above paragraphs demonstrate that it is impossible for one person, or even for a group of people from one department, to apply the RCM process without working with others. The diversity of the information that is needed and the diversity of the people from whom it must be sought mean that it can only be done on the basis of extensive consultation and cooperation, especially between production/operations and maintenance people. The most efficient way to organize this is to arrange for the key people to apply the process in small groups.

13.2 RCM Review Groups

The eight questions that make up the RCM3 process have been considered at length in Chapters 3 to 9. The nature of these questions, especially those dealing with functions, consequence assessments, and the selection of suitable failure management policies, is such that even where historical records exist, they are simply not comprehensive enough to provide all the answers on their own. Yet the questions must still be answered, so the information still has to be obtained from somewhere.

As noted earlier in the chapter, in most cases, "somewhere" turns out to be "someone"—someone with intimate knowledge and experience of the asset under review. However, we have seen in the previous chapter that the nature of the RCM process is such that it is impossible for one person or even for a group of people from one department to apply the RCM process on their own. The diversity of the information that is needed and the diversity of the people from whom it must be sought mean that it can only be done on the basis of extensive consultation and

cooperation, especially between production/operations and maintenance people. The most efficient way to organize this is to arrange for the
key people to apply the process in small groups. This chapter considers
who should participate in a typical RCM review group and what each
group actually does.

Who should participate?

In practice, the places in every group do not have to be filled by exactly
the same people as shown in Figure 13.1. The objective is to assemble a
group that can provide most if not all of the information described in
Section 13.1. These are the people who have the most extensive knowledge and experience of the asset and of the process it is part of. To ensure
that all the different viewpoints are taken into account, this group should
include a cross section of users and maintainers and a cross section of
the people who do the tasks and the people who manage the aforemen-
tioned resources. In general, it should consist of not less than four and
not more than seven people, the ideal being five or six.

Facilitator

Operations Supervisor

Maintenance Supervisor

RCM3 Review Group

Operator

Maintainer

Specialist

FIGURE 13.1   A typical RCM review group

The group should consist of the same individuals throughout the
analysis of any one asset. If the people present at each meeting change, too
much time is lost going over ground that has already been covered for the
benefit of the newcomers. People can be specialists in any of the following:

- Some aspect of the process. These aspects usually tend to be dangerous or environmentally sensitive issues.
- A particular failure mechanism, such as fatigue or corrosion.

- A specific type of equipment, such as hydraulic systems.
- Some aspect of maintenance technology, such as vibration analysis or thermography.
- Suppliers and manufacturers of the assets under review.
- Vendors and suppliers of spare parts (i.e., long lead items).

Unlike other group members, specialists only need to attend meetings at which their specialty is under discussion.

What does each group do?

The objective of each group is to use the RCM process to determine the maintenance requirements of a specific asset or a discrete part of a process. Under the guidance of a facilitator, the group analyzes the context in which the asset is operating and then completes the RCM Information Worksheet as explained in Chapters 4 to 7. (The actual writing is done by the facilitator, so the group members do not have to handle any chapter if they don't wish to.) The members of the group then use the RCM decision diagram shown in Chapter 10 to decide how to deal with each of the failure modes listed on the Information Worksheet. Their decisions are recorded on RCM Decision Worksheets as explained in Chapter 10.

The RCM meetings also provide a very efficient forum for key people to learn how to operate and maintain *new* equipment, especially if one of the vendor's field technicians attends meetings held during the final stages of commissioning. The RCM process provides a framework for such technicians to transfer everything they know about the asset to the other group members in an orderly and systematic fashion. The RCM worksheets enable the organization to capture the information in writing for dissemination to anyone else who needs to know.

The watchword throughout this process is consensus. Each group member is encouraged to participate and contribute whatever he or she can at each stage in the process. Nothing should be recorded until it has been accepted by the whole group. (As discussed in Section 13.3 of this chapter, the facilitator has a crucial role to play in this aspect of the process.)

This work is done at a series of meetings that last for about 3 hours each, and each group meets at an average rate of anywhere from one to five times per week. If the group includes shift workers, the meetings need to be planned with special care.

The asset should be subdivided and allocated to groups in such a way that any one group can complete the entire process in not less than 5 and not more than 15 meetings—certainly no more than 20.

What do participants get out of the process?

The flow of information that takes place at these meetings is not only into the database. When any one member of the group makes a contribution, the others immediately learn three things:

- More about the asset, more about the process that it is part of, and more about what must be done to keep it working. As a result, instead of having five or six people who each know a bit—often a surprisingly little bit—about the asset under review, the organization gains five or six experts on the subject.
- More about the objectives and goals of their colleagues. In particular, maintenance people learn more about what their production colleagues are trying to achieve, while operations people learn much more about how maintenance can—and cannot—help them to achieve it.
- More about the individual strengths and weaknesses of each team member. On balance, much more tends to be learned about strengths than weaknesses, which has a salutary effect on mutual respect as well as mutual understanding.

In short, participants in this process gain a much better understanding of:

- What each group member (themselves included) should be doing
- What the group is trying to achieve by doing it
- How well each group member is equipped to make the attempt

This changes the group from a collection of highly disparate individuals from two notoriously adversarial disciplines (operations and maintenance) into a team.

The fact that the team members have each played a part in defining the problems and identifying solutions also leads to a much greater sense of ownership on the part of the participants. For instance, operators start talking about "their" machines, while maintenance people are much more inclined to offer constructive criticism of "their" schedules.

This process has been described as "simultaneous learning," because the participants identify what they need to learn at the same time as they learn it. (This is much quicker than the traditional approach to training, which starts with a training needs analysis, proceeds through the development of a training program, and ends with the presentation of training courses—a process that can take months.)

One limitation of group learning in this fashion is that unless specific steps are taken to disseminate the information further, the only people who benefit directly are the members of each group. Two ways to overcome this problem are:

- To ensure that anyone in the organization can gain access to the RCM database at any time
- To use the output of the RCM process to develop formal training courses

The RCM meetings also provide a very efficient forum for key people to learn how to operate and maintain new equipment, especially if one of the vendor's field technicians attends meetings held during the final stages of commissioning. The RCM process provides a framework for such technicians to transfer everything they know about the asset to the other group members in an orderly and systematic fashion. The RCM worksheets enable the organization to capture the information in writing for dissemination to anyone else who needs to know.

13.3 RCM3 Facilitators

Section 13.2 of this chapter mentioned that the facilitator has a crucial role to play in the implementation of RCM. The primary function of an RCM facilitator is to facilitate the application of the RCM philosophy by asking questions of a group of people chosen for their knowledge of a specific asset or process, ensuring that the group reaches consensus about the answers, and recording the answers.

Of all the factors that affect the ultimate quality of the analysis, the skill of the facilitator is the most important. This applies both to the *technical quality* of the analysis and to:

- The pace at which the analysis is completed
- The attitude of the participants toward the RCM process

The author trained many RCM facilitators over the years and came to realize that the skills required by the RCM facilitator extends beyond understanding the RCM process only. The facilitator must be able to identify the asset systems that need to be included in the RCM program, the information that is needed to conduct a successful RCM analysis, the people who must participate and how to manage the overall process – inside and outside the meetings. To achieve a reasonable standard, an RCM facilitator has to be competent in 55 key areas. These can be divided into eight main skill sets:

- Performing asset criticality review and asset prioritization (where to focus)
- Developing a risk framework and tolerability thresholds (developing a risk matrix)
- Planning the RCM project (scope and analysis boundary)
- Applying the RCM3 logic (following the process)
- Managing the analysis
- Conducting the meetings
- Time management and implementation planning
- Interacting with people outside the review group—administration, logistics, and managing upward

Performing Asset Criticality Review and Asset Prioritization

Not all assets are equal and not all assets should be treated the same. RCM3 takes time and involve valuable resources as discussed in Section 2 of this chapter, the RCM effort should focus on areas where it matters most. Therefore, not all assets should probably be considered for the RCM analysis process (although some users decide to do so). The question arises, which assets should be included and where to start.

Asset Criticality Assessment. The asset criticality assessment and prioritization process (ACAP) discussed in detail in Chapter 2 (Section 2.5) provides the understanding of which assets should be analyzed using RCM. It is a focusing tool used by the facilitator for selecting the systems and subsystems that pose the biggest threat to the organization (for achieving business goals) and the ones that would benefit the most from a proper RCM analysis. Following the asset prioritization process, the facilitator would be able to recommend the systems that should be included in the RCM project. Planning the analyses is discussed later in this section. Although not a topic for discussion in this book, the organization should have a parallel asset strategy development process (i.e., Maintenance Task Analysis) that can be used for analyzing and developing maintenance programs for less critical asset systems.

Asset Prioritization. The asset prioritization process includes the review of the data and information contained in the Work Management System (WMS) as well as field observations and interviews. Assets are categorized according to their impact on Safety, Environmental and Operations (and any other business objective, i.e., service levels and quality). The facilitator must be able to assist in the development of a framework for assessing the asset criticality and set priorities based on criticality. Based on the outcome of the asset criticality and prioritization process, the facilitator should engage the stakeholders to determine a tolerable level of risk for each asset system.

Following the asset prioritization process, the facilitator will be able to recommend *where to start* and *what assets to include* in the RCM program.

Asset Verification. Asset registers and hierarchies are rarely complete or correct. This may be for many reasons ranging from incorrect data entry or asset modifications and/or replacements. The facilitator must verify that the asset registry contained in the system of record is reflecting what is installed in the field and vice versa. The asset verification process should reveal any anomalies and discrepancies. The information gathered during the verification process is valuable for drafting the operating context discussed in Chapter 3.

Remediation and Corrective Action. Any anomalies and urgent findings should be recorded and remediation or corrective action must be recommended. Variations lead to ineffective implementation and recording of asset information and maintenance program.

Developing a Risk Framework and Tolerability Thresholds (Developing a Risk Matrix)

The concept of risk management is not new to RCM facilitators; it is done intuitively through applying the RCM logic. With RCM3, however, all decisions are based on the risk associated with each failure, and it is essential to have the risk framework and what is categorized as tolerable and intolerable properly defined prior to the start of any RCM3 analysis. The development of the risk matrix as described in Chapter 7 may not be the duty of the facilitator alone, but he or she should be involved in the process.

Planning the RCM Project (Scope and Analysis Boundary)

Facilitators are expected to work independently on RCM projects. As with any moderately resource intensive initiative, the first critical success factor is to ensure that the RCM project is planned as thoroughly as possible before work starts. To ensure that the plan is realistic and to secure a reasonable degree of ownership of the plan, the plan must be prepared in close consultation with the managers and supervisors of the areas affected. In preparing the plan, the following questions need to be answered:

- What equipment will be covered by the project?
- To what extent is equipment duplicated?

- How many RCM meetings are needed to analyze each item?
- How many people at each level will be directly involved in review groups?
- How many different review groups are needed? (The answer to this question is governed by the answers to the previous questions.)
- How many review groups will each facilitator work with at once (in other words, how many review projects will be in progress at once)?
- What other factors might have a bearing on the manning, the start date and the duration of each analysis (such as vacation and other training commitments of key people, parallel initiatives, shutdowns, shift arrangements, etc.)?
- On what basis are the reviews to be audited and by whom is this to be done in each section?

A second objective of this exercise is to establish the objectives of the RCM pilot-projects as clearly as possible and to agree how management would decide whether these had been achieved at the end of each project. Proper planning of the RCM3 project ensures that project risks are identified and effectively managed or mitigated. Over and above the benefits of mitigating the risks, the following benefits can also be achieved:

- Resources are identified, and participants are trained in advance.
- Project scope and analysis boundaries are well defined and agreed upon prior to start of the project.
- Expectations are managed throughout the project with stakeholder involvement at all levels (deliverables, performance criteria, etc.).
- Project timelines are met more often (activities are planned and scheduled).
- Logistics are planned, and review group meetings are scheduled in advance to avoid conflicts.
- Audit meetings are scheduled in advance to ensure proper attendance.
- Any deviation from plan can be addressed appropriately and timely.

Generally, the following decisions are made by the facilitator:

- Identify and select the assets for review.
- Prepare for the meetings.
- Select levels of analysis and define analysis boundaries.
- Identify and select assets for review.

The assets that should be considered for RCM analysis and review are usually selected based on criticality (relevant importance of the system or asset to the business) or on criteria that may include one or more of the following:

- Risks(s) associated with failure (safety, environment, production, customer service, etc.)
- Availability requirements
- Current maintenance program *(being reactive)*
- High maintenance cost
- Unknown systems or assets or inadequate maintenance strategies, etc.

The high-level asset criticality assessment determines what assets should be considered first and which assets will not be included in the RCM program. The asset criticality assessment and asset prioritization process was discussed in the previous section.

Prepare for Meetings. Prior to the first meeting, the facilitator should collect whatever basic information is readily available about the asset/process. This includes flow diagrams, general arrangement drawings, operating manuals, technical history records, if any, and electrical, hydraulic, and pneumatic circuit drawings. (Note that this does not mean that the facilitator should attempt to start the analysis alone, nor should too much time be wasted looking for information.) Collecting the information forms part of the asset verification and validation process discussed in the previous section.

Select Levels of Analysis and Define Boundaries. The equipment to be analyzed by each review group will be identified during the planning phase. However, it may become necessary to group the equipment differently in order to carry out a sensible analysis. This means that the final decision about equipment grouping/levels of analysis is made by the facilitator, who then has to define the boundaries of the analysis accordingly.

Applying the RCM Logic

The facilitator must ensure that the RCM process is applied correctly by the review group. This entails ensuring that all the questions embodied in the RCM process are asked correctly in the correct sequence, that they have been correctly understood by all the group members, and that the group reaches consensus about the answers.

Managing the Analysis

By and large, the following decisions are made by the facilitator and/or the facilitator alone does the work.

Handle Complex Failure Modes Appropriately. Decide when to choose which of the four options listed in Chapter 6, Section 6.7 (pages 164–171), when listing failure modes.

Know When to Stop Listing Failure Modes. Knowing when to stop listing the failure modes that might cause each functional failure is one of the key elements of successful facilitating and requires careful judgment. Moving on too soon to the next functional failure means that critical failure modes may be overlooked or that failure effects are inadequately described. Listing too many failure modes leads to analysis paralysis.

Interpret and Record Decisions with a Minimum of Jargon. As a rule, the facilitator physically records the decisions of the group. In so doing, care must be taken to ensure that all technical terms used will be understood by everyone on the site (including auditors, design engineers, and senior managers).

Recognize When the Group Doesn't Know. The facilitator has to distinguish between uncertainty (the group is not 100% sure, but sure enough to make a viable decision) and ignorance (the group simply doesn't know enough to make a viable decision).

Curtail Attempts to Redesign the Asset in RCM Meetings. Attempts to redesign the asset is the biggest single time waster in RCM review meetings. The facilitator should simply note that redesign is compulsory or may be desirable, and he or she may jot down a suggestion if the answer seems obvious. The redesign process itself should be carried out elsewhere. (This is not to say that the RCM group cannot get involved in the redesign process—the group should—it simply means that it shouldn't do so in the RCM meeting.)

Complete the RCM Worksheets. Whether they are stored manually or electronically, the RCM Information and Decision Worksheets should be completed in a way that is clear and readable. Abbreviations should be avoided, and they should contain a reasonable minimum of spelling mistakes and grammatical errors.

Prepare on Audit File. As discussed in Chapter 11, managers with overall responsibility for each asset need to audit the analyses carried out by the review groups. Before this can be done, the facilitator needs to prepare the RCM worksheets in a clear, coherent fashion. This usually entails binding them into a formal document called the audit file. This file should also contain enough background information—schematic drawings, known failure data, even photographs of the equipment—to enable the auditors to do their job properly.

Enter RCM Data into a Computerized Database. Data are entered into a computer and displayed on a big-screen television or projector screen while the review group observes what is being entered, while it is being entered. The facilitator must reach consensus, and the review group must be satisfied that the facilitator is capturing the information correctly. The facilitator may use an abbreviated style to save time during the meetings. The facilitator must at least be able to type as fast as he or she can write. The software or worksheets used for capturing the data must be easy to use, and everyone in the meeting must be familiar with what is being entered. As mentioned in Section 10.4, the software must be unobtrusive and used to increase the efficiency and effectiveness of the facilitation process and never to drive the RCM analysis itself.

Conducting the Meetings

The following points deal with the way in which the facilitator interacts with participants at meetings on a purely human level.

Set the Scene. At the first meeting of each group, the facilitator must lay out basic meeting norms with the group (issues such as use of names, dress, punctuality, cell phones, etc.) and ensure that every group member understands the scope and objectives of the exercise and why he or she has been asked to take part. At the start of all subsequent meetings, the facilitator should briefly recap what has been done to date and provide a brief agenda for the meeting. The facilitator should also ensure that the members of the group have enough materials (drafts of completed worksheets, etc.) to enable them to keep track of the process.

Set a Good Example. How the facilitator conducts himself or herself in meetings has a profound effect on the way the other group members behave. In particular, the facilitator should set a good example by displaying a positive attitude toward the process, take care to preserve the dignity of group members, and provide positive feedback in response to positive contributions.

Ask the RCM Questions in Order. Once the meetings are under way, the key role of the facilitator is to ask the questions required by the RCM process. It is essential to avoid any tendency to skip questions or to take answers for granted. (In particular, take care not to ignore or overlook questions designed to establish whether any task is worth doing.)

Ensure That Each Question Has Been Correctly Understood. In spite of the fact that they should all have attended a basic RCM training course, group members are not as familiar with the RCM process as the facilitator. As a result, they often misunderstand the questions, especially in the early stages, and the facilitator must be alert to such misunderstandings. Common mistakes were discussed in Section 11.2 of Chapter 11.

Encourage Everyone to Participate. Everyone who has something to contribute should do so. This entails encouraging reticent people to participate, while ensuring that dominant personalities do not take over the meetings to the exclusion of everyone else. Interest can be sustained and participation encouraged by asking group members to do small tasks between meetings such as clarifying technical points (perhaps by calling a vendor, measuring a dimension, checking out a quality standard, etc.).

Answer the Questions. Facilitators should avoid what often becomes a strong temptation to answer the RCM questions directly. However, it is legitimate to clarify doubtful answers by further questioning.

Secure Consensus. One of the most important functions of the facilitator is to ensure that the group reaches consensus. Consensus does not mean that decisions are made by casting a vote. It also does not mean that everyone must agree completely with every decision. It does mean that everyone is prepared to accept the majority view. (If group members simply cannot reach consensus, the facilitator should ask someone whose expertise is respected by the members of the group to counsel them further—and if necessary, to make the final judgment.)

Motivate the Group. As discussed above, one of the most important factors that affect the attitude of the group is the attitude of the facilitator. Other motivational issues that the facilitator may need to deal with are waning enthusiasm, especially if a large number of meetings will be needed to review a big asset, and skepticism, where group members don't believe that their recommendations will be taken seriously by management.

Manage Disruptions Appropriately. All meetings occasionally suffer from disruptions. However, in the case of RCM, the group is trying to do a great deal of work that requires intense concentration, so interruptions can be especially unwelcome. Three areas that usually need special care are digressions, personality clashes, and grievances that are not related to the RCM process.

Coach the Group or Individual Members. It is sometimes necessary for the facilitator to provide formal coaching to individuals or to the group as a whole in some element of the RCM philosophy. However, coaching is inefficient and time-consuming, so it should not be seen as a substitute for formal training in RCM.

Time Management and Implementation Planning

RCM is a resource-intensive process—sufficiently so for management at all levels to be concerned about the amount of time and effort it takes to complete each analysis. Both the resources required to apply RCM and the duration of each project are profoundly affected by the pace at which facilitators conduct meetings and the way they manage their time outside meetings. As a result, facilitators need to develop their time management skills every bit as much as their skills in any other aspect of RCM. The RCM3 process lends itself to more efficient use of time and resources. The last question in the RCM3 process asks: what can be done to manage or reduce tolerable risks in a cost-effective way? From this it is clear that it is not necessary to take every failure mode through the RCM decision logic, tolerable risks could be left alone or further optimized can be achieved outside the RCM review meeting. It is, however, necessary to mention that the failures that pose tolerable risks may still have severe economic consequences and careful consideration not to discuss these is required.

Five overall key measures of time management effectiveness include:

Pace of Working. A number of people are present at each RCM meeting, so the amount of time spent in these meetings has the greatest impact on the total number of labor hours spent on the RCM process.

Slow progress at meetings also means that more meetings need to be held, which could delay the project completion date. As a result, this is the most important of the five measures of time effectiveness. Experienced facilitators should be able to complete six failure modes within 1 hour of analysis time.

Total Number of Meetings Held. The total number of meetings needed to perform a complete analysis should be estimated as part of the RCM project planning phase. A second measure of time effectiveness is to compare the actual number of meetings held with this estimate. However, estimates can themselves be inaccurate, so it is usually acceptable for a facilitator to complete any one analysis within 20% of the estimated number of meetings (with due allowance for the learning process in the case of new facilitators).

Actual Completion Date Versus Target Completion Date. The completion date of each set of meetings should also be determined during the RCM project planning phase. The facilitator should go to great lengths to achieve this date. Completion of the meetings is usually delayed either because the number of meetings required exceeds the estimate or because meetings are not held as planned. If either of these problems occurs, every effort should be made to recover lost ground, if necessary by scheduling extra meetings

Time Spent Preparing for an Audit. As explained earlier, the facilitator needs to prepare an RCM audit file after the meetings have been completed. Since recommendations cannot be implemented until they have also been audited, this step should also be carried out as quickly as possible. An experienced facilitator should be able to have an analysis ready for final audit no more than 2 weeks after the last meeting of the review group.

Time Outside Meetings. Facilitators are also scarce and expensive resources, so they owe it to themselves and to their employers to use their own time as effectively as possible. In the RCM context, this means that the amount of time that facilitators spend on administrative work outside meetings should be about the same as the time spent in the meetings themselves. Figure 13.2 illustrates the typical workflow—from start to finish—for a RCM project.

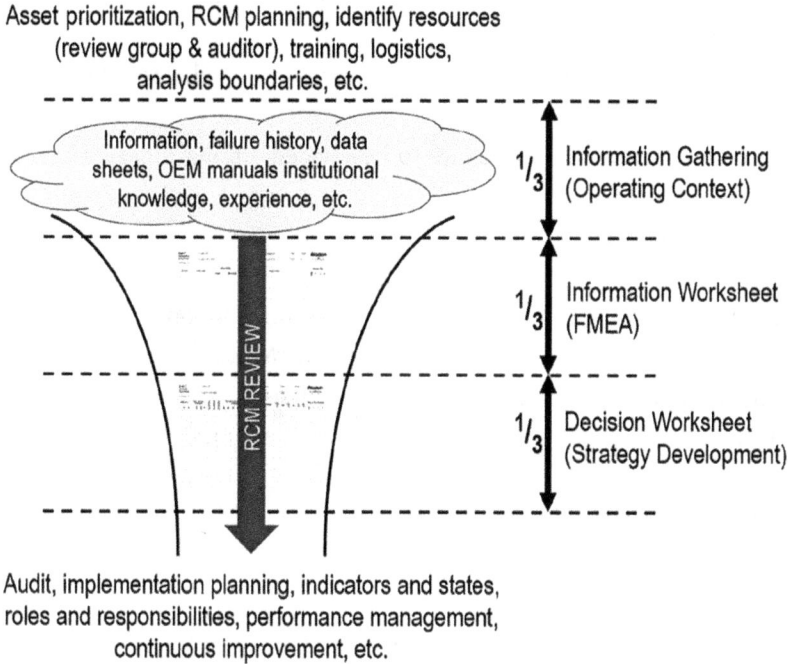

FIGURE 13.2 A typical RCM project workflow

Interacting with People Outside the Review Group— Administration, Logistics, and Managing Upward

This part of this chapter deals with activities where the facilitator interacts with people (usually managers) who are not members of review groups. These interactions involve making decisions, providing information, or getting work done. Who actually does each task may vary from place to place, but regardless of who is supposed to do it, the facilitator still plays a major part in ensuring that it actually gets done. As a result, facilitators tend to be judged on progress in the following areas as much as in any other.

Set Up the RCM Project as a Whole. This consists of the following steps:

- Decide which assets (or which parts of which assets) are to be analyzed using the RCM process.
- Establish the objectives of each analysis, and agree when and how their achievement is to be measured.
- Estimate how many RCM meetings will be needed to review each asset.
- Decide how the assets are to be divided among different review groups.
- Decide who will audit each analysis.

These steps are usually carried out in close consultation with the RCM project manager and the asset manager. If RCM is new to the business unit, this phase also tends to be done with assistance from experienced consultants (especially in estimating the number of meetings).

Plan the Project. Before starting each analysis, each of the following must be planned in detail:

- Decide who is going to participate in each review group.
- Arrange training in RCM for group members and auditors who have not yet been trained.
- Decide when, where, and at what time every meeting is to be held.
- Decide when the analysis will be audited.
- Decide when to hold the top management presentation.

These steps are also usually carried out in consultation with the RCM project manager and the asset manager.

Communicate the Plans. Participants and their bosses should receive written notice of initial plans for training courses and meetings. Any subsequent revisions to these plans should also be communicated in good time. Auditors need to be reminded about forthcoming audits.

Once the meetings are under way, the RCM project manager should ensure that people actually attend planned meetings. Attendance norms should be clearly defined, well publicized, and strictly adhered to.

Set the Meeting Venue. An RCM meeting room should be big enough for people to sit around a table comfortably without touching each other, and it should be reasonably close to the group members' normal workplace. It should also be quiet, reasonably secluded, well lit, and adequately ventilated. The meeting should not be interrupted by phone calls or pagers. A flip chart or whiteboard is usually essential. Whether or not refreshments are provided at meetings depends on organizational norms.

Communicate Urgent Findings. Appropriate managers should be told before the audit about findings or recommendations that may be of special interest to them, or that may need urgent attention (such as serious safety or environment hazards). This ensures that potentially dangerous problems are dealt with quickly, and it also helps to sustain the interest of the people who are providing the resources for the project.

Communicate Progress. Keep management informed about progress against plan. Bring to management's attention problems that you cannot solve yourself and that are impeding or threatening to impede progress, such as sustained absenteeism from meetings, seriously counterproductive behavior, excessive interruptions, etc.

Ensure That RCM Worksheets Are Audited. The facilitator should usually attend audit meetings in person to answer queries, note corrections, and (if required) provide guidance to the auditors on the RCM process (although the auditors must undergo formal training in RCM before attempting to audit an RCM analysis). The facilitator must also ensure that consensus is achieved between the auditors and the review group during the audit process. This entails reporting audit findings back to the group and ensuring that differences are resolved. Finally, the facilitator must update the worksheets to incorporate the results of the audit.

Make a Presentation to Top Management. A short, high-quality summary of at least one major RCM analysis should be presented to the senior managers of each business unit in which the process is applied. It should show how the initial objectives of the analysis have been or will be achieved and what had to be done to achieve them.

Support Implementation. Ensuring that RCM decisions are implemented is usually the overall responsibility of the asset manager, although the facilitator will need to remain involved. The key elements of the implementation process were discussed in Chapter 11.

Sustain a Living Program. After completing each analysis, the facilitator should work with the RCM project manager and the asset managers to set up meetings to reappraise and where necessary update the analysis. These meetings should be held at intervals of 9 to 12 months, and ideally the meetings should be facilitated by the original facilitator. This issue is discussed in more detail in Section 13.5 of this chapter.

Who Should Facilitate?

Facilitators should have a strong technological background, should be highly methodical, and should be natural consensus builders. They can work as facilitators on a full-time or part-time basis. In addition, they should also have a reasonable understanding of the process and of the technology embodied in the assets under review, but they should *not* be experts on either subject. This whole approach is based on the notion that the other group members are the experts in these areas. (It may also explain why process experts and line maintenance managers and supervisors should participate in the process as group members, but they should not do so as facilitators.)

The field in which a facilitator *should*, of course, be an expert is RCM, which means that appropriate training will usually be required. In order to become fully competent in all 55 skill areas, facilitators require at least 10 days of intensive formal practical training before starting to work with groups. Thereafter, most facilitators require further mentoring from a skilled RCM practitioner for a period of a few months after their formal training program. In order to secure the highest possible level of "ownership" of and long-term commitment to the conclusions drawn during the process, the facilitator should also be a full-time employee of the organization that will be operating and/or maintaining the asset in the long term. (This is one of many reasons why it is strongly recommended that outsiders should not be used as RCM facilitators.)

13.4 Implementation Strategies

Broadly speaking, RCM can be applied in one of three ways:

- A task force approach, in which it is used to focus on acute problems only
- A selective approach, in which it is applied on a systematic basis to analyze only those assets that are thought likely to benefit most from the application of RCM
- The comprehensive approach, in which RCM is used to analyze all the equipment on a site

Key elements of each of these approaches are discussed in the following paragraphs.

The Task Force Approach

The quickest and biggest short-term returns are usually achieved when RCM is applied to an asset or process that is suffering from intractable problems that have serious consequences. Such problems can often be solved surprisingly quickly by training a small group (the "task force") to carry out a comprehensive RCM analysis of the affected system. Such groups should consist of members drawn from the same disciplines as the groups described in Section 13.2 of this chapter. They often work full time on the review project until it is complete, and the group is then disbanded.

- The main advantages of this approach are that it is quick, because only one or two groups have to make their way up the RCM learning curve; it is easy to manage, because only a small number of people are involved; and if it is successful—which is usually the case—it can yield substantial returns (in terms of improved plant performance) for a *relatively small investment.*

- The main disadvantages of this approach are that *it does nothing to secure the long-term involvement and commitment* of all the people in the organization to the results, so the results are much less likely to endure; and because it is narrowly focused, *it does little to foster best practice* across the entire organization.

The Selective Approach

In addition to acute problems that might lend themselves to the task force approach, most organizations also have some assets that are more susceptible than others to chronic problems that are difficult to identify. These problems usually manifest themselves as downtime, poor product quality, poor customer service, or excessive maintenance costs. Other areas might be confronted with unacceptable safety or environmental hazards that need to be tackled on a systematic basis.

Given hundreds if not thousands of items to choose from in a large undertaking, it makes sense to start applying a technique with the power of RCM in areas where the worst of these problems are encountered. Once these have been dealt with, a decision is taken about whether RCM will be used to analyze assets with less serious problems, and so on.

The author has found that in most cases, the simplest, quickest, and most effective way to identify where physical assets are causing the most serious problems (especially in terms of failure consequences) is to ask their users. This usually means production or operations managers at all levels.

If the worst problems are not immediately obvious, or if it is not possible to achieve consensus about where to start on an informal basis, then it is sometimes necessary to decide on a more formal basis where RCM should be applied. This can be done in three stages:

- Identify "nonsignificant" assets. These are assets that are not likely to benefit much from the RCM process.

- Rank the assets that are significant in descending order of importance.
- Decide whether to use a "template" approach for very similar assets.

Significant Assets. An asset is judged to be significant if it could suffer from any failure mode that on its own:

- Could threaten safety or breach any known environmental standard
- Would have significant economic consequences

Items are also judged to be significant if they contain hidden functions whose failures would expose the organization to a multiple failure with significant safety, environmental, or operational consequences.

Conversely, for any item to be classified as nonsignificant, we must be sure that:

- None of its failure modes will affect safety or the environment.
- None of its failure modes will have significant operational consequences.
- It does not contain a hidden function whose failure exposes the organization to the risk of a significant multiple failure.

The process of identifying significant items is quick, approximate, and conservative. In other words, if it is not certain that any asset is not significant in the sense defined above, then it should be subjected to a full RCM review. Note that the assessment of significance can be done at any level, with the understanding that this may not be the level at which the RCM analysis is eventually conducted.

When making decisions about significance, note also that the RCM process is applied to any asset in its operating context. This context is a function of the process or system that the asset is part of, so any asset should only be analyzed in the context of a specific process or system (such as a packing line, a rolling mill, or a crane). The selection of significant items should never be based on generic items or components (all pumps, all bearings, all relief valves), because these would necessarily have to be taken out of context.

In the civil aviation industry, a surprisingly high percentage of items can be classified as nonsignificant in the sense described above. However, for 30 years this industry has been designing aircraft specifically to avoid or minimize the consequences of failure, so there is a very high (but still not infallible) level of redundancy built into their assets.

Assets in other industries, however, tend to enjoy a much lower level of redundancy, so a rather higher proportion of items end up being classified as significant, especially if due consideration is given to failures that could affect safety or the environment. This means that most organizations will still be confronted with a large number of items that should be analyzed. If the answer is not self-evident, the next question that needs to be answered systematically becomes "Where do we start?" And the answer is *to rank significant items in order of importance.*

A large number of techniques have been developed that attempt to provide a systematic, usually quantitative basis for deciding what assets are likely to benefit most from the application of analytical processes such as RCM. Sometimes called "criticality assessments," most of these techniques use some variation of a concept known as the "probability/ risk number," or PRN.

A PRN is derived by attaching a numerical value to the probability of failure—or failure rate—of an asset (the higher the probability, the higher the value), and by attaching another value to the severity of the consequences of the failure (again, the more serious the failure, the higher the value). The two numbers are multiplied to give a third, which is the PRN. Assets with the highest PRNs are analyzed first, then those with lower scores, and so on until assets are encountered where the likely return does not justify detailed analysis.

More sophisticated variations of this process build up composite PRNs by attaching different numerical weightings to different categories of failure consequences (typically, high for safety or environmental consequences, intermediate for operational, and lower for direct repair costs). If hard data about historical failure rates and costs are available, these rankings can be further refined using Pareto analysis.

Systematic rankings of this sort can be useful in helping to clarify and build consensus about what assets really matter and about where large, complex systems are particularly vulnerable. However, the criteria and the relative weightings used to assess severity and probability vary widely from company to company, so most criticality assessment processes use scales and values that are unique to specific organizations.

Templating. Another way to reduce the investment in RCM is to use the analysis of one asset as a "template" for another. For reasons that were stressed repeatedly throughout Chapters 3 to 7, this approach can only be applied to assets or processes that are very similar, if not identical, and that are operating in virtually the same context.

When this approach is adopted, an RCM group carries out a comprehensive, *zero-based* analysis of the first item or process in a series of very similar items or processes, and then the group uses this analysis as the basis for a review of the other items in the series. To do this, the group asks if the functions and performance standards of each subsequent item differ in any way from those listed on the worksheets for the *zero-based* item. The differences (if any) are recorded on the worksheets for the second item, and the analysts move on to compare the functional failures in the same way, and so on until they have completed the entire analysis.

If the items are technically virtually identical and the operating context is very similar, this approach can save considerable amounts of time and effort, because in most cases, a substantial proportion of the analysis remains unchanged for the subsequent items.

However, while it is technically appealing, templating can also have quite serious motivational drawbacks. This is because the operators and maintainers of the subsequent assets are asked to accept decisions made by others, which naturally reduces their sense of ownership. In extreme cases, the latter people may even reject the initial analysis out of hand because "it was not invented here." This phenomenon has led some organizations not to use templating at all, but to start all analyses from a zero base.

(Interestingly, this can lead to some quite different maintenance programs as different groups select different methods of dealing with the same failure. One way in which this can occur quite legitimately was explained in Figure 8.23 on page 288.)

Advantages and Disadvantages of the Selective Approach. Typically, organizations that adopt the selective approach apply RCM to between 20% and 40% of their assets.

The main advantage of this approach is that the investment is only made where it will yield quick and (usually) measurable returns. Because RCM is only applied to part of the facility, the overall project is less costly and hence easier to manage than if an entire facility is analyzed.

The main disadvantage of this approach is that it places much greater emphasis on the technical and operational performance of the equipment than on the people on whom the equipment ultimately depends in the long term (the operators and maintainers).

The Comprehensive Approach

Some organizations decide that the benefits of RCM are so great in terms of both improved equipment performance and improved motivation of their people that they decide to use RCM to analyze all the equipment on their sites. The two alternatives that are most often used are:

- To analyze all the assets on the site in one short, intensive campaign. Campaigns of this nature usually last from 6 to 18 months on most sites. Up to 20 or even more groups can be active at once, working under the direction of anywhere from 3 or 4 to 30 or 40 facilitators. As soon as a group completes the analysis of its asset or process, a new group is activated. In this way, the entire campaign is finished quickly, and the organization enjoys the benefits equally quickly. In fact, this is an excellent way to achieve massive and lasting step changes in maintenance performance for companies that need to do so in a hurry.

However, this approach is highly resource-intensive, so it needs a great deal of careful planning and management attention. It should not really be considered if a number of other initiatives are to be undertaken in parallel with RCM.

- A second possibility is still to review all the equipment on the site, but to do so in stages. Perhaps four or five groups are activated at a time, working under the direction of one or two facilitators. On this basis, it could take 5 to 10 years to analyze all the equipment on a large site (3 to 4 on a smaller one). The organization still derives all the benefits of RCM, but it takes much longer to do so. This approach is less disruptive in the short term, but if expectations are not very carefully managed, it could be seen to be dragging on forever and hence could become demotivating. On the other hand, it means that RCM can be applied in parallel with other initiatives and vice versa.

Since the people who could benefit from this approach often substantially outnumber the assets, it is usually necessary to analyze most if not all of the assets so that everyone can take part in the process.

Advantages and Disadvantages of the Comprehensive Approach. The main disadvantages of this approach are that it is slower, because more people have to become familiar with the RCM methodology; and it is more difficult to manage, because many more people are involved.

The main advantage is that it secures much more broadly based long-term ownership of maintenance problems and their solutions. Not only does this improve individual motivation and teamwork, but it also ensures that the results of the exercise are far more likely to endure. (Best practice becomes "part of the way we do things around here.")

Deciding Which Approach to Use

If it is to be applied correctly, RCM requires a substantial commitment of resources. If the comprehensive approach described above is applied, it needs the wholehearted involvement and cooperation of large numbers of people. As a result, it is wise to decide in stages which approach should be used.

Since managers have to commit the resources to RCM, it makes sense to start by giving them the opportunity to learn what RCM is all about, to assess for themselves what resources are required to apply it, and to judge for themselves what potential benefits it offers in their areas of responsibility. The best way to do this is usually to arrange for them to attend an introductory training course.

If the response is favorable, the next step is to run one or two pilot projects. These enable the organization to gain firsthand experience of the dynamics of the whole RCM process, what it achieves, and what resource commitments are needed to achieve it.

However, before undertaking any pilot project, it is essential both to assess the resources required to do it relative to the likely benefits and to plan the project as thoroughly as possible. This should be done in close consultation with the managers of the area where a pilot project is likely to be undertaken. It entails the following steps:

- Confirm the scope of the project and define the objectives (the "now" state and the desired end state).
- Estimate the time needed to review the equipment in each area.
- Identify the project manager and facilitator(s).
- Identify the participants (by title and by name).
- Plan training for the participants and facilitators.
- Plan the date, time, and location of each meeting.

When the pilot project is complete, the participants are in a position to evaluate the results for themselves and to decide whether, where, and how quickly RCM should be applied to the remaining assets in the organization. Chapter 14 explains that RCM yields substantial returns, but that the nature of these returns varies widely from one organization to another. As a result, the best time to decide which approach to adopt is after a small number of pilot projects have been completed and the organization is able to judge for itself what returns RCM offers in relation to what inputs.

13.5 RCM in Perpetuity—a Living Program

The application of RCM leads to a much more precise understanding of the functions of the assets that have been reviewed, and a much more scientific view of what must be done to cause them to continue to fulfill their intended functions. However, the analysis will not be perfect—and never will be perfect—for two reasons:

- The evolution of a maintenance policy is inherently imprecise. Numerous decisions have to be made on the basis of incomplete or nonexistent hard data, especially about the relationships between age and failure. Other decisions have to be made about the likelihood and the consequences of failure modes that haven't happened yet, and that may never happen. In an environment like this, it is inevitable that some failure modes and effects will be overlooked completely, while some failure consequences and task frequencies will be assessed incorrectly.
- The assets and the processes they are a part of will be changing continuously. This means that even parts of the analysis that are completely valid today may become invalid tomorrow.

The people involved in the process will also change. This is partly because the perspectives and priorities of those who take part in the original analysis inevitably change with time, and it is partly because people simply forget things. In other cases, people leave and their places are taken by others who need to learn why things are as they are. All these factors mean that the validity of the RCM database and people's attitudes toward it will inevitably deteriorate if no attempt is made to prevent this from happening.

One way to do this is to use the RCM process to analyze all significant unanticipated failure modes that occur after the initial analysis has been completed. This is usually done by convening an ad hoc group that uses RCM to determine the most effective way of dealing with the failure. The results of their deliberations should be woven into the RCM database for the affected asset. The ad hoc group itself should include as many as possible of the people who carried out the original analysis.

A second—and much surer—way to ensure that RCM databases remain current in perpetuity is to ask the original groups to review the database for "their" asset on a formal basis once every 9 to 12 months. Such a review meeting need not last for more than one afternoon. Specific questions that should be considered include the following:

- Has the *operating context* of the equipment changed enough to change any of the decisions made during the initial analysis? (Examples include a change from a single shift operation to double shifting, or vice versa.)
- Have any *performance expectations* changed enough to necessitate revisions to the performance standards recorded on the RCM worksheets?
- Since the previous meeting, have any *failure modes* occurred that should be recorded on the Information Worksheets?
- Should anything be added to or changed in the descriptions of *failure effects?* (This applies especially to the evidence of failure and estimates of downtime.)
- Has anything happened to cause anyone to believe that *failure consequences* should be assessed differently? (Possibilities here include changes in environmental regulations and changed perceptions about acceptable levels of risk.)
- Is there any reason to believe that any of the *tasks* selected initially are not, in fact, technically feasible or worth doing?
- Has any evidence emerged that suggests that the *frequency* of any task should be changed?
- Has anyone become aware of a *proactive technique* that could be superior to one of those selected previously? (In most cases, "superior" means "more cost-effective," but it could also mean "technically superior.")
- Is there any reason to suggest that a task or tasks should be *done* by someone other than the person selected originally?
- Has the asset been *modified* in a way that adds or subtracts any functions or failure modes or that changes the technical feasibility of any tasks? (Special attention should be paid to control systems and protective systems.)

If such reviews are carried out on a regular basis, they only take a small fraction of the time and effort that are needed to set up the database to begin with, but these reviews ensure that the organization continues to enjoy the benefits of the original exercise in perpetuity. These benefits are discussed in more detail in Chapter 14.

13.6 How RCM Should *Not* Be Applied

If it is applied correctly, RCM yields results very quickly. However, not every application of RCM yields its full potential. Some even achieve little or nothing. In the author's experience, some of the main reasons why this happens are technical in nature, but the majority are organizational. The most common are discussed in the following paragraphs.

The Analysis Is Performed at Too Low a Level

The problems that arise if an RCM analysis is performed at too low a level were listed in detail in Section 6.7 of Chapter 6. Most important among these are that the analysis takes far longer than it should, it results in a massive increase in paperwork, and the quality of the decisions deteriorates. As a result, people start finding the process tedious and lose interest, it costs much more than it should, and it does not achieve as much as it could.

Too Hurried or Too Superficial an Application

This is usually the result of insufficient training or too heavy an emotional investment in the status quo on the part of key participants. It often results in a set of tasks that are almost the same as they were to begin with.

Too Much Emphasis on Failure Data

There is often a tendency to overemphasize the importance of data such as MTBFs and MTTRs. This issue is discussed at length in Chapter 12. Such data are nearly always overemphasized at the expense of properly defined and quantified performance standards, the thorough evaluation of failure consequences, and the correct use of data such as P-F intervals.

Asking a Single Individual to Apply the Process

One of the least effective ways to apply RCM is to ask a single individual to apply the process on his or her own. In fact, no matter how much effort a single individual applies to the development of a maintenance program (whether using RCM or any other technique), the resulting schedules nearly always die when they reach the shop floor, for two main reasons:

- **Technical validity.** No one individual can possibly have an adequate understanding of the functions, the failure modes and effects, and the failure consequences of the assets for which his or her program is being developed. This leads to programs that are usually generic in nature, so people who are supposed to do them often see them as being incorrect if not totally irrelevant.
- **Ownership.** People on the shop floor (supervisors and craftspeople) tend to view the schedules as unwelcome paperwork that appears from some ivory tower and disappears after it is signed off. Many of them learn that it is more comfortable just to sign off the schedules and send them back than it is to attempt to do them. (This leads to inflated schedule completion rates, which at least keeps the planners happy.) The main reason for the lack of interest is undoubtedly sheer lack of ownership.

The only way around the problems of technical invalidity and lack of ownership is to involve shop-floor people directly in the maintenance strategy formulation process, as discussed earlier in this chapter. Done correctly, not only does this produce schedules with a much higher degree of technical validity than anything that has gone before, but it also produces an exceptionally high level of ownership of the final results.

Using the Maintenance Department on Its Own to Apply RCM

In many organizations, an almost impenetrable divide still exists between the maintenance and production functions. This often leads the maintenance people in such organizations to try to apply RCM on their own. In fact, as Chapter 2 made clear, maintenance is all about ensuring that assets continue to function to standards of performance required by their users. We have seen that the users are nearly always production or operations people. If these people are not closely involved in helping to define functions and performance standards, two problems usually arise:

- The maintenance people do it for them. In the author's experience, this nearly always leads to large numbers of inaccurate function statements and performance standards, and consequently it leads to distorted or inappropriate programs designed to preserve those functions.
- There is little or no buy-in to the maintenance program on the part of the users, who, after all, are the customers of the maintenance service. This, in turn, means that users understand less clearly why it is in their own interests to release machines for essential maintenance, and also why operators need to be asked to carry out certain maintenance tasks.

In addition to defining what they want the asset to do, users also have a vital contribution to make to the rest of the strategy formulation process. As explained in Section 13.1 of this chapter, by participating in the FMEA, they learn a great deal about failure modes caused by human error and hence what they must do to stop breaking their machines. They also play a key role in evaluating failure consequences, and they have invaluable personal experience of many of the most common warnings of failure. All this is lost if they do not participate in the process.

In short, from a purely technical point of view, it is rapidly becoming apparent that it is virtually impossible to set up a viable, lasting maintenance program in most industrial undertakings without involving the users of the assets. (This focus on the user—or customer—is, of course, the essence of TQM.) If their involvement can be secured at all stages in the process, that notorious barrier rapidly starts to disappear, and the two functions start to work, often for the first time ever, as a genuine team.

Asking Manufacturers or Equipment Vendors to Apply RCM on Their Own

A universal feature of traditional asset procurement is the insistence that the equipment manufacturer should provide a maintenance program as part of the supply contract for new equipment. Apart from anything else, this implies that manufacturers know everything that needs to be known to draw up suitable maintenance programs.

In fact, as explained on page 158, equipment manufacturers usually possess surprisingly little of the information needed to draw up truly context-specific maintenance programs. They also have other agendas when specifying such programs (not least of which is to sell spares). What is more, either they are committing the users' resources to doing the maintenance (in which case they don't have to pay for it, so they have little interest in minimizing it), or they may even be bidding to do the maintenance themselves (in which case they may be keen to do as much as possible).

This combination of extraneous commercial agendas and ignorance about the operating context means that maintenance programs specified by manufacturers often embody a high level of overmaintenance (sometimes ludicrously so) coupled with massive overprovisioning of spares.

Most maintenance professionals are aware of this problem. However, despite this awareness, most of us still persist in demanding that manufacturers provide these programs, and then we accept that the programs must be followed in order for warranties to remain valid (and so bind ourselves contractually to doing the work, at least for the duration of the warranty period).

None of this is meant to suggest that manufacturers mislead us deliberately when they put together their recommendations. In fact, they usually do their best in the context of their own business objectives and with the information at their disposal. If anyone is at fault, we are really the ones—the users—for making unreasonable requests of organizations that are not in the best position to fulfill them.

A small but growing number of users solve this problem by adopting a completely different approach to the development of maintenance programs for new assets, by involving the manufacturers' field technicians in a user-driven RCM analysis, as discussed on page 158.

In this way, the user gains access to the most useful information that the manufacturer can provide, while still developing a maintenance program directly suited to the context in which the equipment will actually be used. The manufacturer may lose a little in up-front sales of spares and maintenance but will definitely gain all the long-term benefits associated with improved equipment performance, lower through-life costs, and a much better understanding of the real needs of the customer. A classic win-win situation.

Using Outsiders to Apply RCM

It is wise to steer clear of the temptation to use third parties to formulate maintenance strategies. In this context, they suffer from most of the shortcomings that apply to single individuals, maintenance departments on their own, and manufacturers/equipment vendors, as discussed above. In addition, most outsiders know little about the dynamics of the organization for which the schedules are being written, such as the operating context of each asset, the risks that the organization is prepared to tolerate, and the skills of the operators and maintainers of the assets. This often results in generic analyses that contain many more assumptions than if the analyses were facilitated by informed insiders. What is more, after the initial analyses have been completed, outsiders more often than not move on to other organizations. After they have gone, there is often no one left with a sufficiently strong sense of ownership of the analyses and their outcomes to ensure that they stay alive in the sense discussed in Section 13.5 of this chapter.

Finally, the fact that most outsiders are usually working under contract introduces commercial constraints that can distort the RCM process if they are not managed very carefully indeed. In particular, the need to finish contracts on time and on budget creates additional time pressures that can cause too many decisions to be taken too quickly. These could have devastating consequences years, even decades, after the contracts are complete.

On the other hand, if RCM is applied by properly trained insiders, their own jobs—indeed their own lives—often quite literally depend on the long-term validity of each analysis. As a result, they will naturally be more inclined (and less constrained) to take whatever extra time is needed to ensure that all reasonably foreseeable risks are dealt with appropriately.

Using Computers to Drive the Process

Chapter 10 mentioned that computerized databases should be used to store and sort the information generated by the RCM process. However, as with so much in the world of information technology, it is easy to succumb to the temptation to go beyond what computers should be used for and to focus on their apparently "nice-to-have" uses.

For instance, it is tempting to computerize RCM algorithms such as the main decision diagram on page 356. This is often done by creating a screen that asks, say, question 1 and setting up the system so that an answer of no brings up another screen (or window) that asks question HS, while an answer of yes leads to a new screen that asks questions ES, ER, and so on. This is done in the utterly mistaken belief that a succession of screens will somehow speed up or streamline the process. In fact, there is simply no way that referring to a succession of 12 to 20 screens is quicker than reading a single sheet of paper, so using a computer in this fashion slows the process down.

Using a computer inappropriately to drive the process can also have a strong negative influence on perceptions of RCM. Too much emphasis on a computer means that RCM starts being seen as a mechanistic exercise in populating a database, rather than a means to explore the real needs of the asset under review. For this reason, the author agrees with Smith (1993) when he says that there is no "software code to do the engineering thinking for us" and that the computer "doesn't replace the need for solid engineering know-how and judgement." In short, RCM is *thoughtware*, not software. Although we recognize this, we also recognize the importance of good software to capture the information and support the process.

Conclusion

These comments suggest that the surest way to achieve most if not all of the positive benefits of RCM is to apply the process at the right level, and to do so on a formal basis using groups of properly trained people who represent the operations and maintenance functions and who have an intimate firsthand knowledge of the equipment under review.

13.7 Building Skills in RCM3

RCM provides a common framework that enables people from diverse backgrounds to achieve consensus about a wide range of highly technical issues. However, this process itself embodies many concepts that are new to most people. They need to learn what these are and how they fit together before they can use the process successfully. (Some people who have been steeped in traditional approaches to maintenance also need to unlearn a great deal.)

The best way to ensure that large numbers of people acquire the relevant skills quickly is to provide suitable training. The most appropriate mix of courses for people at different levels is as follows:

- **Maintainers and operators.** A course in the basic principles of RCM. Such a course should incorporate a variety of case studies and practical exercises that enable delegates to gain an appreciation of how the theory works in practice.

- **Maintenance managers, engineers, operations managers, supervisors, and senior technicians**. A course that covers the same ground as the course for craftspeople and operators, but that also explains what must be done to manage the implementation of RCM.
- **Facilitators.** Facilitators should be introduced to RCM in an introductory course such as the one described above, and then the facilitators should undergo at least 10 more days of intensive formal training before starting to work with groups. Thereafter, most facilitators require further mentoring from a skilled RCM practitioner for a period of a few months after their formal training program, before they become fully competent in all 55 of the key skill areas listed in Section 13.3 of this chapter.

(For a description of a comprehensive array of training and other support services that meets all the above requirements, go to http://www.aladon.com.)

CHAPTER 14

What RCM Achieves

The application of RCM results in three tangible outcomes that were discussed at length in Chapter 11:

- Maintenance schedules to be done by the maintenance department
- Revised operating procedures for the operators of the asset
- A list of areas where one-time changes must be made to the design of the asset or the way in which it is operated in order to deal with situations where the asset cannot deliver the desired performance in its current configuration

Two other less tangible outcomes that were mentioned in Chapter 11 are that participants in the process learn a great deal about how the asset works and that they also tend to function better as teams.

Achieving all these outcomes requires a great deal of time and effort, especially if RCM is applied as described in Chapter 11. However, if RCM is applied correctly, it yields returns that far outweigh the costs involved. Most applications pay for themselves in a matter of months, although some have paid for themselves in 2 weeks or less. The wide variety of ways in which RCM pays for itself are discussed at length in Section 14.3 of this chapter. In order to place this discussion in perspective, we first need to consider different ways in which it is possible to measure the performance of the maintenance function.

Maintenance performance can be considered from two quite distinct viewpoints. The first focuses on how well maintenance ensures that assets continue to do what their users want them to do. This is usually referred to as "maintenance effectiveness," and it is likely to be of most interest to the users or "customers" of the maintenance service. The second viewpoint concentrates on how well maintenance resources are being used. This is referred to as "maintenance efficiency." It is usually of more interest to managers who are directly responsible for maintenance. These two issues are considered separately in the next two sections of this chapter.

14.1 Maintenance Effectiveness

Chapters 2 and 4 emphasized that the objective of maintenance is to ensure that any physical asset continues to fulfill its intended functions to the standards of performance desired by the user. As a result, any assessment of how well maintenance is achieving its objectives must entail an assessment of how well the assets are continuing to fulfill their functions to the desired standard. This is influenced, in turn, by three issues:

- "Continuity" can be measured in several different ways.
- Users have different expectations of different functions.
- Individual assets can have more than one and often several functions, as explained in Chapter 2.

These issues are considered in more detail in the following paragraphs.

Different Ways of Measuring Maintenance Effectiveness

The primary function of any highly mechanized and fully loaded manufacturing facility is to produce at least as many units of salable product as it was expected to produce when it was built. ("Fully loaded" means that it is operating 7 days per week/24 hours per day and that there is a ready market for every unit that the facility can produce.) In this context, any failure that reduces output results in lost sales.

In cases like these, the simplest overall measure of the operational performance of the facility as a whole is total output per period.

If the facility is not producing what the users—usually the owners—feel it should be producing on a regular basis, they will not be satisfied until the situation is put right. At least until then, the users will be inclined to judge effectiveness in terms of total output against targets. This should be recognized when setting up any system for tracking maintenance effectiveness.

All is not necessarily well if overall output is on target. A facility that is producing the right number of units could still be experiencing problems that affect safety, product quality, operating costs, environmental integrity, customer service, and so on, so these also need to be measured and dealt with appropriately.

There are a great many ways in which we can measure how effectively an asset is fulfilling its functions. Five of the most common are:

- **How often it fails.** This is the most widely understood meaning of the term "reliability." It is usually measured by the mean time between failures or the failure rate.
- **How long it lasts.** This is usually thought of as its "life" or its "lifespan," at the end of which the item under consideration fails and is either rebuilt or discarded and replaced with a new one. Strictly speaking, this phenomenon should be described as "durability."
- **How long it is out of service when it does fail.** This is usually referred to as "downtime" or "unavailability." It measures how much of the time the item is incapable of fulfilling a stated function to the satisfaction of the user, in relation to the amount of time the user would like it to be capable of doing so. Unavailability (or the converse, availability) is usually expressed as a percentage.
- **How likely it is to fail in the next period.** This is assuming that it has survived to the beginning of that period. We have seen that this is the conditional probability of failure. This could perhaps be described as a measure of dependability, if only to distinguish it from the other three variables. One common variation of this measure is the "B10 life." Chapter 12 explained that this is usually

measured from the moment the item is put into service and is the period before which not more than 10% of the items can be expected to fail. (In other words, the conditional probability of failure in the stated period is 10%.)

- **How efficient it is.** In common business usage, the term "efficiency" actually has two quite distinct meanings. The first measures output relative to input, while the second measures how well something is performing against how well it should be performing.

For instance, in a power station, energy efficiency measures the amount of energy exported in relation to the amount of energy released by the fuel. Depending on the technology used (coal, gas, combined cycle, etc.), this usually ranges from about 35% to about 58%. However, if a station that should average 40% energy efficiency is only averaging 38%, it will be exporting 95% of the energy that it should be exporting. In the context of this book, the first (40%) measure is a functional performance standard. As explained in Chapter 3, this is used to judge whether the item has failed. The second (95%) measure is used to judge the effectiveness with which the organization is achieving the desired performance on an ongoing basis.

Efficiency also refers to pace of working, and it does so in two ways—how fast an asset should work relative to the pace at which it could work (desired performance versus initial capability) and how fast it actually works relative to the pace at which it should work (actual performance versus desired performance). We have seen that desired performance must be less than initial capability because allowance must be made for deterioration. So in the context of this chapter, efficiency compares the pace at which an asset actually works with the pace at which it should work, not with the pace at which it could work.

Efficiency-type measurements can also apply in a slightly different fashion to the consumption of maintenance consumables (such as lubricating oil and hydraulic oil) and process consumables (such as solvents and reagents used in chemical plants and in the extraction of minerals).

All five of these measures are valid. It is simply a matter of deciding which is the most appropriate in the context under consideration.

For example, if a turbogenerator set has the lowest energy costs per unit of output among all those used by an electric utility, it is likely that its users would want it to generate (base load) power for as much of the time as possible. In terms of this function, the most appropriate measure of maintenance effectiveness is availability. (The operators may occasionally choose to run the set at less than full load. They may even choose to shut it down completely from time to time for purely operational reasons. Slowdowns or shutdowns of this nature affect the utilization of the asset as opposed to its availability. In essence, availability measures what percentage of time the machine is available to fulfill its primary performance requirement, while utilization measures how much it actually fulfills it.)

On the other hand, the generator set might only be used periodically to satisfy peak demands for power (peak loads). In this case, the primary concern of the users will be that the generator comes on stream as soon as it is required, so a primary measure of effectiveness will be how often it does so (or conversely, how often it fails to do so, expressed by a failure rate).

When measuring safety, performance is usually measured in terms of number of days or number of labor hours worked between lost time incidents (or fatalities). This is a form of mean time between failures. Similar measures are used for environmental incidents.

On the product quality front, a scrap rate of, say, 4% can be seen as a measure of unavailability, in the sense that while a machine is producing scrap, it is not available to produce first-grade product. (A scrap rate of 4% corresponds to a yield of 96%.) Scrap rates can also be expressed as 20 parts per million, which is another way of expressing a failure rate. Both are valid measures of maintenance effectiveness, especially in highly mechanized or automated processes.

Different Expectations

Every function has associated with it a unique set of continuity (reliability and/or durability and/or availability and/or dependability) expectations.

For instance, two of the functions associated with the bodywork of a car are "To isolate the occupants of the car from the elements" and "To look acceptable." Most car owners expect the bodywork to be able to fulfill the first function throughout the expected life of the car (unless the car is a convertible or unless they open a door or a window). On the other hand, everyone knows that cars get dirty—and hence start "to look unacceptable"—in the space of a few days or weeks. So in the first case we have a continuity expectation that might be measured in hundreds of thousands of miles or decades, while in the second case the continuity expectation is measured in hundreds of miles or days.

This issue is complicated by the fact that the loss of nearly every function can be caused by more than one failure mode—in fact, sometimes by dozens. Each failure mode has associated with it a specific failure rate (or MTBF), and each will take the function out of service for an amount of time that is specific to that failure mode. As a result, the continuity characteristics of any function will actually be a composite of the continuity characteristics of all the failure modes that could cause the loss of that function.

For instance, take the function "To look acceptable," which was mentioned above. In addition to the accumulation of dirt, this function could be lost due to rust or corrosion, fading of the paintwork, external damage (sideswiped in a parking lot), and vandalism, among others. It should also be apparent that some of these failure modes have little or nothing to do with maintenance. For instance, external damage is mainly a function of how this car—or the other vehicle involved—is operated, although design may play a small part by adding rubbing strips to reduce damage and/or by making it easier and cheaper to replace damaged panels. The probability of vandalism is also a function of where the car is used (the operating context), so it is almost completely beyond the control of the designer and the maintainer. The rate of accumulation of dirt is a function of where and when a car is used (road conditions and climatic conditions), and it is managed by a suitable maintenance program (washing the car). Corrosion and fading of paintwork can be influenced substantially at the design stage (although yet again the operating context—climatic conditions and the provision of shelter—and to some extent maintenance activities—polishing and chassis washing—can play a part in moderating the severity and frequency of these failures).

This example leads to two important conclusions:

- We need a thorough understanding of all the failure modes that are likely to cause each loss of function in order to be able to design, operate, and maintain an asset in such a way that the effectiveness expectations that we have of each function will be achieved.
- It is unreasonable to hold the maintainer of an asset alone accountable for the achievement of any continuity (reliability/availability/durability/dependability) targets for any asset or any function of any asset. The achievement of these targets is also a function of how the asset is designed, built, and operated. Accountability for achieving the associated targets should be divided jointly between the people responsible for all these functions. (In other words, maintenance effectiveness as it is being defined in this chapter is not only a measure of the effectiveness of the maintenance department. It also measures how effectively all the people associated with the asset are playing their part in doing whatever is necessary to ensure that the asset continues to do what its users want it to do.)

Different Functions

Perhaps the most important point about measuring the effectiveness of maintenance activities is the fact that every asset has more than one and sometimes dozens of functions. As explained above, a unique set of continuity expectations is associated with each function. This means that if an asset has 10 functions, then the effectiveness with which the asset is being maintained can be measured in (at least) 10 different ways.

For instance, let us consider how maintenance effectiveness might be measured by the owner of a typical suburban gas station. For the purpose of this example, the asset is a storage and pumping system used for gasoline. In this system, unleaded gasoline is stored in an underground tank with a capacity of 50,000 gallons. It is periodically filled by a road tanker to a level of 48,000 gallons. An upper-level switch in the tank switches on a local warning light if the tank has been filled to a level of 48,500 gallons, and another switches on another warning light in the main office if the level drops to 5,000 gallons. A low-level alarm sounds in the office if the tank level drops to 2,000 gallons, and a local ultimate high-level alarm sounds if the tank level reaches 49,000 gallons. The tank is double skinned to ensure that gasoline is contained in the event of a leak in the inner skin. A level indicator indicates the fuel level in the tank.

The tank supplies gasoline to five pumps. Each pump is switched on and off by pressing and releasing a handle in the nozzle. The nozzle also incorporates a pressure switch that trips the pump when the vehicle fuel tank is filled to the tip of the nozzle. A flowmeter measures the amount of fuel delivered each time the pump is activated and displays the volume and value of the fuel delivered to the customer. This meter is zeroed each time the nozzle is returned to its cradle.

(This system embodies additional secondary functions that deal with access onto and into the tank, drains, venting, valving, ease of use by a customer, other protection, appearance, and so on. These would also be listed in a real-life situation. However, for the purpose of this example, we only consider the functions described above.) On this basis, a list of functions might read:

- To pump between 25 and 40 gallons/minute of gasoline to the vehicle
- To indicate the volume and value of fuel delivered to customer to within 0.03% of the actual volume/value
- To shut off the pump when required by customer or when customer's fuel tank is full
- To contain the gasoline
- To store between 2,000 and 48,000 gallons of gasoline
- To switch on a warning light in the main office if the tank level drops to 5,000 gallons

- To switch on a local warning light if the tank level reaches 48,500 gallons
- To sound an alarm in the main office if the tank level drops below 2,000 gallons
- To sound an alarm if the tank level reaches 49,000 gallons
- To contain the contents of the tank in the event of a leak
- To indicate the level of fuel in the tank to within 0.05% of the actual level

When assessing the maintenance effectiveness of this system, the owner of the gas station will have different criteria for each of the above functions. For instance:

- **Function 1: To pump between 25 and 40 gallons/minute of gasoline to the vehicle.** This function can fail in three ways with three quite different sets of consequences, so each functional failure needs to be considered on its own merits:
  - **Functional failure A: Fails to pump at all.** Obviously, if a pump isn't working, it cannot be used to pump gasoline. However, there are five pumps in the station, so the level of availability required depends on the pattern of demand. For example, the station owner may tell us that he "hardly ever" has all five pumps in use at once—so seldom that we can ignore the possibility. He might also tell us that four pumps tend to be in use simultaneously for a total of not more than 1 hour a day, and then never for more than about 10 minutes at a time. If each pump has an average availability of 95%, two pumps will be out of service simultaneously for no more than 2% of the time. In other words, four pumps would be available 98% of the time, while there is a demand for four pumps 4% of the time. Under these circumstances, only a tiny fraction of customers would need to wait for gasoline, and then not for very long. This might tempt the owner to accept an availability of 95%. (If he regularly had five or more customers wanting to buy gas at the same time,

he would expect a much higher availability. But it may cost him somewhat more to achieve it, especially if he has to pay a premium for rapid response when calling out technicians to deal with failures.)

– **Functional failure B: Pumps less than 25 gallons/minute.** Some regular customers might find slow pumps sufficiently irritating to take their business elsewhere, especially if there are faster alternatives nearby. Consequently, the owner is likely to want any pump that wasn't failed completely to pump at the required rate "all the time—or at least, as close to all the time as you can make it." This might turn out to mean 99.8% of the time that the pump is not otherwise out of action—another form of availability.

– **Functional failure C: Pumps more than 40 gallons/minute.** If the pump pumps too fast, it is likely to generate sufficient back pressure to keep tripping the "tank-full" pressure-sensing mechanism in the nozzle. Customers would have to learn to throttle back the filling rate by not depressing the handle so much, which many regulars might also find irritating enough to cause them to take their business elsewhere. As a result, the owner is likely to say that he wouldn't want this failed state to occur "too often." He might then quantify this expectation as a failure rate—say, not more than once in 50 years on any one pump.

• **Function 2: To indicate the volume and value of fuel delivered to customer to within 0.03% of the actual volume/value.** This function can fail in two ways:

– **Functional failure A: Indicates that more than 0.03% less fuel has been delivered than actual.** If this happens, the station owner appears to be selling less fuel than he is actually selling, so he loses money. The failure becomes apparent after a while, because the ratio of fuel sold to fuel received will start to decline. Nevertheless, the owner would probably still seek a low failure rate—say, not more than one in 1,000 years on any one pump. (If the indicator fails completely, it shows that nothing has been delivered. If this happens, one lucky customer might get a free tank of fuel; then the station manager would shut down the affected pump until the problem is rectified.)

- **Functional failure B: Indicates that more than 0.03% more fuel has been delivered than actual.** If this happens and it comes to the attention of either the customers or the trading standards authorities (probably both), the station owner would be in serious trouble. Many of his customers would regard him as a crook and take their business elsewhere. The authorities would probably fine him and, depending on the severity of the discrepancy, might even revoke his license to trade (thus putting him out of business). Whatever else happens, his standing in the community would take a beating. The severity of these consequences would lead him to seek a very low failure rate—say, once in 50,000 years on any one pump. (Whether this is achievable or not is another issue entirely.)

- **Function 3: To shut off the pump when required by customer or when customer's fuel tank is full.** This function can fail in three ways:
  - **Functional failure A: Fails to shut off when required by the customer.** If the pump carries on pumping after the customer releases the handle, the back-pressure sensor should shut it off when the tank is full. As a result, the customer will end up with much more gas in the tank than he or she wanted. This would almost certainly lead to a row about how much should be paid for, and it would end in the possible loss of a customer. As a result, the station owner would probably require a fairly low failure rate—say, once in 1,000 years on any one pump.
  - **Functional failure B: Fails to shut off when tank is full.** Many customers rely on the sensor to tell them when the tank is full. If it fails to do so, the pump should shut off when the customer releases the handle. However, it is likely that the tank will overflow onto the shoes of the customer before he or she is able to

react, leading to a lot of unpleasantness and perhaps a demand for compensation. This too would lead the owner to expect a low failure rate—say again, once in 1,000 years for any one pump.

- **Functional failure C: Both local switches unable to switch off pump.** If the sensor and the handle both fail to shut off the pump, it will carry on pumping gasoline all over the forecourt until the electric supply is shut off at the main circuit breaker. This would create a nasty fire hazard, so the owner would expect a very low failure rate—say, once in 1,000,000 years. (This is attainable if each switch independently achieves 1 in 1,000.)

- **Function 4: To contain the gasoline.** When asked about this function, the station owner might say something like, "We have had one leak in the gasoline system in the last 10 years—and that was one too many." Here the user is measuring effectiveness in terms of a failure rate. When pressed, he might accept a rate of, say, one in 500 years for a "small" leak, which he might choose to define as less than 5 gallons per hour. (It is highly unlikely that anyone would measure containment in terms of availability, because, say, 99% availability means that the system would be leaking 1% of the time—about 800 hours out of 10 years. Even 99.9% still means that it would leak for 80 hours. Clearly this is nonsense.)

- **Function 5: To store between 2,000 and 48,000 gallons of gasoline.** This function can also fail in three ways, each of which must again be considered separately:

  - **Functional failure A: Level drops below 2,000 gallons.** Based on normal patterns of demand, fresh supplies of gasoline are ordered when the tank level approaches 5,000 gallons, and we are told that they are nearly always delivered before the level reaches 2,000 gallons. If the level in the tank drops much below 2,000 gallons, there is a greatly increased chance that the tank will empty, causing the station to lose business. As a result, the station manager expedites the delivery if the level drops to 2,000 gallons (as indicated by the low-level alarm). He says he needs to expedite deliveries about once a year, which he says is "just about acceptable." Here he is again judging effectiveness in terms of a failure rate. (Note that this failed state is caused by increased demand and/or slow delivery. It has nothing to do with the maintenance department in the classical sense. Nonetheless, dealing with this failure can be seen as maintenance because we are seeking to "cause the business to continue.")

– **Functional failure B: Level rises above 48,000 gallons.** The level in the tank is only likely to rise above 48,000 gallons if the delivery driver is not paying attention to the tank level indicator when filling the tank or if the level indicator itself has failed. In both cases the warning light comes on at 48,500 gallons. We are told that this happens "about once every 6 months"—another failure rate that the people involved might say they accept.

– **Functional failure C: Tank contains something other than gasoline.** The tank can only contain something other than gasoline if it is filled with something else, probably diesel. If this happens, customers could fill their tanks with the wrong fuel and cause serious damage to their engines. The station owner figures that the resulting bad publicity and claims for damages could put him out of business, so he would rather this didn't happen at all. When reminded that "never" is an unattainable ideal, he might decide to accept a failure rate of, say, once in 100,000 years.

• **Function 6: To switch on a local warning light in the main office if the tank level drops to 5,000 gallons.** The station manager usually logs the level in all the fuel tanks every day in order to track consumption, and he orders more fuel when levels approach 5,000 gallons. The low-level warning light serves as a reminder if the level indicator fails or if there is a sudden surge in demand between readings. This light is needed about once every 2 years (M_{TED} = 2 years). If it does not work when needed, the low-level alarm sounds when the level drops to 2,000 gallons. If an initial order is placed at this late stage, the tank will almost certainly run dry, and the station will be out of gasoline for several hours. The owner says he will accept a mean time between occurrences of this multiple failure (M_{MF}) of 400 years. In the light of this expectation, the formula on page 232 tells us that the maximum unavailability that the station can tolerate for the low-level warning light is M_{TED}/M_{MF} = 2/400 = 0.5%. This means that the low-level alarm is being maintained effectively if its availability remains above 99.5%.

- **Function 7: To switch on a local warning light if the tank level rises to 48,500 gallons.** The high-level warning light is backed up by an audible alarm, so following similar logic to the above example, the owner might come to the conclusion that he will accept an availability of 97.5% for this warning light.

- **Function 8: To sound an alarm in the main office if the level in the tank drops below 2,000 gallons.** If the level in the tank drops to 2,000 gallons and the low-level warning does not sound, the delivery is not expedited. We are told that under these circumstances, there is a 50% chance that the tank will run dry before the tanker arrives, and the station would be out of gasoline for about 1 hour on average under such circumstances. This leads the station owner to conclude that he will not accept this multiple failure (level drops below 2,000 gallons while low-level alarm is failed) more than "once in a hundred years" (M_{MF} = 100 years). As discussed above, M_{TED} is 1 year, so the station can tolerate a maximum unavailability for the low-level alarm of M_{TED}/M_{MF} = 1/100 = 1%. In the light of this objective, the low-level alarm is being maintained effectively if its *availability* remains above 99%.

Similar logic would be followed to determine availabilities for functions 9, 10, and 11 in the above example. It would also be used to establish effectiveness measures for the functions of this system that were not included in the above list. However, for the functions discussed, the effectiveness expectations of the gas station owner can be summarized as shown in Figure 14.1.

Function	Failed State	Measure of Effectiveness		Comments
		Availability	MTBF	
1	A	≥ 95%		Each pump
	B	≥ 99.8%		Each pump
	C		≥ 50 years	Each pump
2	A		≥ 1,000 years	Each pump
	B		≥ 50,000 years	Each pump
3	A		≥ 1,000 years	Each pump
	B		≥ 1,000 years	Each pump
	C		≥ 1,000,000 years	Each pump
4	A		≥ 500 years	System
5	A		≥ 1 year	Tank
	B		≥ 6 months	Tank
	C		≥ 100,000 years	Tank
6	A	≥ 99.5%		L/L Warning Light
7	A	≥ 97.5%		H/L Warning Light
8	A	≥ 99%		L/L Alarm

FIGURE 14.1   Expectation for effectiveness

The example in the figure illustrates several important points about the measurement of maintenance effectiveness:

• When measuring maintenance performance, we are not measuring equipment effectiveness; we are measuring functional effectiveness. The distinction is important, because shifting emphasis from the equipment to its functions helps people—maintainers in particular—to focus on what the equipment does rather than what it is.
• Even quite simple assets have a surprisingly large number of functions. Each of these functions has a unique set of performance expectations. Before it is possible to develop a comprehensive maintenance effectiveness reporting system, we need to know what all these functions are, and we must be prepared to establish what the user thinks is acceptable or otherwise in each case.

This means that it is not possible to list a single continuity statement for an entire asset, such as "To fail not more than once every 2 years" or "To last at least 11 years." We need to be specific about which function must not be lost more than once every 2 years or must not fail for at least 11 years (or more precisely, which functional failure must not occur more than once every 2 years or which functional failure must not occur before 11 years).

- There is often a tendency to focus too heavily on primary functions when assessing maintenance effectiveness. This is a mistake, because in practice apparently trivial secondary functions often embody bigger threats to the organization if they fail than primary functions. As a result, every function must be considered when setting up maintenance effectiveness measures and targets.

For instance, the primary functions listed for the gasoline system are to pump and to store fuel (functions 1 and 5, respectively). However, two of the highest expectations of the owner centered on two secondary functional failures—2B (a failure that could put him out of business) and 3C (a failure with serious safety implications).

Multiple Performance Standards and the OEE. If a function embodies multiple performance standards, it is tempting to try to develop a single composite measure of effectiveness for the entire function. For instance, the primary function of a machine performing a conversion operation in a manufacturing facility usually incorporates three performance standards:

- It must work at all.
- It must work at the right pace.
- It must produce the required quality.

The effectiveness with which it continues to meet each of these expectations is measured by availability, efficiency, and yield. This suggests that a composite measure of the effectiveness with which this machine is fulfilling its primary function on an ongoing basis could be determined by multiplying these three variables, as follows:

Overall effectiveness = availability × efficiency × yield

For instance, the primary function of a milling machine might be:

To mill 101 ± 1 workpieces per hour to a depth of 1 inch ± 1/8 inch

If this machine is out of action completely for 5% of the time, its availability is 95%. If it is only able to produce 96 pieces per hour when it is running, its efficiency is 96%. If 2% of its output are rejects, its yield is 98%. Applying the above formula gives an overall effectiveness of $0.95 \times 0.98 \times 0.96 = 0.894$, or 89.4%.

This particular composite measure is sometimes referred to as "overall equipment effectiveness," or OEE. Composite measures of this sort are popular because they allow users to assess maintenance effectiveness at a glance. They also seem to offer a basis for comparing the performance of similar assets (so-called benchmarking). However, these measures actually suffer from numerous drawbacks, as follows:

- The use of three variables in the same equation implies that all three have equal weighting. This may not be the case in practice.

  For instance, in the milling machine example above, the workpiece may have a work-in-process value of $200 at that point in the process. The organization might be making a gross profit of $100 on a finished product sale price of $500. This means that 1% downtime or 1% loss of efficiency costs the company one sale per hour—a lost profit of $100 per hour. On the other hand, 1% scrap means that the organization has to write off 1 workpiece per hour, representing $200 worth of work in process in addition to $100 lost profit—a total loss of $300 per hour. Consequently, the machine in the above example is losing:

  $$(5 \times 100) + (4 \times 100) + (2 \times 300) = \$1,500 \text{ per hour}$$

- due to downtime, slow running, and rejects. However, an identical machine producing the same product might suffer from 4% downtime, run at 98% of its rated speed, and produce 4% scrap. In this case the overall effectiveness would be 0.96 × 0.98 × 0.96 = 0.903, or 90.3%. This is apparently a better performance than the first machine. However, this machine is losing:

$$(4 \times 100) + (2 \times 100) + (4 \times 300) = \$1,800 \text{ per hour}$$

- which is actually a significantly worse performance than that of the first machine!

- It is possible for many assets to operate too fast as well as too slowly. Overspeeding an asset would increase the OEE as defined above, which means that it is possible to obtain an apparent improvement in overall performance by forcing the asset to operate in a failed state.

  For instance, a primary performance standard of the milling machine was that it should produce 101 ± 1 workpieces per hour. The "+ 1" means that if the machine produces more than 102 units per hour, it is in a failed state (perhaps because it starts going faster than a bottleneck assembly process, leading to a buildup of work in process, or because going too fast causes the milling cutter to overheat and damage the workpieces, or because it leads to excessive tool wear). However, if it operates at 103 workpieces per hour, the apparent efficiency is 102%. This increases the overall equipment effectiveness as defined above, at a time when the machine is actually in a failed state. This is clearly nonsense.

- The OEE as defined above only relates to the primary function of any asset. This is misleading, because as in the case of the gasoline storage system, every asset—machine tools included—has many more functions than the primary function, and each of these will have their own unique performance expectations. Consequently, the OEE is not a measure of overall effectiveness at all, but only a measure of the effectiveness with which the primary function of the asset is being fulfilled.

- Finally, for the reasons discussed earlier, truly user-oriented maintenance enterprises need to turn their attention away from equipment effectiveness and toward functional effectiveness. So, if measures of this sort must be used, it is much more accurate to refer to them as measures of primary functional effectiveness (PFE) rather than overall equipment effectiveness.

Conclusion

The two most important conclusions to emerge from this section of the chapter are that:

- When evaluating the contribution that maintenance is making to the performance of any asset, the effectiveness with which each function is being fulfilled must be measured on an ongoing basis. This, in turn, requires a crystal-clear understanding of all the functions of the asset, together with an equally clear understanding of what is meant when it is said to be failed.
- The ultimate arbiter of effectiveness is the user (whose expectations must, in turn, be realistic). What users expect will vary—quite legitimately—from function to function and from asset to asset, depending on the operating context.

14.2 Maintenance Efficiency

As mentioned at the beginning of this chapter, maintenance efficiency measures how well the maintenance function is using the resources at its disposal. The large number of ways in which this can be done are generally well understood, so they are only discussed briefly in this part of this chapter for the sake of completeness.

Efficiency measures can be grouped into four categories. These are *maintenance costs, labor, spares and materials,* and *planning and control.*

Maintenance Costs

The costs referred to in this part of this chapter are the direct costs of maintenance labor, materials, and contractors, as opposed to the indirect costs associated with poor asset performance. The latter issues were discussed in Section 14.2 of this chapter.

In many industries, the direct cost of maintenance is now the third-highest element of operating costs, behind raw materials and either direct production labor or energy. In some cases, it has risen to second or even first place. As a result, controlling these costs has become a top priority.

Some industries offer scope for substantial reductions in direct maintenance costs, especially those whose processes embody mature or stable technologies and/or that have a large legacy of second-generation thinking embodied in their maintenance practices. However, in other industries, especially those that are newly mechanizing or automating their processes at a significant rate, the sheer volume of maintenance work to be done is often growing at such a pace that maintenance costs are likely to rise in absolute terms over the next 10 years or so. As a result, take care to evaluate the pace and direction of technological change before committing to substantial long-term reductions in total maintenance costs.

The most common ways in which maintenance costs are measured and analyzed are as follows:

- Total cost of maintenance (actual and budgeted):
  - For the entire facility
  - For each business unit
  - For each asset or system
- Maintenance cost per unit of output
- Ratio of parts to labor expenditure

Labor

The cost of maintenance labor typically amounts to between one-third and two-thirds of total maintenance costs, depending on the industry and overall wage levels in the country concerned. In this context, maintenance labor costs should include expenditure on contract labor (which is often incorrectly grouped under "spares and materials" because it is bought out). When considering maintenance labor, it is also wise not to make the common mistake of treating maintenance work done by operators as a zero cost "because the operators are there anyway." In using operators for this work, the organization is still committing resources to maintenance, and the cost should be acknowledged accordingly.

Common ways of measuring and analyzing maintenance labor efficiency include the following:

- Maintenance labor cost (total and per unit of output)
- Time recovery (time performing specific tasks as a percentage of total time paid for)
- Overtime (absolute hours and as a percentage of normal hours)
- Relative and absolute amounts of time spent on different categories of work (proactive tasks, default actions and modifications, and subsets of these categories)
- Backlog (by number of work orders and by estimated hours)
- Ratio of expenditure on maintenance contractors to expenditure on full-time maintenance employees

Spares and Materials

Spares and materials usually account for the portion of maintenance expenditure that does not come under the heading of "labor." How well they are managed is usually measured and analyzed in the following ways:

- Total expenditure on spares and materials (total and per unit of output)
- Total value of spares in stock
- Stock turns (total value of spares and materials in stock divided by the total annual expenditure on these items)
- Service levels (percentage of requested stock items that are in stock at the time the request is made)
- Relative and absolute values of different types of stocks (consumables, active spares, "insurance" spares, dead stock)

Planning and Control

How well maintenance activities are planned and controlled affects all other aspects of maintenance effectiveness and efficiency, from the overall utilization of maintenance labor to the duration of individual stoppages. Typical measures include:

- Total hours of predictive/preventive/failure-finding maintenance tasks issued per period
- The above hours as a percentage of total hours
- Percentage of the above tasks completed as planned
- Planned hours worked versus unplanned hours
- Percentage of jobs for which the time was estimated
- Accuracy of estimates (estimated hours versus actual hours for jobs that were estimated)

Some of these efficiency measures are useful for making immediate decisions or initiating short-term management action (expenditure against budgets, time recovery, schedule completion rates, backlogs). Others are more useful for tracking trends and comparing performance with similar facilities in order to plan longer-term remedial action (maintenance costs per unit of output, service levels, and ratios in general). Together, they are a great help in focusing attention on what must be done to ensure that maintenance resources are used as efficiently as possible.

Maintenance efficiency is also quite easy to measure. The issues that it addresses are usually under the direct control of maintenance managers. For these two reasons, there is often a tendency for these managers to focus too much attention on efficiency and not enough on maintenance effectiveness. This is unfortunate, because the issues discussed under the heading of maintenance effectiveness usually have a much greater impact on the overall physical and financial well-being of the organization than those discussed under the heading of maintenance efficiency. Truly customer-oriented maintenance managers direct their attention accordingly. As the next section of this chapter explains, the greatest strength of RCM is the extent to which it helps them to do so.

14.3 What RCM Achieves

The use of RCM3 helps to fulfill all the Fourth-Generation Maintenance expectations. The principle benefit of RCM is that it pays for itself, additional benefits are summarized in the following paragraphs starting with safety and environmental integrity.

Greater Safety and Environmental Integrity

RCM contributes to improved safety and environmental protection in the following ways:

- The *systematic review of the safety and environmental implications of every evident failure before considering operational issues* means that safety and environmental integrity become—and are seen to become—top maintenance priorities.
- From the technical viewpoint, *the decision process dictates that failures that could affect safety or the environment must be dealt with in some fashion*—it simply does not tolerate inaction; as a result, tasks are selected that are designed to reduce all equipment-related safety or environmental hazards to an acceptable level, if not eliminate them completely. The fact that these two issues are dealt with by groups that include both technical experts and representatives of the likely victims means that they are also dealt with realistically.
- The structured approach to protected systems, especially the concept of the hidden function and the orderly approach to failure-finding, leads to substantial improvements in the maintenance of protective devices. *This greatly reduces the probability of multiple failures that have serious consequences.* (This is perhaps the most powerful single feature of RCM. Using it correctly significantly lowers the risk of doing business.)
- Involving groups of operators and maintainers directly in the analysis makes them much more sensitive to the real hazards associated with their assets. This makes them *less likely to make dangerous mistakes and more likely to make the right decisions when things do go wrong.*

- The *overall reduction in the number and frequency of routine tasks* (especially invasive tasks that upset basically stable systems) reduces the risk of critical failures occurring either while maintenance is under way or shortly after start-up.

  This issue is particularly important if we consider that preventive maintenance played a part in four of the worst accidents in industrial history (Bhopal, Chernobyl, Piper Alpha, and Texas City). One was caused directly by a proactive maintenance intervention that was currently under way (cleaning a tank full of methyl isocyanate at Bhopal). On Piper Alpha, an unfortunate series of incidents and oversights might not have turned into a catastrophe if a crucial relief valve had not been removed for preventive maintenance at the time.

The most common way to track performance in the areas of safety and environmental integrity is to record the number of incidents that occur, typically by recording the number of lost-time accidents per million labor hours in the case of safety, and the number of excursions (incidents where a standard or regulation is breached) per year in the case of the environment. While the ultimate target in both cases is usually zero, the short-term target is always to better the previous record.

To provide an indication of what RCM has achieved in the field of safety, Figure 14.2 shows the number of accidents per million takeoffs recorded each year in the commercial civil aviation industry over the period of development of the RCM philosophy (excluding accidents caused by sabotage, military action, or turbulence). The percentage of these crashes that were caused by equipment failure also declined. Much of the improved reliability is, of course, due to the use of superior materials and greater redundancy, but most of these improvements were driven, in turn, by the realization that maintenance on its own could not extract the required level of performance from the assets as they were then configured. As explained in Chapter 12, this shifted attention from a heavy reliance on fixed-time overhauls in the 1960s

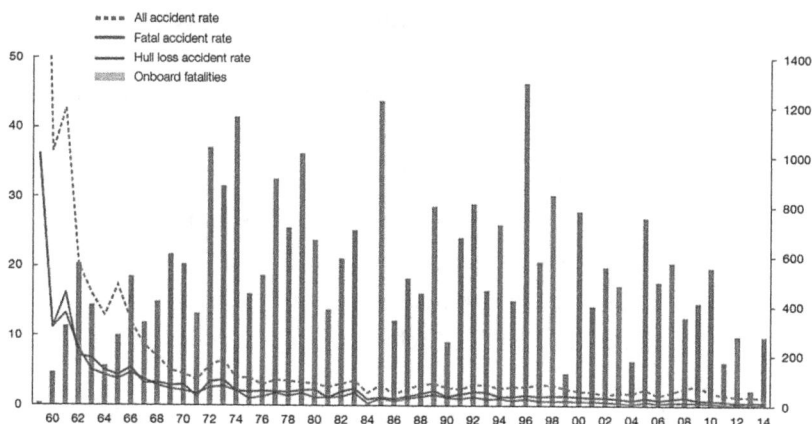

FIGURE 14.2 Safety in the civil aviation industry. (*Source:* C. A. Shifrin, "Aviation Safety Takes Center Stage Worldwide," *Aviation Week & Space Technology*, Vol. 145, No. 19, pp. 46–48.)

to doing whatever is necessary to avoid or eliminate the consequences of failures, be it maintenance or redesign (the cornerstone of the RCM philosophy). It also reduced the number of crashes that might otherwise have been caused by inappropriate maintenance interventions.

Higher Plant Availability and Reliability

The scope for performance improvement clearly depends on the performance at the outset. For example, an undertaking that is achieving 95% availability has less improvement potential than one that is currently only achieving 85%. Nonetheless, if it is correctly applied, RCM achieves significant improvements regardless of the starting point.

Plant performance is, of course, improved by reducing the number and the severity of unanticipated failures that have operational consequences. The RCM process helps to achieve this in the following ways:

- The *systematic review of the economic risks associated with every failure* that has not already been dealt with as a safety hazard, together with the stringent criteria used to assess task effectiveness, ensures that only the most effective tasks are selected to deal with each failure mode.

- The emphasis placed on on-condition tasks helps to ensure that *potential failures are detected before they become functional failures.* This helps reduce economic risks in three ways:
  - Problems can be rectified at a time when stopping the machine will have the least effect on operations.
  - It is possible to ensure that all the resources needed to repair the failure are available before it occurs, which shortens the repair time.
  - Rectification is only carried out when the assets really need it, which extends the intervals between corrective interventions. This, in turn, means that the asset has to be taken out of service less often.

- By relating each failure mode to the relevant failed state, the Information Worksheet provides a tool for *quick failure diagnosis*, which leads, in turn, to *shorter repair times*.

- The previous example suggests that greater emphasis on on-condition maintenance reduces the frequency of major over-hauls, with a corresponding long-term increase in availability. In addition, a comprehensive list of all the failure modes that are reasonably likely, together with a dispassionate assessment of the relationship between age and failure, reveals that there is often no reason at all to perform routine overhauls at any frequency. This leads to a reduction in previously scheduled downtime without a corresponding increase in unscheduled downtime.

  For instance, RCM enabled a major integrated steelwork to eliminate all fixed-interval overhauls from its steelmaking division. In another case, the intervals between major overhauls of a stationary gas turbine on an oil platform were increased from 25,000 to 40,000 hours without sacrificing reliability.

- In spite of the above comments, it is often necessary to plan a shutdown or an overhaul for any of the following reasons:
  - To prevent a failure that is genuinely age-related
  - To rectify a potential failure
  - To rectify a hidden functional failure
  - To carry out a modification

In these cases, the disciplined review of the need for preventive or corrective action that is part of the RCM process leads to *shorter shutdown* worklists, which leads, in turn, to shorter shutdowns. Shorter shutdowns are easier to manage and hence more likely to be completed as planned.

- Short shutdown worklists also lead to fewer start-up failure problems when the plant is started up again after the shutdown, because it is not disrupted as much. This too leads to an overall increase in reliability.
- RCM provides an opportunity for those who participate in the process to learn quickly and systematically how to operate and maintain the new plant. This enables them to avoid many of the errors that would otherwise be made as a result of the learning process, and as well it enables them to ensure that the plant is maintained correctly from the outset.
- At least four organizations with whom one of the authors has worked in the United States achieved what each organization described as "the fastest and smoothest start-up in the company's history" after applying RCM to new installations. In each case, RCM was applied in the final stages of commissioning. The companies concerned are in the manufacturing, chemical, and water utility business.
- Superfluous functions are eliminated, and hence so are superfluous equipment and failures. As mentioned in Chapter 4, it is not unusual to find that between 5% and 20% of the components of a complex plant are utterly superfluous but can still disrupt the plant when they fail. Eliminating such components leads to a corresponding increase in reliability.
- By using a group of people who know the equipment best to carry out a systematic analysis of failure modes, it becomes possible to identify and eliminate chronic failures that otherwise seem to defy detection, and to take appropriate action.

Improved Product Quality

By focusing directly on product quality issues, RCM does much to improve the yield of automated processes.

Greater Maintenance Efficiency (Cost-Effectiveness)

RCM helps to reduce, or at least to control the rate of growth of, maintenance costs in the following ways.

Less Routine Maintenance. Wherever RCM has been correctly applied to an existing fully developed preventive maintenance system, it has led to a reduction of 40% to 70% in the perceived routine maintenance workload. This reduction is partly due to a reduction in the number of tasks, but mainly it is due to an overall increase in the intervals between tasks. It also suggests that if RCM is used to develop maintenance programs for new equipment or for equipment that is currently not subject to a formal preventive maintenance program, the routine workload would be 40% to 70% lower than if the maintenance program were developed by any other means.

Note that in this context, "routine" or "scheduled" maintenance means any work undertaken on a cyclic basis, be it the daily logging of the reading on a pressure gauge, a monthly vibration reading, an annual functional check of a temperature switch, or a five-yearly fixed-interval overhaul. In other words, it covers scheduled on-condition tasks, scheduled restoration, scheduled discard tasks, and scheduled failure-finding.

For example, RCM has led to the following reductions in routine maintenance workloads when applied to existing systems:

- A 50% reduction in the routine maintenance workload of a confectionery plant
- A 50% reduction in the routine maintenance requirements of the 11-kV transformers in an electrical distribution system
- An 85% reduction in the routine maintenance requirements of a large hydraulic system on an oil platform
- A 62% reduction in the number of low-frequency tasks that needed to be done on a machining line in an automotive engine plant
- A 70% reduction in the number of PM tasks on an electric shovel resulting in a 4% increase in machine availability

- More than 50% reduction in scheduled maintenance for a waste-water treatment plant, mainly due to built-in redundancy and over-maintenance

Note that the reductions mentioned above are only reductions in perceived routine maintenance requirements. In many PM systems, fewer than half of the schedules issued by the planning office are actually completed. This figure is often as low as 30% and sometimes even lower. In these cases, a 70% reduction in the routine workload will only bring what is issued into line with what is actually being done, which means that there will be no reduction in actual workloads.

Ironically, the reason why so many traditionally derived PM systems suffer from such low schedule completion rates is that much of the routine work is perceived—correctly—to be unnecessary. Nonetheless, if only a third of the prescribed work is being done in any system, that system is wholly out of control. A zero-based RCM review does much to bring situations like these back under control.

Better Buying of Maintenance Services. Applying RCM to maintenance contracts leads to savings in two areas.

First, a clear understanding of failure consequences allows buyers to specify response times more precisely—even to specify different response times for different types of failures or different types of equipment. Since rapid response is often the costliest aspect of contract maintenance, judicious fine-tuning in this area can lead to substantial savings.

Second, the detailed analysis of preventive tasks enables buyers to reduce both the content and the frequency of the routine portions of maintenance contracts, usually by the same amount (40%–70%) as any other schedules that have been prepared on a traditional basis. This leads to corresponding savings in contract costs.

Less Need to Use Expensive Experts. If field technicians employed by equipment suppliers attend RCM meetings as suggested in Chapter 13, the exchange of knowledge that takes place leads to a quantum jump in the ability of the maintainers employed by the users to solve difficult problems on their own. This leads to an equally dramatic drop in the need to call for (expensive) help thereafter.

Clearer Guidelines for Acquiring New Maintenance Technology. The criteria used to decide whether a proactive task is technically feasible and worth doing apply directly to the acquisition of condition monitoring equipment. If these criteria are applied dispassionately to such acquisitions, a number of expensive mistakes can be avoided.

Most of the Items Listed under "Higher Plant Availability and Reliability." Most of the items listed in the previous section of this chapter also improve maintenance cost-effectiveness. How they do so is summarized below:

- Quicker failure diagnosis means that less time is spent on each repair.
- Not only does detecting potential failures before they become failed states mean that repairs can be planned properly and hence carried out more efficiently, but it also reduces the possibility of the expensive secondary damage that could be caused by the functional failure.
- The reduction or elimination of overhauls together with shorter worklists for the shutdowns that are necessary can lead to very substantial savings in expenditure on parts and labor (usually contract labor).
- The elimination of superfluous functions and equipment also means the elimination of the need either to prevent it from failing in a way that interferes with production or to repair it when it does fail.
- Learning how the plant should be operated together with identifying chronic failures leads to a reduction in the number and severity of failures, which leads to a reduction in the amount of money that must be spent on repairing them.
- The most spectacular case of this phenomenon encountered concerned a single failure mode caused by an incorrect machine

adjustment (operator error) in a large process plant. It was identified during an RCM review and was thought to have cost the organization using the asset just under $1 million in repair costs alone over a period of 8 years. It was eliminated by asking the operators to adjust the machine in a slightly different way.

Longer Useful Life of Expensive Items. By ensuring that each asset receives the bare minimum of essential maintenance—in other words, the amount of maintenance needed to ensure that what it can do stays ahead of what the users want it to do—the RCM process does much to help ensure that just about any asset can be made to last as long as its basic supporting structure remains intact and spares remain available.

As mentioned on several occasions, RCM also helps users to enjoy the maximum useful life of individual components by selecting on-condition maintenance in preference to other techniques wherever possible.

Greater Motivation of Individuals. RCM helps to improve the motivation of the people who are involved in the review process in a number of ways. First, a clearer understanding of the functions of the asset and of what they must do to keep it working greatly enhances their competence and hence their confidence.

Second, a clear understanding of the issues that are beyond the control of each individual—in other words, of the limits of what each person can reasonably be expected to achieve—enables people to work more comfortably within those limits. (For instance, no longer are maintenance supervisors automatically held responsible for every failure, as so often happens in practice. This enables them—and those about them—to deal with failures more calmly and rationally than might otherwise be the case.)

Third, when each group member knows that he or she played a part in formulating the goals, in determining what should be done to achieve them, and in deciding who should do what to achieve them, these group members feel a strong sense of ownership.

This combination of competence, confidence, comfort, and ownership means that the people concerned are much more likely to want to do the right job right the first time.

Better Teamwork. The ways in which the highly structured RCM approach to maintenance problem analysis and decision making contributes to team building were summarized in Chapter 13. Not only does this approach foster teamwork within the review groups themselves, but it also improves communication and cooperation between:

- Production or operations departments and the maintenance function
- Management, supervisors, technicians, and operators
- Equipment designers, vendors, users, and maintainers

A Maintenance Database. The RCM Information and Decision Worksheets provide a number of additional benefits:

- **Adapting to changing circumstances.** The RCM database makes it possible to track the reason for every maintenance task right back to the functions and the operating context of the asset. As a result, if any aspect of the operating context changes, it is easy to identify the tasks that are affected and to revise them accordingly. (Typical examples of such changes are new environmental regulations, changes in the operating cost structure that affect the evaluation of operational consequences, or the introduction of new process technology.) Conversely, it is equally easy to identify the tasks that are not affected by such changes, which means that time is not wasted reviewing these tasks.

  In the case of traditionally derived maintenance systems, such changes often mean that the whole maintenance program has to be reviewed in its entirety. As often as not, this is seen as too big an undertaking, so the system as a whole gradually falls into disuse.

- **Providing an audit trail.** Rather than prescribing specific tasks at specific frequencies, more and more modern safety legislation is demanding that the users of physical assets must be able to produce documentary evidence that their maintenance programs are built on rational, defensible foundations. The RCM worksheets provide this evidence—the audit trail—in a coherent, logical, and easily understood form.

- **Creating more accurate drawings and manuals.** The RCM process usually means that manuals and drawings are read in a completely new light. People start asking "What does it do?" instead of "What is it?" This leads them to spot a surprising number of errors that may have gone unnoticed in as-built drawings (especially process and instrumentation drawings). This happens most often if the operators and craftspeople who work with the machines are included in the review teams.

- **Reducing the effects of staff turnover.** All organizations suffer when experienced people leave or retire and take their knowledge and experience with them. By recording this information in the RCM database, the organization becomes much less vulnerable to these changes.

    For example, a major automotive manufacturer was faced with a situation where a plant was to be relocated and most of the workforce had chosen not to move with the equipment to the new site. However, by using RCM to analyze the equipment before it was moved, the company was able to transfer much of the knowledge and experience of the departing workers to the people who were recruited to operate and maintain the equipment in its new location.

- **Introducing expert systems.** The information on the Information Worksheet in particular provides an excellent foundation for an expert system. In fact, many users regard this worksheet as a simple expert system in its own right, especially if the information is stored in a simple computerized database and sorted appropriately.

An Integrative Framework. All the issues discussed above are part of the mainstream of maintenance management, and many are already the target of improvement programs. A key feature of RCM is that it provides an effective step-by-step framework for tackling *all* of them at once and for involving everyone who has anything to do with the equipment in the process.

With RCM3, organizations are putting risk and reliability management mainstream with organizational management systems. Compliance to international standards for risk (ISO 31000) and physical asset management (ISO55000) is now possible.

CHAPTER 15

A Brief History of RCM

15.1 The Experience of the Airlines

In 1974, the United States Department of Defense commissioned United Airlines to prepare a report on the processes used by the civil aviation industry to prepare maintenance programs for aircraft. The resulting report was entitled "Reliability-Centered Maintenance."

Before reviewing the application of RCM in other sectors, the following paragraphs summarize the history of RCM up to the time of publication of the report by Nowlan and Heap (1978). The italicized paragraphs quote extracts from their report.

The Traditional Approach to Preventive Maintenance

The traditional approach to scheduled maintenance programs was based on the concept that every item on a piece of complex equipment has a "right age" at which complete overhaul is necessary to ensure safety and operating reliability. Through the years, however, it was discovered that many types of failures could not be prevented or effectively reduced by such maintenance activities, no matter how intensively they were performed. In response to this problem, airplane designers began to develop design features that mitigated failure consequences—that is, they learned how to design airplanes that were "failure tolerant." Practices such as the replication of system functions, the use of multiple engines and the design of damage tolerant structures greatly weakened the relationship between safety and reliability, although this relationship has not been eliminated altogether.

Nevertheless, there was still a question concerning the relationship of preventive maintenance to reliability. By the late 1950s, the size of the commercial airline fleet had grown to the point at which there was ample data for study, and the cost of maintenance activities had become sufficiently high to warrant a searching look at the actual results of existing practices. At the same time the Federal Aviation Agency, which was responsible for regulating airline maintenance practices, was frustrated by experiences showing that it was not possible to control the failure rate of certain unreliable types of engines by any feasible changes in either the content or frequency of scheduled overhauls. As a result, in 1960 a task force was formed, consisting of representatives from both the FAA and the airlines, to investigate the capabilities of preventive maintenance.

The work of this group led to the establishment of the FAA/Industry Reliability Program, described in the introduction to the authorizing document as follows:

> *"The development of this program is towards the control of reliability through an analysis of the factors that affect reliability and provide a system of actions to improve low reliability levels when they exist. In the past, a great deal of emphasis has been placed on the control of overhaul periods to provide a satisfactory level of reliability. After careful study, the Committee is convinced that reliability and overhaul time control are not necessarily directed at associated topics. . . ."*

This approach was a direct challenge to the traditional concept that the length of time between successive overhauls of an item was an important factor in controlling its failure rate. The task force developed a propulsion-system reliability program, and each airline involved in the task force was then authorized to develop and implement reliability programs in the area of maintenance in which it was most interested. During this process, a great deal was learned about the conditions that must exist for scheduled maintenance to be effective. Two discoveries were especially surprising:

Scheduled overhaul has little effect on the overall reliability of a complex item unless the item has a dominant failure mode. There are many items for which there is no effective form of scheduled maintenance.

The History of RCM Analysis

The next step was an attempt to organize what had been learned from the various reliability programs and develop a logical and generally applicable approach to the design of preventive maintenance programs. A rudimentary decision-diagram technique was devised in 1965, and in June 1967 a paper on its use was presented at the AIAA Commercial Aircraft Design and Operations Meeting. Subsequent refinements of this technique were embodied in a handbook on maintenance evaluation and program development, drafted by a maintenance steering group formed to oversee development of the initial program for the new Boeing 747 airplane. This document, known as MSG-I, was used by special teams of industry and FAA personnel to develop the first scheduled maintenance program based on the principles of reliability-centered maintenance. The Boeing 747 maintenance program has been successful.

Use of the decision-diagram technique led to further improvements, which were incorporated two years later in a second document, MSG-2: Airline Manufacturer Maintenance Program Planning Document.

MSG-2 was used to develop the scheduled maintenance programs for the Lockheed 101 and the Douglas DCJ0 airplanes. These programs have also been successful. MSG-2 has also been applied to tactical military aircraft; the first applications were for aircraft such as the Lockheed S-3 and P-3 and the McDonnell F4J. A similar document prepared in Europe was the basis for the initial programs for such aircraft as the Airbus Industrie A-300 and the Concorde.

The objective of the techniques outlined in MSG-1 and MSG-2 was to develop a scheduled maintenance program that assured the maximum safety and reliability of which the equipment was capable and also provided them at the lowest cost. As an example of the economic benefits achieved with this approach, under traditional maintenance policies the initial program for the Douglas DC-8 airplane required scheduled overhaul for 339 items, in contrast to seven such items in the DC-10 program. One of the items no longer subject to overhaul limits in the later programs was the turbine propulsion engine. Elimination of scheduled overhauls for engines led to major reductions in labor and materials costs, and also reduced the spare-engine inventory required to cover shop maintenance by more than 50%. Since engines for larger airplanes then cost more than US$1 million each, this was a respectable saving.

As another example, under the MSG-1 program for the Boing 747, United Airlines expended only 66,000 manhours on major structural inspections before reaching a basic interval of 20,000 hours for the first heavy inspections of this airplane. Under traditional maintenance policies it took an expenditure of more than 4 million manhours to arrive at the same structural inspection interval for the smaller and less complex Douglas DC-8. Cost reductions of this magnitude are of obvious importance to any organization responsible for maintaining large fleets of complex equipment. More important:

> • *Such cost reductions are achieved with no decrease in reliability. On the contrary a better understanding of the failure process in complex equipment has actually improved reliability by making it possible to direct preventive tasks at specific evidence of potential failures.*

Although the MSG-1 and MSG-2 documents revolutionized the procedures followed in developing maintenance programs for transport aircraft, their application to other types of equipment was limited by their brevity and specialized focus. In addition, the formulation of certain concepts was incomplete. For example, the decision logic began with an

evaluation of proposed tasks, rather than an evaluation of the failure con-
sequences that determine whether they are needed, and if so, their actual
purpose. The problem of establishing task intervals was not addressed,
the role of hidden-function failures was unclear, and the treatment of
structural maintenance was inadequate. There was also no guidance on
the use of operating information to refine or modify the initial program
after the equipment entered service or the information systems needed for
effective management of the ongoing program.

All these shortcomings, as well as the need to clarify many of the
underlying principles, led to analytic procedures of broader scope and
their crystallization into the logical discipline now known as Reliability-
Centered Maintenance.

15.2 The Evolution of RCM3

We have seen in Chapter 1 that since the release of the Nowlan and Heap
report in 1978 when the process was first called Reliability-Centered
Maintenance (RCM). RCM become very popular. Many attempts by
different people and organizations have been made to industrialize
RCM but the most notably was the development of RCM2™ by John
Moubray and the application thereof by the worldwide Aladon net-
work since 1991. The development of the the SAE Standard JA1011,
Evaluation Criteria for RCM Processes, and the associated guide SAE
JA1012, *A Guide to the RCM Standard* led many organizations to follow
the standard giving RCM credibility.

The author was introduced to RCM in the late 1990s and has spent
most of his time working exclusively in applying RCM2 to assets in
almost all industries known to people, ranging from mining, water and
wastewater utilities, aviation, transport, manufacturing, nuclear, and
power generation and distribution.

The author was trained and mentored by the late John Moubray and
worked with some of the senior and most experienced Aladon Network
members for more than 20 years. The idea to include a *risk-based*
approach for failures impacting safety and environment was introduced

by John Moubray at a meeting with Aladon Network members in 2003. As mentioned in Chapter 1, work started on the *risk-based RCM* in 2004 but only really picked up momentum in 2014 when the author and Theuns Koekemoer came together and used the *ground-breaking work of John Moubray* as the basis to develop RCM3. The development took 4 years of hard work and many revisions, mostly around the decision logic and how risk management changes the way people think about maintenance. RCM3 is a dramatic departure from the RCM process described in the SAE Standard.

RCM3 adds a new dimension to how maintenance and risk management strategies are defined and even though RCM3 is different, it still complies fully with the SAE JA1011 RCM Standard and actually surpasses the standard by extending the functionality to align with the newer ISO standards for Risk- and Asset Management while incorporating all of the valid RCM methodology steps.

Reliability management has become highly specialized and, with the introduction of new standards and technology, RCM3 places reliability mainstream with organization management systems. RCM3 moved closer to directly influence and contribute to other business processes.

Comparison Between RCM2 and RCM3

RCM Element	RCM2 (SAE JA 1011)	RCM3 Highlights	Reason for additions / changes	Improvements and advantages
Operating Context	Mentions and considers the Operating Context (OC) throughout the process. Operating Context is considered when failure modes are identified and when failure management strategies are developed (Failure consequences are different when OC varies). SAE JA1011[1] mentions the OC as important but not necessary a requirement.	The Operating Context is the **FIRST** step and question that must be answered in RCM3: *What are the conditions under which the equipment is expected to operate?* The OC is not just important, but essential for developing a sensible and defensible risk management program.	The Operating Context *must* be defined prior to the FMEA, listing functions and performance standards, failed states, failure modes and failure effects is all based on the OC. The inherent risk posed by each failure and everything impacting the performance of an asset, are influenced by and derived from the OC.	Defining the Operating Context is undeniable the first step of the RCM process and all assumptions and decisions are based on the OC, making RCM3 compliant and exceeding requirements of SAE standard. Risk assessment and risk management must be performed within the context – according to ISO 31000[2] (ISO Standard for Risk Management). True optimization is only possible when the OC is defined.
Functions	Requires the definition of Primary and Secondary Functions with associated standards of performance • Performance standards should be defined (where possible) • Specific about the definition of functions for protective devices	Requires the definition of Primary and Secondary Functions with associated standards of performance • Performance standards should be defined (where possible) • Specific about the definition of functions for protective & detective devices • Expands Secondary Functions to include cleanliness, regulations, regulatory requirements, recycle / repurpose / reuse	The expectations of modern equipment have changed with the changing expectations of the people who own and operate the equipment. Expectation further changed with new advanced technologies and innovation, through interconnectivity, mobility and predictive technology. Rising pressure from governments and societies with regards to sustainability and environmental integrity, places higher demands on reusable energy and focus on sustainable operations.	Requirements for asset performance now includes elevated consideration for sustainability and environmental integrity. The focus is now more on what the equipment's role is in society as a whole rather than a siloed view of the organization who owns and operates the equipment. The performance standard for initial capability (inherent reliability) is now drawing the attention to defect elimination and longer asset life while meeting regulations and regulatory requirements.
Functional Failures	Functional failures are acknowledged as "failed states": • General failed state • Total failure • Partial failure	Now defined as "Failed State" and acknowledges the differences between: • General failed state • Failing state • Failed state • End state (as part of the failure process)	The general failed state, the failing state (process of failing), partial failure (failed state where equipment no longer meets performance criteria) are now clearly defined and distinguished from the end state (total failure). The RCM3 process deals with all possible failures at the appropriate level.	Agreement between different disciplines (e.g. engineering, operations and maintenance) can be reached much faster and therefore the process of identifying appropriate risk management strategies, is much quicker (saving time and money). The new definition encourages the use of new maintenance techniques and technology.
Failure Modes	Defines a Failure Mode as the event that causes the Functional Failure /	Defines a Failure Mode as a 'cause' and 'mechanism' that causes the Failed	RCM review groups (and facilitators) are forced to define at least one or	Consistent and improved root cause failure identification is now possible.

[1] SAE JA1011 - Evaluation Criteria for Reliability-Centered Maintenance (RCM) Processes, August 1999
[2] ISO 31000 - Risk management - Principles and guidelines, 2009

RCM Element	RCM2 (SAE JA 1011)	RCM3 Highlights	Reason for additions / changes	Improvements and advantages
	Failed State. The facilitator / review group must constantly be reminded of the correct level of detail (not to describe failure effects / symptom of the failure).	State. This allows the facilitator / review group to identify root causes consistently and with the correct level of detail. The failure mechanism also ties in with the degradation mechanisms (terminology used in RBI).	more failure mechanisms for each failure mode ensuring the level of detail is sufficient and appropriate for developing risk management strategies that are both technically feasible and worth doing. Failure causes are events causing the failed states while failure mechanisms are the conditions leading to the failures (e.g. corrosion, normal wear and tear, etc.).	even for inexperienced facilitators. Templating of like type equipment is easier to perform, more information is carried over. Integration with other risk-based approaches (e.g. RBI) are now easy to achieve. Root causes are identified and treated and no longer the symptoms associated with failures.
Failure Effects	Failure Effect is defined as one statement (one paragraph that describes what will happen if the failure mode occurs and nothing was done to prevent it). It requires the facilitator to record the physical effects of each failure by asking the following questions: • What evidence (if any) that the failure has occurred? • In what ways (if any) it poses a threat to safety or the environment? • In what ways (if any) it affects production or operations? • What physical damage (if any) is caused by the failure? • What must be done to repair it?	Like RCM2, Failure Effects are described if no maintenance is being performed and no attempt is made to prevent them, but the effects are now separated in three levels: Local Effect, Next Higher-Level Effect and End Effect RCM3 also describes Potential Worst-Case Effect (where protection is also in a failed state – allowing for true zero-base analysis). It does so by asking the following questions: • When is the failure most likely to occur? • How often the failure would occur if no attempt is made to prevent it? • What evidence (if any) that the failure has occurred? • In what ways (if any) it poses a threat to safety or the environment? • In what ways (if any) it affects production or operations? • What physical damage (if any) is caused by the failure? • What must be done to repair it? • Does it cause any secondary damage? • What is the revenue loss (if any)?	Separating the effect description makes it possible to distinguish more easily between the specifics of complex failure effects. Reporting on failure effects (assessing the consequence severity) to different levels in the organization is more granular and less time is spent during the analysis and the subsequent analysis audit meetings. Describing failure effects are far easier and the separation allows the different disciplines in the review group (engineering, operations and maintenance) to focus on their areas of expertise and knowledge. The first question now truly considers the Operating Context and when failures are more likely to occur. (e.g. storm events, start-up, take-off or landing, following maintenance intervention, etc.).	Easier and more comprehensive templating at equipment type level (Local Effect descriptions included in the analysis template). Indicators easier to define (clear difference between what operator / maintenance personnel sees vs. what management wants to see). Potential worst case analysis now possible describes multiple failure conditions separate and with appropriate level of detail. The focus is on increasing the reliability of the protected function/system as a first priority. True zero-base analysis now possible without considering protective systems to mitigate inherent risk. Using consequence definitions as defined in the organizations risk framework, allows everyone to relate and understand the effects of failure and the risk it poses. It is now possible to quantify inherent risk and develop risk mitigation strategies for intolerable risks.

RCM Element	RCM2 (SAE JA 1011)	RCM3 Highlights	Reason for additions / changes	Improvements and advantages
Consequences vs. Risk	Considers the consequences of failure and categorizes them in four categories: Safety/Environmental, Operational, Non-Operational and a single category of hidden failure consequences. All failures are treated and consequences are evaluated based on the four categories.	Considers evident Physical and Economic Risks and separates the Hidden Risks in two categories: Hidden Physical and Hidden Economic Risks. Physical Risks are risks impacting health, safety or the environment while Economic Risks impact operational capability and financial well-being.	Assessing and managing risks allow the review team to distinguish between tolerable and intolerable risks as defined by the organizations risk framework. Not all risks are intolerable and therefore not all failures need to be analyzed – saving time and valuable resources.	This is valuable in high risk environments. Improved integrity and improved planning for testing protective devices are possible. The focus is on the devices that could impact safety vs. the economic impact of the same.
Inherent Risk	Follows a subjective approach to risk management and addresses risk only when failures (or multiple failures) impact safety or the environment. RCM2 is a process to determine what must be done to an asset system to preserve its functions (while minimizing or avoiding failure consequences).	RCM3 addresses risk directly and the risk management approach is based on ISO 31000 Standards for Risk Management. RCM3 is the process used to determine what must be done to an asset system to preserve its functions while minimizing the risks associated with failures to a tolerable level. RCM3 further considers a probabilistic risk assessment at component level when compulsory redesigns or one-time changes are required. Every reasonably likely failure mode is assessed and quantified in terms of its inherent risk. Less likely failure modes are considered based on inherent risk.	Inherent risk is quantified in relative terms as if no maintenance is being performed and if protection associated with failure is unavailable (zero-base). RCM3 is aligned with ISO Standards for Asset Management and Risk (ISO 55000[3] and ISO 31000). RCM3 considers risk mitigation through addressing the probability and the consequence severity both as proactive risk management strategies. This provides more ways to proactively deal with intolerable risk and more decisions are made (fewer compulsory redesigns). For tolerable risks, the default risk mitigation strategies (no scheduled maintenance, spare part policies, etc.) are true default actions. Further risk reduction (for tolerable risk) may be considered provided it can be achieved in a cost-effective manner. This truly makes risk management strategies feasible and worth doing.	The revised risk achieved through the new RCM3 decision process, demonstrates the impact of risk mitigation - both on cost and risk exposure. It allows for proper and formal assessment to determine requirement for one-time changes (redesigns) based on the relative risk. Risk is quantified in relative terms and less compulsory redesign decisions are made – this allows the review group to make more decisions (less open-ended results) and it leads to a more defensible failure management program. Once risk management strategies have been defined, especially for failure modes posing intolerable risks, it is possible to determine the risk and financial impact of the recommendations. Less likely failure modes are evaluated based on the real risk they pose, leading to realistic asset management strategies.

[3] ISO 55000 – ISO 55000 - Asset management - Overview, principles and terminology, 2014

RCM Element	RCM2 (SAE JA 1011)	RCM3 Highlights	Reason for additions / changes	Improvements and advantages
Decision Diagram	The RCM2 Decision Diagram treats all Hidden Functions the same (single approach). The decision logic considers predictive and preventive maintenance tasks as proactive failure management strategies and failure finding, redesigns and no scheduled maintenance as default actions. A combination of tasks is also seen as a default action and consequence mitigation is achieved primarily through optimizing protective devices (protected functions). For any proactive maintenance task (PM) to be considered, the PM must be both technically feasible (according to the failure characteristics) and worth doing (reduces the consequences to an acceptable level).	RCM3 incorporates additional criteria to identify *Hidden Physical* and *Economic Risks*. A true zero-base analysis is only possible if protection related to the failure under consideration is ignored. Focus is placed on reliability of the protected function first. Failure-finding intervals are optimized through increasing reliability of the protected function (when applicable) as the primary concern. Dependency on protective devices are reduced. The *worth doing* criteria for different risk criteria is significantly different from the RCM2 decision logic. Any *Physical Risk* must be reduced to a tolerable level. *Economic Risks* are considered (first) and not cost only. The mitigation strategy must reduce intolerable operational risk (now quantified) in order to be considered. The RCM3 process leads to more defensible risk mitigation.	For hidden failures having an intolerable *physical risk*, risk thresholds are used to determine the failure-finding intervals. For hidden failures having an intolerable *economic risk*, the cost of doing failure- finding is compared to the cost of the multiple failure and intervals are optimized based on cost. *Functional checks* designed for protective devices that fail (not fail-safe devices) are now included (where applicable). The focus in RCM2 could be (and has been) misinterpreted as being biased towards protective devices present in the system (especially standby and redundant equipment), which resulted in "No Scheduled Maintenance" decision for the protected function. This meant that the risk to the organization is drastically increased during repair time when the protected function failed (risk of multiple failures).	The criteria for *Hidden Economic Risks* determine the optimum interval for failure finding (providing highest availability) at the lowest cost. The cost of the failure-finding task must still be acceptable to the user, otherwise a one-time change may be considered to reduce the overall cost of multiple failures (where applicable). Improved integrity through functional testing for protective systems that fail (based on risk tolerance). The RCM3 decision diagram focus on the protected function as a priority. The need for a protective device and failure-finding intervals are only considered AFTER the integrity of the protected function has been addressed. These decisions are all risk based.
SAE JA 1011/1012 International RCM Standard	RCM2 complies fully with the minimum requirements of the SAE JA 1011 and SAE JA 1012 RCM standards.	RCM3 complies fully with the minimum requirements of the SAE JA 1011 and SAE JA 1012 Standards and goes beyond these requirements. RCM3 aligns with ISO 55000 and ISO 31000 Management Systems.	To align and integrate RCM with recognized and adopted International Management Systems. To mainstream RCM with International Asset Management Systems.	RCM3 now aligns with new and emerging standards making the results easier to defend. International standards and management systems are rarely challenged. RCM3 will become the new standard.

Asset Hierarchies and Functional Block Diagrams

Plant Registers and Asset Hierarchies

Most organizations own, or at least use, hundreds if not thousands of physical assets. These assets range in size from small pumps to steel rolling mills, from aircraft carriers to office blocks. The assets may be concentrated on one small site or spread over thousands of square miles. Some of these assets will be mobile; others will be fixed.

Before any organization can apply RCM—a process used to determine what must be done to ensure that any physical asset continues to do whatever its users want it to do—it must know what these assets are and where they are. In all but the smallest and simplest facilities, this means that a list of all the plant, equipment, and buildings that are owned or used by the organization, and that require maintenance of any sort, must be prepared. This list is known as the asset register. Prioritizing the assets is especially useful in determining which assets should be considered for RCM analysis. The assets should be prioritized in order of relative risk they pose to achieving the business objectives.

The register should be designed in a way that makes it possible to keep track of the assets that have been analyzed using RCM, those that have yet to be analyzed, and those that are not going to be analyzed. (The asset register is also needed for other aspects of maintenance management, such as the planning and scheduling of routine and nonroutine maintenance tasks, history recording, and maintenance cost allocation. As a result, it should be set up and the associated numbering systems designed in such a way that it can be used for all these purposes.)

Chapter 8 explained that RCM can be applied at almost any level in a hierarchy. It also suggested that the most appropriate level is the level that leads to a reasonably manageable number of failure modes per function. "Appropriate" levels become much easier to identify if the plant register is set up as a hierarchy that makes it possible to identify any system or any asset at any level of detail, down to and including individual components ("line replaceable items") or even spare parts.

The truck on page 00 provides one example of such a hierarchy. Figure AII.1 shows another example covering a typical hierarchy for a water and wastewater utility.

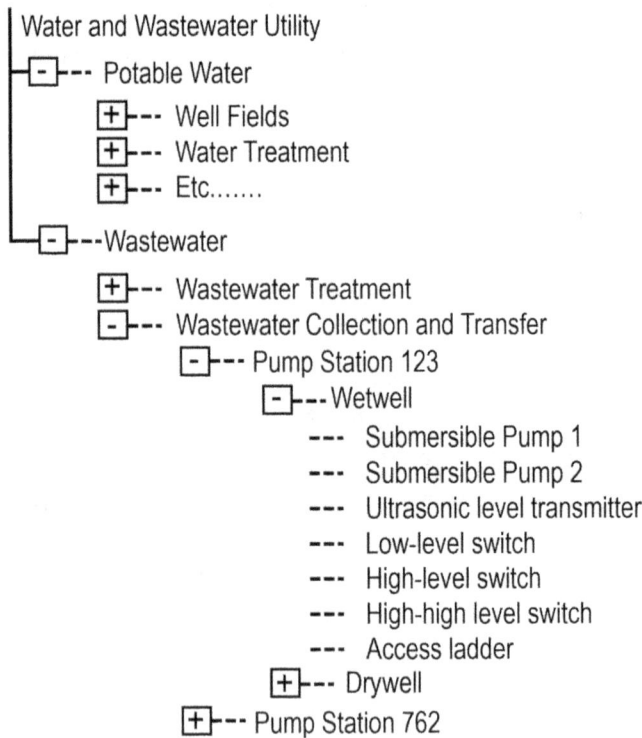

```
Water and Wastewater Utility
 └[-]--- Potable Water
          [+]--- Well Fields
          [+]--- Water Treatment
          [+]--- Etc.......
     └[-]---Wastewater
              [+]--- Wastewater Treatment
              [-]--- Wastewater Collection and Transfer
                      [-]--- Pump Station 123
                              [-]---Wetwell
                                    --- Submersible Pump 1
                                    --- Submersible Pump 2
                                    --- Ultrasonic level transmitter
                                    --- Low-level switch
                                    --- High-level switch
                                    --- High-high level switch
                                    --- Access ladder
                              [+]--- Drywell
                      [+]--- Pump Station 762
```

FIGURE AII.1   Asset hierarchy for Water Utility

Functional Hierarchies and Functional Block Diagrams

It is possible to develop a hierarchy showing the primary functions of each of the assets in the asset hierarchy. Figure AII.2 shows how this might be done for the asset hierarchy presented in Figure AII.1.

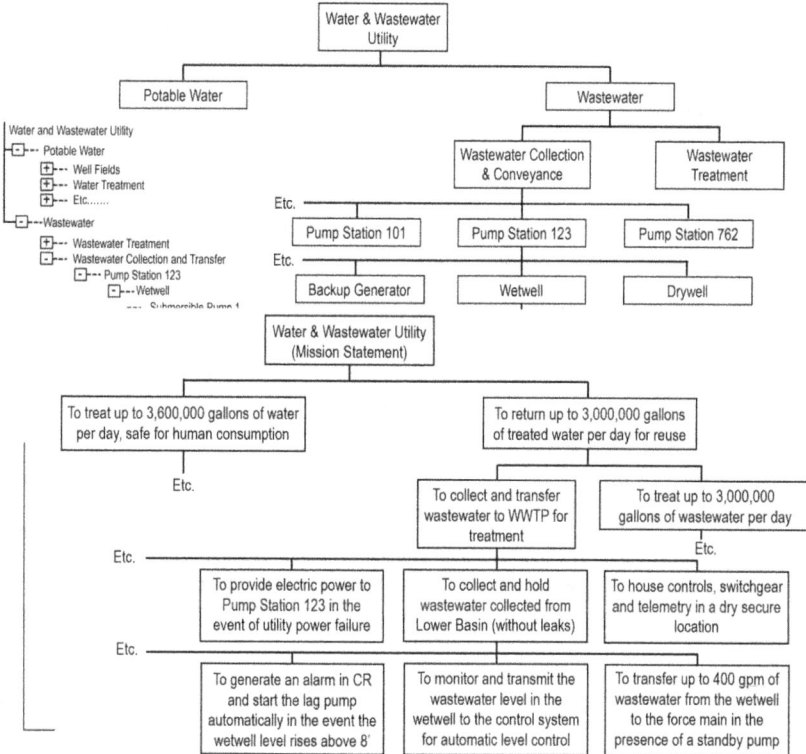

FIGURE AII.2

Variations of the functional hierarchy in Figure AII.2 are used to show the relationships between functions at the same level. These are usually known as "functional block diagrams," and they can be used to depict the relationships in a number of different ways. For instance, Smith (1993) defines a functional block diagram as "a top-level representation of the major functions that the system performs." On the other hand, Blanchard and Fabrycky (1990), who prefer the term "functional flow diagram," suggest that these diagrams can be prepared at many different levels. Smith tends to use the diagrams to show the movement of materials, energy, and control signals through and between different elements of a system, whereas Blanchard and Fabrycky use them to depict the movement of a single asset through different mission phases (such as an aircraft moving from start-up to taxi, takeoff, climb, cruise, descent, landing, and so on).

A functional block diagram for the boiler house in Figure AII.1 shows that the coal flows from the coal handling plant to the two boilers, and the residue flows to the ash handling plant. It also shows what materials and services flow across the system boundaries. This is illustrated in Figure AII.3, which goes on to show a more detailed functional block diagram for one of the boilers. A more complex version of these diagrams could also be used to show what control and indication signals pass across the system boundaries.

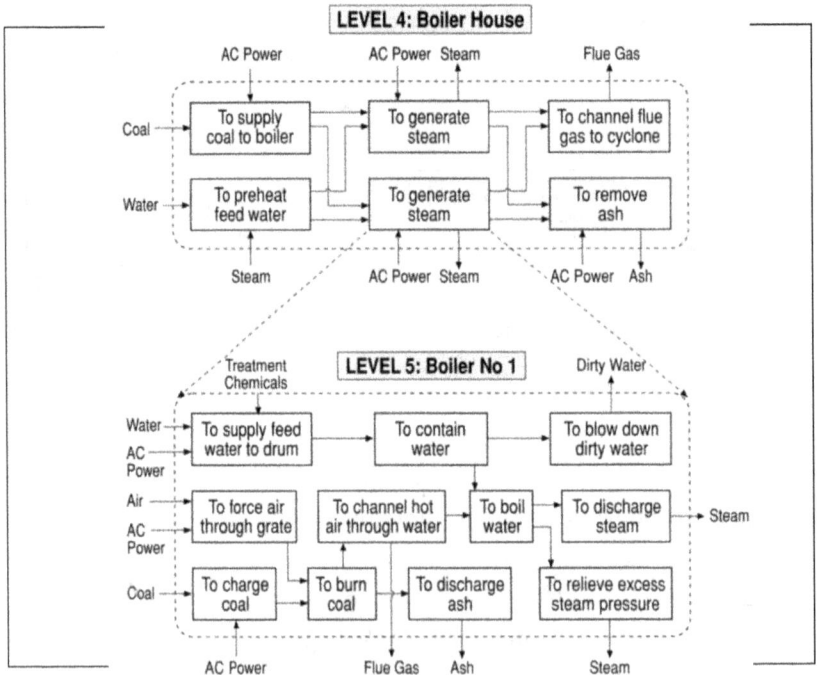

FIGURE AII.3

Functional hierarchies and functional block diagrams are an essential part of the equipment design process, because design starts with a list of desired functions, and designers have to specify an entity (asset or system) that is capable of fulfilling each functional requirement.

As mentioned in Chapter 4, functional block diagrams can also be of some help when RCM is applied to facilities where the processes or the relationships between them are not intuitively obvious. These tend to be large, poorly accessible, very complex, monolithic structures such as naval vessels, combat aircraft, and the less accessible parts of nuclear facilities.

However, in most other industrial applications (such as thermal power stations, food and automotive manufacturing plants, offshore oil platforms, petrochemical and pharmaceutical plants, and vehicle fleets), there is usually no need to draw up functional block diagrams before embarking on an RCM project, for the following reasons:

- In most industries, the relationships between different processes are usually well enough understood by the participants in RCM review groups to make these diagrams unnecessary.

  For example, the boiler house operators and maintainers would be fully aware of the fact that coal, water, and air go into the boiler at one end and that steam, flue gas, and ash (and occasionally dirty water) come out of the other. Most of them would probably regard the notion that these simple facts should be drawn in a diagram at best as a waste of time. As discussed at length in Chapter 4, the real challenge is not to identify the simple and usually obvious relationships between processes; instead, it is to define the desired performance relative to the initial capability for all the key elements of each system and then to define what must be done to ensure that the system continues to deliver the desired performance.

  In cases of uncertainty, equipment is usually accessible enough for it to be easy to go and see what goes where, and if not, the required information can be extracted from a set of process and instrumentation drawings. In fact, a good set of P&IDs nearly always eliminates the need for functional block diagrams entirely as a precursor to the application of RCM. In such cases, block diagrams add significantly to the time, effort, and cost of the RCM process while adding nothing to its value.

- Functional block diagrams only identify primary functions at each level, so they only tell part of the story. (For instance, nearly all the assets at the fourth level and below in Figure AII.1 have a secondary containment function. This cannot be shown in a functional block diagram without making it unmanageably cumbersome.)

- As explained in Section 4.3 of Chapter 4, the principal functions of the assets in the hierarchy above the level chosen for the analysis should be summarized in suitably worded operating context statements. These statements are written only for those assets that are relevant to the analysis in question. As a result, time is not wasted defining the functions of assets that are not germane to the asset under consideration. (If large numbers of assets are analyzed, these high-level context statements evolve into a de facto functional hierarchy for the entire organization—one that is far more detailed than a crude, single-statement-per-asset diagram.)

- Assets at or below the level chosen for the analysis are dealt with as part of the normal RCM process. Section 6.7 of Chapter 6 showed that the functions of lower-level assets are either listed as secondary functions in the main analysis or dealt with as failure modes, or in the case of exceptionally complex subsystems, broken out for separate analysis.

  For instance, the example of the truck shown in Figure 6.16 in Chapter 6 showed how a blockage in the fuel line could simply be treated as a failure mode of either the engine or the drive system, without needing a separate function statement for the fuel system or the fuel line.

Functional block diagrams tend to be of most value to outsiders seeking to apply RCM on behalf of equipment users. Because they are outsiders, they need these diagrams—usually prepared at the expense of the owners of the assets—to improve their own understanding of the processes that they are about to analyze. The best way to avoid this expense is not to employ outsiders as analysts in the first place, but rather to train people who have a reasonable firsthand working knowledge of the plant as RCM facilitators.

System Boundaries

When applying RCM to any asset or system, it is, of course, important to define clearly where the system to be analyzed begins and where it ends. If a comprehensive asset hierarchy has been drawn up and a decision taken to analyze a particular asset at a particular level, then the system usually automatically encompasses all the assets below that system in the asset hierarchy. The only exceptions are subsystems that are judged to be so insignificant that they will not be analyzed at all, or very complex subsystems that are set aside for separate analysis.

Care is needed with control loops that consist of a sensor in one system that sends a signal to a processor in a second system, which, in turn, activates an actuator in a third. Chapter 6 explained that this issue can often be dealt with either by conducting the analysis at a high enough level to ensure that the system encompasses the entire loop or by analyzing control systems separately (after the controlled systems have been analyzed). However, sometimes this is not practical, in which case a decision must be made about which system will encompass the control loop in its entirety.

Care is also needed to ensure that assets or components right on the boundaries do not fall between the cracks. This applies especially to items like valves and flanges.

It is wise not to be too rigid about boundary definitions, because as understanding grows during the RCM process, perceptions about what should or should not be incorporated in the analysis frequently change. This means that boundaries may need to be extended to incorporate some subsystems, others may be dropped, and yet others that are included initially may be set aside for later analysis.

(Again, the strongest exponents of rigid boundary definitions tend to be external contractors seeking to apply RCM on behalf of end users, because system boundaries must be defined precisely in order to define the commercial scope of the contracts. The fact that the analysis is the subject of a formal contract means that boundaries have to be defined much more precisely—and much more rigidly—than is necessary from a purely technical point of view. The upshot is that either contracts of this type have to be renegotiated every time a boundary needs to be changed, or the boundary is not moved when it should be, resulting in a suboptimal analysis. The best way to avoid the time and cost associated with these commercial maneuverings is to avoid contracting out this aspect of maintenance policy formulation altogether.)

Human Error

Chapter 6 mentioned that a great many equipment failures are caused by human error. It went on to say that if a specific human error is considered to be a credible reason why a functional failure could occur, then that error should be included in the FMEA. However, human error is an enormous subject in its own right. The purpose of this appendix is to provide a brief summary of the major categories of human error and to suggest how they might be dealt with within the framework of RCM.

Principal Categories of Human Error

When considering the interaction between people and machines, Blanchard et al. (1995) group the main factors into four categories:

- Anthropometric factors
- Human sensory factors
- Physiological factors
- Psychological factors

Nearly every human error can be traced to a failure or a problem that has occurred in at least one of these four areas. As a result, we review them briefly in the first part of this appendix, before looking in more detail at the fourth category.

Anthropometric Factors

Anthropometric factors are those that relate to the size and/or strength of the operator or maintainer. Errors occur because a person (or part of a person, such as a hand or arm):

- Simply cannot fit into the space available to do something
- Cannot reach something
- Is not strong enough to lift or move something

If a failure is occurring or is reasonably likely to occur for any of these reasons, it is highly unlikely that a proactive maintenance task will be found to deal with it. Note also that if a human error occurs for one of these reasons, the human error is not the root cause. The failure mode is actually poor design, and the resulting human error is a failure effect.

If the consequences are such that something must be done about a failure that is occurring for anthropometric reasons, the only viable course of action is likely to be redesign. This will nearly always involve reconfiguring the asset in such a way that it becomes more accessible or easier to move. In this context, Figure AIII.1 shows some dimensions that are considered by the U.S. Navy to be adequate for reasonable human access in confined spaces.

Human Sensory Factors

Human sensory factors concern the ease with which people can see, hear, feel, and even smell what is going on around them. In the case of operators, this tends to apply to the visibility and legibility of instruments and control consoles. For maintainers, it relates to the visibility of components in the nooks and crannies of complex systems. The volume and variability of background noise levels also affect the ability of both operators and maintainers to discern what is happening to their equipment.

Note again that if errors are occurring or thought to be likely for these reasons, the human error is not the root cause but is the effect of some other failure. The remedies also usually entail redesigning the asset (making things easier to see, reducing noise levels).

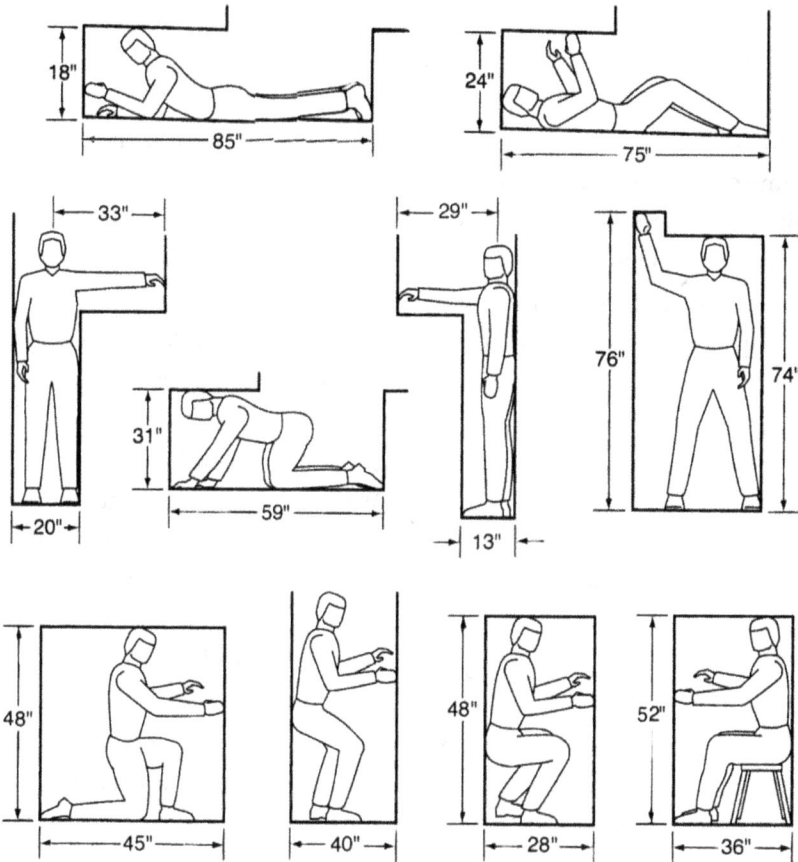

FIGURE AIII.1   Where people fit. (From NAVSHIPS 94234, *Maintainability Design Criteria Handbook for Designers of Shipboard Electronic Equipment*, U.S. Navy, Washington, D.C.)

Physiological Factors

The term "physiological factors" refers to environmental stresses that affect human performance. The stresses include high or low temperatures, loud or irritating noises, excessive humidity, high vibration, exposure to toxic chemicals or radiation, or simply working for too long—especially at a physically or mentally demanding task—without an adequate break.

Sustained exposure to these stresses leads to reduced sensory capacity, slower motor responses, and reduced mental alertness. These are all manifestations of (human) fatigue, and all greatly increase the chances that the people concerned will make a slip, lapse, or mistake (these three terms are defined in the next section of this appendix).

If errors occur or are thought to be likely for any of these reasons, humans are once again not the root cause, but the error is the effect of some other failure. Again, if the consequences warrant it, the remedy is likely to be some form of one-off change. Either the design of the physical environment can be changed in such a way that the error-inducing stresses are reduced (for instance, by reducing temperatures or by providing hearing protection), or operating procedures could be changed in a way that gives overstressed people a chance to recover (longer, more frequent, or more carefully timed rest breaks).

Another environmental stress factor is a relentlessly hostile or adversarial organizational climate. While this does not necessarily have a physiological effect, it can lead to an increased predisposition toward psychological errors. In many cases, it boils down to excessive and inappropriate use of high-task/low-relationship leadership styles. Unfortunately, there is not much that RCM can do about this problem.

However, what RCM can do is alleviate—if not eliminate—the hostile relationship that so often exists between maintenance and operations people, as explained on page 000. This makes people less inclined to blame each other for errors and more inclined to find solutions.

Psychological Factors

The three sets of factors discussed so far all relate to external phenomena that cause humans to make an error. As a result, these factors are relatively easy to identify and to deal with (although doing so may sometimes be expensive). A far more complex and challenging category consists of errors that find their roots in the psyches of the humans themselves. As a result, these psychological factors are discussed in more detail here.

Psychological Errors.   Reason (1991) divides the psychological categories of human error into those that are unintended and those that are intended. An unintended error is one that occurs when someone does a task that he or she should be doing but does it incorrectly ("does the job wrong").

An intended error occurs when someone deliberately sets out to do something, but what he or she does is inappropriate ("does the wrong job"). Reason divides these two categories further as follows:

- Unintended errors are subdivided into slips and lapses.
- Intended errors are subdivided into mistakes and violations.

These categories are illustrated in Figure AIII.2 and are discussed briefly in the following paragraphs.

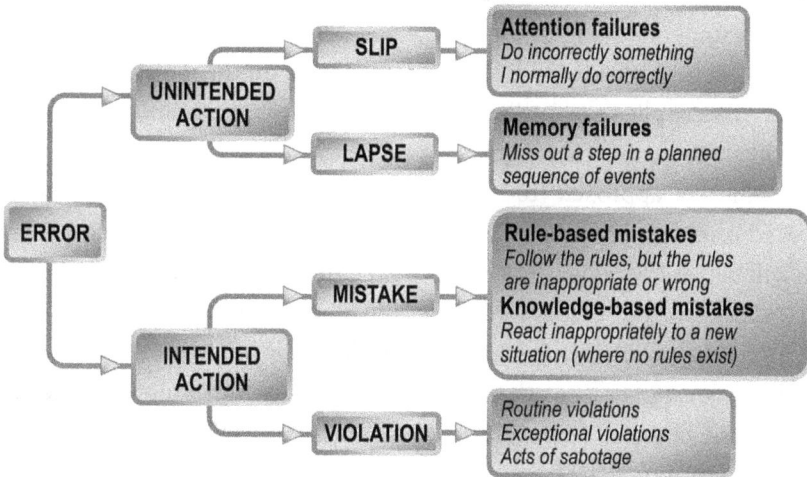

FIGURE AIII.2

Slips and Lapses. Slips and lapses are also known as skill-based errors. They occur when somebody who is fully qualified to do a job—and who may even have done it correctly many times in the past—does the job incorrectly. Slips occur when somebody does something incorrectly (for instance, if an electrician wires a motor incorrectly, causing it to run backward). Lapses occur when someone misses a key step in a sequence of activities (for instance, if a mechanic leaves a tool behind in a machine he or she was working on or forgets to fit a key component while reassembling it).

These errors usually happen because the person concerned was distracted, preoccupied, or simply absent-minded. As a result, the errors are usually unpredictable, although their likelihood increases if the person is working in a physically hostile environment or if the task is exceedingly complex. However, if the environment is reasonably benign and the task is fairly simple, then this category of human errors is perhaps the only one where it is fair to describe the error as the root cause of the failure.

The possibility of a great many slips and lapses can be reduced if operators and maintainers are involved directly in the RCM process (especially the FMEA). This leaves them with a much broader and deeper understanding of the effects and consequences of their actions, which, in turn, results in greater motivation to do the job right the first time. This applies especially to tasks where the consequences of failure are likely to be most serious.

Another approach (to remedy slips that occur during assembly) is based on the assumption that if something can be installed incorrectly, it will be. The remedy is to go back to the drawing board.

This is the essence of the Japanese concept of poka-yoke ("mistake-proofing"). Ideally this philosophy should be applied to original designs rather than retrofitted to existing assets, because it is usually cheaper to build in good practice initially than to modify out bad practice later.

Mistakes

Mistake 1: Rule-Based Mistakes

Rule-based mistakes occur when people believe that they are following the correct course of action when doing a task (in other words, applying a "rule"), but in fact, the course of action is inappropriate. Rule-based mistakes are further subdivided into misapplication of a good rule and application of a bad rule.

In the first case, under a given set of conditions, a person selects a course of action that seems appropriate, usually because it has been successful in dealing with similar conditions in the past—hence the term "good rule." However, some subtle variations on this occasion mean that the course of action, undertaken deliberately, is wrong.

For instance, a protected system might be set up in such a way that excessive pressure should cause an alarm to sound and a warning light to illuminate. However, a situation might arise where the alarm is failed, the pressure increases, and the light comes on. The absence of the alarm may lead the operator to believe that the warning light on its own is only a false alarm, especially if it has a history of spurious failures. In this case, the operator may choose to take no action until the light is repaired—a course of action that has been appropriate in the past. On this occasion, however, it is not the right thing to do.

The application of a bad rule means just what it says. The normally chosen or prescribed course of action is just plain wrong.

A classic example of a bad rule is a maintenance program that schedules items for fixed-interval overhauls in order to deal with failure modes that conform to failure pattern E or F (see Figure 2.5 or 12.1). In the case of pattern F especially, an action designed to improve reliability will, in fact, make it worse, by upsetting a stable system and inducing startup failures.

In these cases, the root cause of the failure is the rule itself or the process by which it is selected. If the rule is promulgated or selected by someone other than the person who performs the task—in other words, if the person doing the task is only following orders—then the mistake is really the effect of another failure.

The RCM process helps to reduce the possibility of misapplying good rules in two ways:

- The thorough analysis of failure effects, especially what could happen if a hidden function is in a failed state when it is needed, means that people are less likely to jump to inappropriate conclusions when the situation does arise (especially if they have been involved in the RCM process).
- By focusing attention on the functions and maintenance of protective devices, the RCM process greatly reduces the probability that these devices will be in a failed state in the first place.

The chances of bad habits developing are also reduced if care is taken during the FMEA to identify failure modes that give rise to spurious alarms and to take steps subsequently to reduce them to a minimum. (In cases where the frequency and/or the possible consequences of a false alarm warrant it, the most appropriate remedy usually entails redesign.)

RCM helps to reduce the possibility of applying bad rules, because the whole RCM process is all about defining the most appropriate rules for maintaining any asset.

Of course, care must be taken to ensure that the rules of RCM itself are not applied badly. This is best done by ensuring that everyone involved in the application of RCM is adequately trained in the underlying principles.

Mistake 2: Knowledge-Based Mistakes

Knowledge-based mistakes occur when someone is confronted with a situation that has not occurred before and that has not been anticipated (in other words, one for which there are no rules). In situations like this, the person has to make a decision about an appropriate course of action, and a mistake occurs if this decision is wrong.

In practice, the author has found that a common problem that occurs in this context is a belief on the part of senior managers and engineers that "I know; therefore my company knows." In fact, if a crisis occurs late at night when all the senior people are off-site, the requisite knowledge is useless if it is not in the mind of the person who has to take the first steps to deal with the crisis.

This suggests that the first and most obvious way to avoid knowledge-based mistakes is to improve the knowledge of the people who have to make the decisions. In most cases, these are the operators and maintainers. Operators and maintainers are likely to make appropriate decisions more often if they clearly understand the way the system works (its functions), the things that can go wrong (functional failures and failure modes), and the symptoms of each failure (failure effects). As mentioned several times in Chapters 4, 13, and

14, this understanding is hugely enhanced if operators and maintainers are involved directly in the RCM process. The most important findings can be disseminated subsequently to people who do not participate in the analysis by incorporating the findings into training programs.

If necessary, the possibility of knowledge-based mistakes can also be reduced by designing (or redesigning) systems in ways that:

- Minimize complexity, so that there is less to know.
- Minimize novelty, because new and alien technologies put people at the bottom of the learning curve, where mistakes are most likely to happen.
- Avoid tight coupling. This means designing systems in such a way that if failures do occur, consequences develop slowly enough to give people time to think and hence more opportunity to make the right decisions.

Violations.   A violation occurs when someone knowingly and deliberately commits an error. Violations fall into three categories:

- **Routine violations.** For instance, when people make a habit of not wearing items of protective clothing (such as hard hats) despite rules that clearly state that they should
- **Exceptional violations.** For instance, if someone who usually wears a hard hat knowingly rushes outside without the hat on "because I couldn't find it and didn't have time to look for it"
- **Sabotage.** When someone maliciously causes a failure

The remedy for routine and exceptional violations usually consists of appropriate enforcement of the rules by management. However, once again, involvement in the RCM process gives people a clearer understanding of the need for safety procedures and the risks they are running if they violate them. The management of sabotage is beyond the scope of this book.

Conclusion

The most important conclusions to emerge from this appendix are that:

- Not all human errors are necessarily the fault of the person who made the error. In many cases, the error is forced either by external circumstances or by inappropriate rules. So if blame is to be allocated for any error, care must be taken to identify the real source.
- Human error is at least as common a reason why equipment fails to do what its users want it to do as deterioration, if not more so. As a result, it should be dealt with as part of the RCM process, either as a failure mode when it is a root cause or as a failure effect when it consists of inappropriate responses to other failures.
- In the industrial context, it is only possible to come to grips with human errors if the people involved in committing the errors are involved directly in identifying them and developing appropriate solutions.

A Continuum of Risk

Chapter 5 suggested that it might be possible to produce a schedule of tolerable risks which combines safety risks and economic risks in one continuum. It suggested that this might be made possible by combining Figures 5.2 and 5.14 in some way.

Figure 5.14, repeated as Figure AIV.1, showed what an organization might decide that it can accept for one event that has economic consequences only.

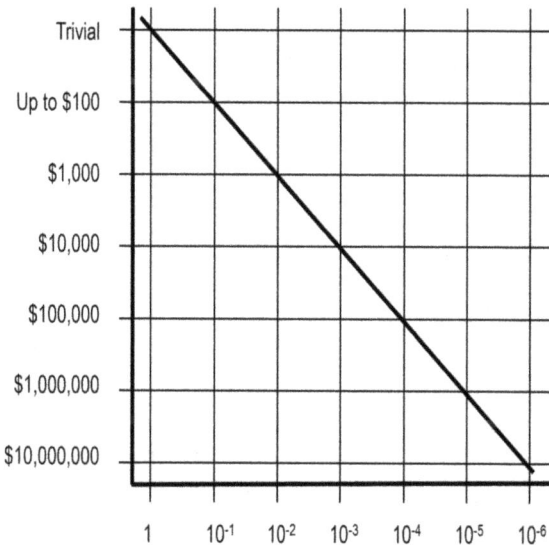

FIGURE AIV.1   Tolerability of economic risk

Tolerability of economic risk Figure 5.2 depicted what one individual might be prepared to tolerate in a specific situation from any event which could prove fatal in that situation, as summarized in Figure AIV.1.

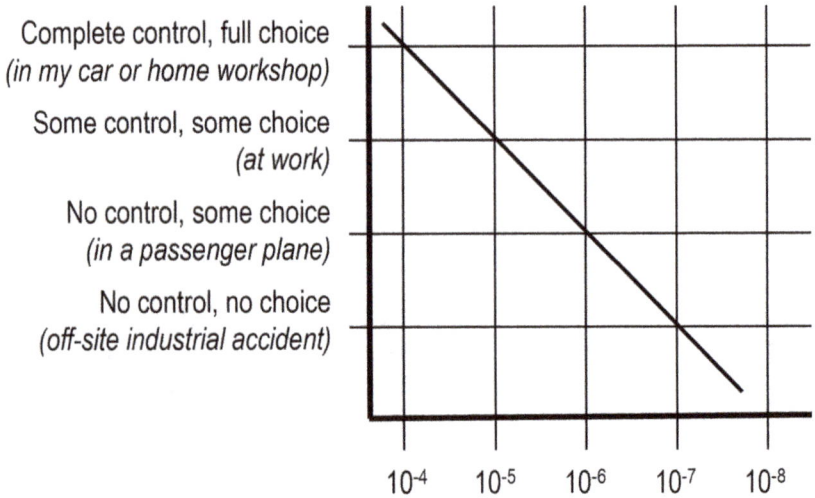

Complete control, full choice
(in my car or home workshop)

Some control, some choice
(at work)

No control, some choice
(in a passenger plane)

No control, no choice
(off-site industrial accident)

10^{-4} 10^{-5} 10^{-6} 10^{-7} 10^{-8}

FIGURE AIV.2 Tolerability of fatal risk

In fact, these two charts cannot be combined as they stand, because Figure AIV.1. is based on the probability of a single event while Figure AIV.2 depicts what one individual might consider to be tolerable for any event.

However, with respect to the latter, Part 3 of Chapter 5 went on to show that it is possible to use what one individual tolerates from any event in a given situation as a basis for deciding what probabilities apply to each event which could place him or her at risk in that situation, as follows:

The first step is to convert what one person tolerates to an overall figure for an entire site. In other words, if I tolerate a probability of 1 in 100,000 (10^{-5}) of being killed at work in any one year and I have 1,000 coworkers who all share the same view, then we all accept that on average 1 person per year on our site will be killed at work every 100 years—and that person may be me, and it may happen this year.

The next step was to translate the probability which myself and my coworkers are prepared to tolerate that any one of us might be killed by any event at work into a tolerable probability for each single event (failure mode or multiple failure) which could kill someone.

For example, continuing the logic of the previous example, the probability that any one of my 1,000 coworkers will be killed in any one year is 1 in 100 (assuming everyone on the site faces roughly the same hazards).

Furthermore, if the activities carried out on the site embody 10,000 events which could kill someone, then the average probability that each event could kill one person must be reduced to 10^{-6}. This means that the probability of an event which is likely to kill ten people must be reduced to 10^{-1}, while the probability of an event that has a 1 in 10 chance of killing one person must be reduced to 10^{-5}. On a site that is divided into several areas and where each area is further divided into several sections, this process of subdividing acceptable risk could be carried out in stages, as shown in Figure AIV.3.

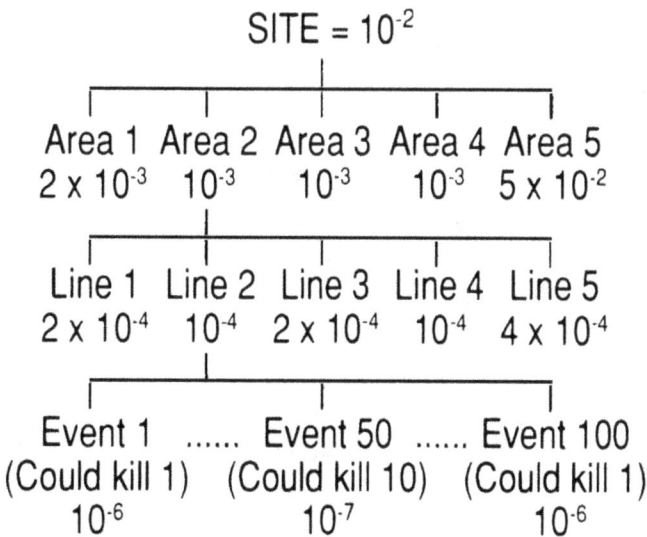

$$\text{SITE} = 10^{-2}$$

Area 1	Area 2	Area 3	Area 4	Area 5
2×10^{-3}	10^{-3}	10^{-3}	10^{-3}	5×10^{-2}

Line 1	Line 2	Line 3	Line 4	Line 5
2×10^{-4}	10^{-4}	2×10^{-4}	10^{-4}	4×10^{-4}

Event 1	......	Event 50	......	Event 100
(Could kill 1)		(Could kill 10)		(Could kill 1)
10^{-6}		10^{-7}		10^{-6}

FIGURE AIV.3   From whole site to one event

In the example shown, an 'event' is either:

- *a single failure mode* (as defined in the FMEA) which on its own has lethal consequences. The probability allocated to this type of event defines the 'tolerable level' which is referred to when the RCM process asks the question "Does this task reduce the probability of the failure to a tolerable level?" See page 102.
- *a multiple failure* where a protected system fails and the protective device which should have rendered the system non-lethal is itself in a failed state. The probability allocated to this type of event defines the 'tolerable level' which is referred to when the RCM process asks, "Does this task reduce the probability of the multiple failure to a tolerable level?" See page 127. It is also the probability used to establish M_{MF} when setting failure finding intervals. See page 169.

In complex systems, it is likely that an approach similar to a fault tree analysis would be used to allocate probabilities (see Andrew and Moss[1993]). However, in this case, we work *downwards* from a top-event probability (the probability of a fatal accident anywhere on the site) to establish objectives for each safety-oriented proactive task and to determine failure-finding task intervals, rather than *upwards* to determine a top-event probability based on an existing maintenance program.

A detailed examination of fault trees is beyond the scope of this book. The purpose of this appendix is only to suggest how it might be possible to convert risks which individual members of society might be prepared to tolerate (another manifestation of 'desired performance') into meaningful information which can be used to establish a maintenance program designed to deliver that performance.

The process described above can be used to produce a graph showing the probabilities of a single fatal event at work which would flow from the risks which one individual is prepared to accept, on the assumption that his or her judgement is accepted by everyone else on the site. This is illustrated in Figure AIV.4. Note that in the next four graphs, the X-axis represents the probability of any one event occurring in any one year, (or more accurately, the annual failure rate.)

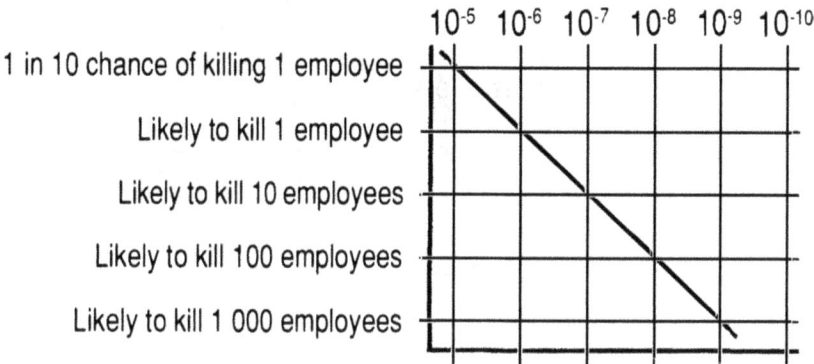

FIGURE AIV.4   Tolerability of one lethal event where I have some control and some choice

The same process could be applied to the situation in which the likely victims have no control but some choice about exposing themselves to the risk. The example in Figure AIV.2 suggests that an airline passenger might be a typical example of someone in this situation. From the maintenance viewpoint, such people are likely to be users of mass transport systems, or people visiting large buildings (shops, offices, sports stadiums, theatres, and so on). In general, these people could be called 'customers'.

In this case, if they all tolerate the same risk as the individual in Figure AIV.2 (and there are the same number of potentially life-threatening events inherent in the system), the process of apportioning risk used in Figure AIV.3 could lead to the single-event probabilities shown in Figure AIV.5

Similar reasoning applied to the no control/no choice scenario might yield the single event probabilities shown in Figure AIV.6. (In practice, most individuals are likely to tolerate an even lower probability of being killed for this reason than is shown in Figure AIV.2 the so-called

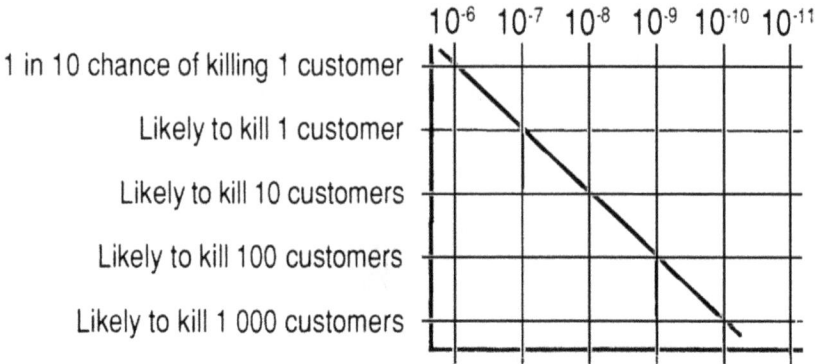

10⁻⁶ 10⁻⁷ 10⁻⁸ 10⁻⁹ 10⁻¹⁰ 10⁻¹¹

1 in 10 chance of killing 1 customer

Likely to kill 1 customer

Likely to kill 10 customers

Likely to kill 100 customers

Likely to kill 1 000 customers

FIGURE AIV.5   Tolerability of one lethal event where I have no control and some choice

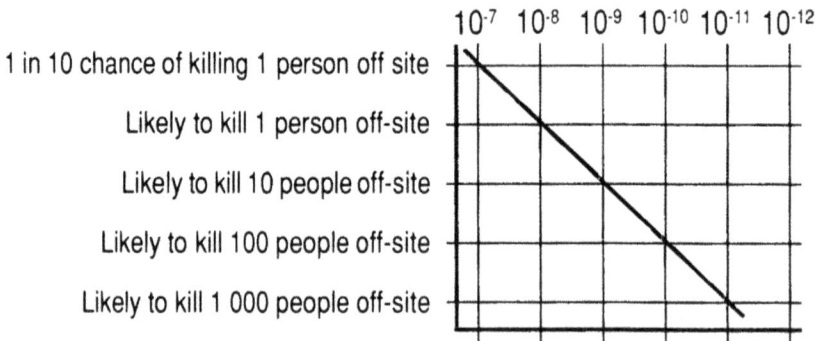

10⁻⁷ 10⁻⁸ 10⁻⁹ 10⁻¹⁰ 10⁻¹¹ 10⁻¹²

1 in 10 chance of killing 1 person off site

Likely to kill 1 person off-site

Likely to kill 10 people off-site

Likely to kill 100 people off-site

Likely to kill 1 000 people off-site

FIGURE AIV.6   Tolerability of one lethal event where I have no control and no choice

"dread" factor. However, in most facilities, fewer events would be likely to have off-site consequences, so the probability for each event might end up about the same.)

Once tolerable probabilities have been determined for single events as shown in Figures AIV.1, AIV.4, AIV.5 and AIV.6, it is of course possible to combine them into a single 'continuum of risk', as shown in Figure AIV.7.

Please note once again that these figures are not meant to be prescriptive and do not necessarily reflect the views of the author or any

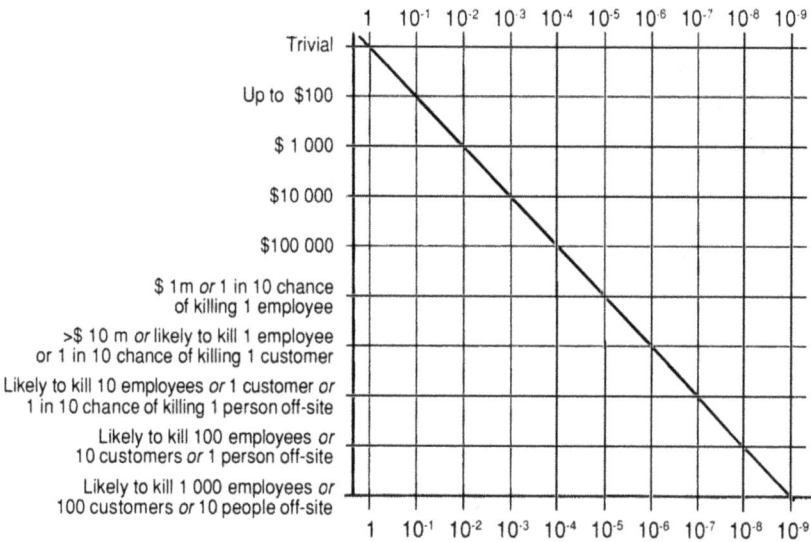

FIGURE AIV.7   A "continuum of risk"

other organization or individual as to what should or should not be tolerable.

Figure AIV.7 is also not intended to imply that 1 employee is worth $10 million. That figure represents a point at which two different value systems happen to coincide. The financial risks which your organization is willing to tolerate and the personal risks which your employees and customers (and society as a whole in the case of no control/no choice hazards) are prepared to tolerate may lead to a completely different set of figures in your operating context.

The key point is that the criterion upon which the whole RCM philosophy is based is what is *tolerable*, not what is practicable or what is a current industry norm (although these may coincide). Part 3 of Chapter 7 suggested that the people who are both morally and practically in the best position to decide what is tolerable are the likely victims. These are the shareholders and their management representatives in the case of financial risks, and employees, customers and the managers who have to clear up afterwards (and bear the responsibility) in the case of personal risks. As mentioned above, this appendix shows one way in which it may be possible to turn informed consensus about tolerable risk into a framework for setting targets for maintenance programs designed to deliver it.

Finally, please bear in mind that the approach outlined in this appendix is not intended to be prescriptive. If you have access to a different framework which satisfies all the parties involved, then by all means use it.

Index

A

ACAP
  benefits, 47–48
  overview, 46–47
actuarial analysis
  complexity, 423
  general information, 422
  reporting failure, 424–425
  Resnikoff conundrum, 425–427
  sample size and evolution, 423
age-related failure, 257–260
  and preventive maintenance, 260–261
  scheduled restoration and scheduled discard
    tasks, 261–269
  See also deterioration
airlines, and RCM, 515–519
Aladon Network, 1–2, 15
appearance, secondary functions, 94
applied stress, increase in, 126–131
applying RCM
  asking a single individual to apply, 474
  asking manufacturers or equipment vendors to
    apply, 476–477
  incorrectly, 473–479
  in perpetuity, 471–473
  too hurried or superficial an application, 473
  at too low a level, 473
  with too much emphasis on failure data,
    473–474
  using computers to drive the process, 478–479
  using outsiders to apply, 477–478
  using the maintenance department on it own to
    apply, 475–476
  See also implementation
asset criticality and asset prioritization. See ACAP
asset criticality assessment, 449
asset prioritization, 449
asset registry, 48
asset verification, 48–49, 449–450
assets
  maintainable and not maintainable, 80–81
  physical assets, 18, 26, 56, 390
  significant assets, 465–467
auditing, 54, 380–381
  preparing an audit file, 454
  providing an audit trail, 513
  what the audit entails, 382–386

when to do the audit, 381–382
who should do the audit, 381
automation, 18
AV process. See asset verification
availability, 51
  and failure-finding tasks, 317–322
  as an RCM achievement, 505–507
  required availability of hidden functions,
    231–233
  See also reliability

B

Basson, Marius, 1–2, 16
batch processes, 63
bathtub curve, 421–422
behavior-based safety, 143–144
best practices, and standardization, 21
buffer stock, 67–68
butterfly effect, 294

C

capability of people, changes to, 391
cause, vs. effect, 148–149
centers of excellence (COEs), and standardization,
  21
chaos theory, 294
cleanliness, secondary functions, 94
comfort, secondary functions, 93–94
commissioning, 420
complexity, 272–273, 423, 428
compliance, legal and regulatory, 215
condition monitoring, 282–283
consequence severity, 115
  auditing, 384–385
  failure effects and, 149–158
  and proactive maintenance, 204–206
consequences, and proactive maintenance,
  204–206
containment, secondary functions, 93
contamination, 125
  secondary functions, 94
control, secondary functions, 92–93
corporate social responsibility (CSR), 21–22
critical assets, 48
customer service, 215
cusum chart, 284

D

data
 entering into a computerized database, 454–455
 summary of key maintenance decision-support data, 430–431
 uses of in formulating maintenance policies, 429–435
data centers, 16
database, as an RCM achievement, 56–57
decision diagrams, 255, 331, 338
 framework of decision processes, 356
 for preliminary assessment of a proposed modification, 343–347
Decision Worksheet
 combination of tasks, 367, 368
 completing, 374–376, 454
 duration of the task, 372–373
 failure consequences, 360–361
 initial task interval, 370–371
 no scheduled maintenance, 369
 one-time change recommendations, 373
 one-time changes, 367, 368
 optimizing existing protective devices, 366–367
 overview, 357–360
 proactive detection, 364–366
 proactive maintenance, 363–364
 proactive risk management strategies, 361–362
 proposed risk management strategy, 370
 quantity of tradespeople required to do the task, 372
 revised risk ranking, 373–374
 storing in a database, 376–378
 unit of measure, 371
 who must do the PM, 371–372
defect elimination, 22–23, 334–335, 404
 See also functional failures
defects, reporting, 403–404
defining failure, 427–428
desired performance, increase in, 126–131
deterioration, 123–124, 256–260
 final stages of, 291–294
 normal deterioration mechanisms, 139
 See also age-related failure
Digitalist Magazine, key factors changing modern businesses, 14–15
dirt, 125
disassembly, 125
downtime, 18, 51
 and the net P-F interval, 279
 vs. repair time, 155–156
drawings, 513

E

economic risk, 209, 214–221, 222
 RCM strategy selection process for, 44
 routine maintenance and hidden functions posing operational risk, 241–242
 strategies dealing with, 350
economic-life limits, 264, 265
economy, secondary functions, 96–97
effect, cause vs., 148–149
efficiency, secondary functions, 96–97
eight questions, 29, 101, 183
 and answers, 437–443
elapsed-time planning, 400–401

eliminating defects, 22–23, 334–335, 404
end effect, 152, 153, 177–178, 179
engineering solutions, 312
engineers, 480
environmental consequences, and one-time changes, 338–339
environmental integrity
 as an RCM achievement, 55, 503–505
 secondary functions, 91
environmental issues, 18
environmental responsibility, 21–22
environmental standards, 64–65
 See also standardization
equipment, manufacturers or vendors, 158–160
equipment training, 57
ESCAPERS categories, 99
evident economic risk, 36
evident failures, 207–210, 313
 and tolerable risk, 222–225
evident functions, 207–210
 substituting for the hidden function, 340
evident physical risk, 35
evident tolerable risk, 36
evolution of RCM3, 519–520
experts, 509–510

F

facilitators, 53
 applying RCM logic, 453
 assessing risk, 203
 building skills, 480
 conducting meetings, 455–457
 developing a risk framework and tolerability thresholds, 450
 general information, 447–448
 interacting with people outside review group, 459–462
 managing the analysis, 453–455
 performing asset criticality review and asset prioritization, 448–450
 planning the RCM project, 450–453
 time management and implementation planning, 457–459
 who should facilitate, 462–463
failed states, 22–23, 32, 103–114
 auditing, 383
 describing, 113–114
 and hidden functions, 243–244
 See also functional failures
fail-safe protective devices, 226–227, 247–248
failure
 calculating probability of a multiple failure, 233–237
 complexity, 272–273
 consequence severity, 204–206
 defined, 102
 defining, 427–428
 evident failures, 207–210
 hidden failures, 208–209
 non-age-related, 269–274
 operational consequences, 214–221
 partial, 104–105
 performance standards and, 103–111
 potential failures and on-condition maintenance, 274–276
 protected functions, 225

reporting, 424–425
start-up, 419–421
statistically random failures, 272
strategic framework summary, 249–251
upper and lower limits, 105–108
variable stress failure, 269–272
failure development period, 277
  See also P-F intervals
failure effects, 151
affecting production or operations, 154–156
auditing, 384
and consequence severity, 33–34, 149–158
defined, 149–150
how to record, 171–171
physical damage, 156
Resnikoff conundrum, 162–163
revenue loss, 157–158
secondary damage, 156–157
sources of information about, 158–163
threat to safety or the environment, 154
what must be done to repair the failure, 157
failure mechanisms, 138
failure mode, 32–33
failure modes
analyzing, 118–122
auditing, 383–384
behavior-based safety, 143–144
categories of, 122–132
causation, 133–134, 135–137
cause versus effect, 148–149
consequences, 145–147
defined, 115–118
details, 132–149
deterioration, 123–124
dirt and contamination, 125
disassembly, 125
evidence failure mode has occurred, 151–153
failure mechanisms, 138
falling capability, 123–126
generic lists of, 160
and hidden functions, 243–244
how often failure mode occurs if nothing done
    to prevent it, 151
how to record, 171–181
human errors, 125–126, 141–142
identifying the root cause, 140–142
incorrect process or packaging materials, 130–131
increase in desired performance or applied
    stress, 126–131
initial incapability, 131–132
and intolerable risk, 211
latent causes, 143
lubrication failures, 124–125
multiple failure modes in a single protective
    device, 324
normal deterioration mechanisms, 139
and the operating context, 149
physical failures, 141
probability, 144–145
reasonable likelihood, 147–148
Resnikoff conundrum, 162–163
root cause of failure, 134–138
secondary damage, 156–157, 225
sources of information about, 158–163
sudden, unintentional overloading, 129–130
sustained, deliberate overloading, 127–128
sustained, unintentional overloading, 128–129

technical history records, 161
when the failure mode is most likely to occur,
    150–151
failure patterns, 38–39
A, 260, 261, 263, 264, 421–422
age-related failure, 260
B, 259, 260, 261, 263, 406–409
C, 260, 263, 414–418
D, 273–274, 418
E, 273–274, 290, 319, 409–414
F, 273–274, 418–421
overview, 269, 405
failure rate, 188
failure-finding intervals
availability and reliability, 317–320
calculating using availability and reliability only,
    320–321
informal approach to setting, 328
methods for calculating, 322–326, 329
practicality of, 329–330
required availability, 321–322
sources of data for calculations, 327–328
failure-finding tasks, 41–42, 313, 442
causing failure, 324–326
in the Decision Worksheet, 364–365
for hidden functions with economic
    consequences, 330
multiple failures and, 314–315
technical aspects of, 315–317
technical feasibility of, 330–333
when failure-finding is not suitable, 332–333
falling capability, 123–126
fatigue, 140
Federal Aviation Agency, Industry Reliability
    Program, 516
FFI. *See* failure-finding intervals
finance, 17
fire hoses, 231
  See also protective devices
First Industrial Revolution, 11
flow processes, 63
FMEAs, 144–145
equipment manufacturers/vendors as sources of
    information for, 158–160
generic lists of failure modes, 160
and the operating context, 149
Resnikoff conundrum, 162–163
technical history records, 161
Fourth Industrial Revolution, 13–23
Fourth Industrial Revolution, The (Schwab), 14
Fourth-Generation Maintenance, 311–312
functional block diagrams, 88
functional checks, 41–42
in the Decision Worksheet, 365–366
functional failures, 32, 103–114
and on-condition tasks, 289–290
gauges and indicators, 110
maintenance managers, 111–112
and the operating context, 111
production managers, 111
safety officers, 111, 113
  See also failed states
functions, 31
auditing, 383
defined, 78–79
dependent primary functions, 89
how to list, 99–100

functions (*Cont.*)
 multiple independent primary functions, 88–89
 performance standards, 79–86
 primary functions, 87–89
 secondary functions, 90–99
 serial primary functions, 89
 superfluous, 98–99
 types of, 86–99

G

globalization, 21

H

health care, 17
Heinrich, Herbert W., 143
Heinrich's Law, 143
hidden economic risk, 35
hidden failures, 208–209, 313, 440
 and on-condition tasks, 302
 detecting, 240
 and one-time changes, 339–342
 preventing, 239–240
 and protective devices, 226–231
 RCM strategy selection process for, 43–44
 risks associated with, 226–249
hidden functions, 208–209
 deciding on a maintenance strategy, 242–248
 duplicating, 341
 failed states and failure modes, 243–244
 failure-finding tasks for hidden functions with
  economic consequences, 330
 making evident by adding another device,
  339–340
 normal circumstances, 246–247
 operating crew, 246
 and protective devices, 245
 and the question of time, 244–245
 required availability of, 231–233
 routine maintenance and hidden functions
  posing a physical risk, 237–241
 routine maintenance and hidden functions
  posing operational risk, 241–242
 substitute a more reliable hidden device, 341
hidden physical risk, 35
high-frequency maintenance schedules, 394–395,
 396–400
history of RCM, 515–520
human errors, 125–126, 141–142
human senses, inspection techniques on, 286–287

I

implementation, 54
 choosing an approach to, 469–470
 comprehensive approach, 468–469
 defect elimination, 404
 key steps, 379–380
 maintenance schedules, 392–403
 one-time changes, 389–391
 RCM audit, 380–386
 reporting defects, 403–404
 risk management strategy task descriptions,
  386, 388
 selective approach, 464–468
 task force approach, 463–464

work packages, 391–394
 See also applying RCM
incorrect application of RCM, 473–479
independent protection layers (IPLs), 248–249
Industrial Revolution
 first, 11
 fourth, 13–23
 second, 12
 third, 12–13
information worksheet, 99–100
 completed example, 181–182
 completing, 454
 describing failed states, 113–114
 and failure modes, 119
 general information, 164
 how to record failure effects, 171–181
 how to record failure modes, 171–174
 levels of analysis, 164–171
 listing functions, 99–100
inherent reliability, 337–338
inherent risk, 34, 53, 200–204
 calculation, 151
initial task interval, 370–371, 386
innovation, 23
international standards, 58
 See also standardization
intolerable risk, 36–37, *198*, 205
 defined, 253
 and failure modes, 211
 and proactive failure management strategies,
  254–255
ISO 9000, 388–389
ISO 31000, 184, 389
ISO 55000, 58, 389

K

Koekemoer, Theuns, 2, 520

L

labor, 500–501
lead time to failure, 277
 See also P-F intervals
levels of analysis, 164–171
 auditing, 383
life limits, 264–265
linear P-F curves, 294–298
loading, 63–64
local effects, 152, 175
logistics
 and repair times, 68
 and skills availability, 69
 spare part optimization, 353
low-frequency maintenance schedules, 394–395,
 400–403
lubrication failures, 124–125

M

maintainers, 479
maintenance
 applying RCM in perpetuity, 471–473
 challenges, 26–28
 cost effectiveness of, 30
 costs, 18, 499–500
 database, 512–513

first-generation, 11
fourth-generation, 13–23
Fourth-Generation Maintenance, 311–312
labor, 500–501
new developments, 24–26
new expectations and reality, 23–24
new techniques, 24, 25
no scheduled maintenance, 42, 351–352, 369
overview, 7–10
planning and control, 502
and RCM, 28–29
relationship of safety and reliability, 10
routine maintenance and hidden functions
posing a physical risk, 237–241
routine maintenance and hidden functions
posing operational risk, 241–242
second-generation, 11–12
spare part optimization, 42, 353
spares and materials, 501
and start-up failures, 420
third-generation, 12–13
Third-Generation Maintenance, 311
See also on-condition maintenance; preventive
maintenance; proactive maintenance
maintenance effectiveness, 482
different expectations, 485–487
different functions, 487–499
measuring, 482–485
maintenance efficiency, 499–502
as an RCM achievement, 508–514
maintenance managers, 111–112, 480
maintenance policies
actuarial analysis in establishing, 422–429
uses of data in formulating, 429–435
maintenance schedules, 54
done by operators, 395–396
high-frequency, 394–395, 396–400
implementation, 392–394
low-frequency, 394–395, 400–403
and quality checks, 396
Maintenance Task Analysis (MTA), 48
maintenance technology, guidelines for acquiring,
510
man-failure, 143
manuals, 513
manufacturers, 158–160
market demand, and seasonal and daily
fluctuations, 70
Mean Time Between Failures (MTBF), 51–52, 328,
432–433
and random failures, 412
Mean Time To Failures (MTTF), 51–52
mean time to repair (MTTR), 68
meetings, facilitating, 455–457
mistake-proofing. See poka-yoke
motivation of individuals, 57, 511–512
Moubray, John, 1–2, 15, 312, 519–520
multiple failure, 313–314
calculating probability of, 233–237
and failure-finding, 314–315

N

National Institute for Occupational Safety and
Health (NIOSH), National Traumatic
Occupational Fatalities (NTOF) data, 9–10
National Safety Council, 9

net P-F interval, 278–280
See also P-F intervals
next-higher-level effect, 152, 153, 176–177
no scheduled maintenance, 42, 351–352, 369
nonlinear P-F curves, 291–294
non-operational consequences, 222–225
and on-condition tasks, 302–303
and one-time changes, 342–347
See also tolerable risk
non-operational risk, 441
normal circumstances, 246–247
Nowlan and Heap report, 1, 16, 272, 343,
425–426

O

OEE, 496–499
on-condition maintenance, potential failures and,
274–276
on-condition tasks, 40–41, 276, 277, 305, 441
condition monitoring, 282–283
determining if tasks are worth doing, 302–303
inspection techniques on human senses,
286–287
the P-F interval and operating age, 290–291
potential and functional failures, 289–290
primary effects monitoring, 285–286
product quality variation, 283–285
selecting the right category, 287–289
technical feasibility of, 281
one-time changes, 42, 54, 212, 240–241, 333–334,
442
applied proactively, 338–347
defect elimination, 334–335
design and maintenance, 336–338
implementing, 389–391
poka-yoke, 335–336
redesign and modification, 334–336
operating context, 30, 53
batch and flow processes, 63
defined, 59
documenting, 70–75
failure modes and, 149
functional failures and, 111
logistics and repair times, 68
market demand and seasonal and daily
fluctuations, 70
operating environment and environmental
standards, 64–65
overview, 59–62
quality standards, 64
raw material supply, 70
recycling, reuse, or repurposing, 70
redundancy, 63
and risk, 185
safety hazards, 66
safety standards, regulations, and regulatory
requirements, 65–66
shift arrangements, 66–67
skills availability and logistics, 69
spares and stocking policies, 68–69
utilization and loading, 63–64
work in progress and buffer stock, 67–68
operating costs, 215
operating crew, 246
operating environment, 64–65
operating performance, improved, 55

operational consequences, 214–217
  avoiding, 217–221
  and on-condition tasks, 302–303
  and one-time changes, 342–347
operational risk, 440
operations managers, 480
operators, 479
outcomes of RCM3 analysis, 53–54
  general information, 481–482
  higher plant availability and reliability, 505–507
  maintenance effectiveness, 482–499
  maintenance efficiency, 499–502
  safety and environmental integrity, 503–505
output, 215
overall equipment effectiveness. *See* OEE
overloading, 127–130
ownership, 474

P

packaging materials, incorrect, 130–131
patterns of failure. *See* failure patterns
Payment Card Industry (PCI) Data Security
  Standard (DSS), 17
performance standards, 31, 50–52
  absolute, 83
  and failure, 103–111
  functions, 79–86
  gauges and indicators, 110
  multiple, 82
  multiple performance standards and the OEE,
    496–499
  qualitative, 82–83
  quantitative, 82
  upper and lower limits, 84–86, 105–108
  variable, 83–84, 108–110
  See also standardization
P-F curves, 2745
  linear P-F curves, 294–298
  nonlinear P-F curves, 291–294
  and random failures, 413
P-F intervals, 276–277
  and condition monitoring, 283
  consistency, 280–281
  and the final stages of deterioration, 293
  how to determine, 298–302
  net P-F interval, 278–280
  and operating age, 290–291
physical assets
  changes to, 390
  dependence on, 18
  longer useful life of, 56
  management of, 26
physical risks, 209, 210–214, 439–440
  RCM strategy selection process for, 44
  routine maintenance and hidden functions
    posing a physical risk, 237–241
  strategies dealing with, 350–351
pilot projects, 470
planned maintenance optimization (PMO), 48
planning RCM, 50
plant operations, changes to, 390
plant performance, as an RCM achievement, 505–507
poka-yoke, 143, 335–336
potential failure
  and on-condition maintenance, 274–276
  and on-condition tasks, 289–290

potential worst-case effect, 152, 153, 178–181
preventive maintenance
  age-related failure and, 260–261
  traditional approach to, 515–517
primary effects monitoring, 285–286
primary functional effectiveness (PFE), 499
PRN, 466–467
proactive detection
  failure-finding tasks, 364–365
  functional checks, 365–366
proactive maintenance
  predictive and preventive maintenance tasks,
    363–364
  and safety, 211–213
  technically feasible and worth doing, 204–206
proactive management, 119
  combination of tasks, 307–312
  failure-finding as a proactive risk management
    strategy, 331–332
  proactive tasks, 303–306
  treating intolerable risk, 254–255
proactive tasks, 206
probability, 188
  calculating probability of a multiple failure,
    233–237
probability/risk number. *See* PRN
process, incorrect, 130–131
product quality, 215
  as an RCM achievement, 507
  variation, 283–285
production managers, 111
project workflow, 459
protected functions, 225
  preventing failure of, 238–239
protection layers, 248–249
protective devices
  fail-safe, 226–227, 247–248
  hidden failures and, 226–231
  multiple failure modes in a single protective
    device, 324
  not fail-safe, 227–231
  optimization of, 313–314, 366–367
  primary functions, 245
  secondary functions, 94–96, 245
PWCE. *See* potential worst-case effect

Q

quality checks, maintenance schedules and, 396
quality standards, 64
  See also standardization

R

raw material supply, 70
RCM
  achievements, 55–57
  applying in perpetuity, 471–473
  applying incorrectly, 473–479
  applying the RCM process, 45–46
  history of, 515–520
  as an integrative framework, 514
  and maintenance, 28–29
  overview, 1–3
  planning, 50
  strategy selection process, 43–45
RCM analysis, history of, 517–519

RCM2, 1–2, 519
RCM3
  and the changing world of maintenance, 7–10
  eight questions, 29, 101, 183, 437–443
  evolution of, 519–520
  overview, 2–3
RCMII (Moubray), 15, 312
reactive maintenance, 119
reasonable likelihood, 147–148
recycling, 70
  secondary functions, 97
redesign. *See* one-time changes
redundancy, 46, 63
regulatory requirements, 65–66
  secondary functions, 98
reliability, 51–52, 99
  comparing, 412–413
  and failure-finding tasks, 317–321
  inherent reliability, 337–338
  as an RCM achievement, 505–507
  and safety, 10
  See also availability
Reliability-Centered Design (RCD), 347–348
reliability-centered maintenance. *See* RCM
*Reliability-Centered Maintenance (RCMII, Second
    Edition)* (Moubray), 15, 312
Reliability-Centered Spares (RCS), 353
renewable strategies, 22
repair costs, and the net P-F interval, 279
repair times, 68
  vs. downtime, 155–156
  excluding from protective devices' availability,
    319–320
reporting defects, 403–404
repurposing, 70
  secondary functions, 97
residual risk, 54
Resnikoff, H. L., 163
Resnikoff conundrum, 162–163, 425–427
reuse, 70
  secondary functions, 97
revenue loss, 157–158
review groups, 52–53, 162, 443–444
  determining inherent risk, 200–204
  what each group does, 445–446
  what participants get out of the process,
    446–447
  who should participate, 444–445
revised operating and maintenance procedures, 54
revised risk, 54
  auditing, 386
risk
  assessment, 188
  auditing, 385
  defined, 34–35, 103
  economic risk, 209, 214–221, 222, 241–242
  evident economic risk, 36
  evident physical risk, 35
  evident tolerable risk, 36
  hidden economic risk, 35
  hidden physical risk, 35
  inherent risk, 34, 53, 201–204
  intolerable risk, 36–37, *198*, 205, 211, 253
  managing, 184–195
  non-operational risk, 441
  operating context, 185
  operational risk, 440

physical risks, 209, 210–214, 439–440
  reduction of as an RCM achievement, 55–56
  revised or residual risk, 54
  tolerable risk, 37, 42–43, 191–194, *198*, 205,
    209, 222–225, 253, 443
  who should evaluate, 194–195
risk assessment
  qualitative, 198
  quantitative, 197–198
  subjective, 197
risk avoidance, 19–20
risk management
  default actions for tolerable risk decisions,
    42–43
  identifying risk management strategies,
    190–194
  managing physical and economic risks, 19–20
  overview, 34–37
  for physical risks, 184–195
  proactive strategies, 37–42
risk management strategies
  combination of tasks, 306–312
  in the Decision Worksheet, 361–362, 370
  failure-finding as a proactive risk management
    strategy, 331–332
  proactive failure management strategies,
    254–255
  proactive strategies and technical feasibility,
    253–254
  proactive tasks, 303–306
  RCM strategy selection process, 43–45
  task descriptions, 386–389
risk matrix, 195–204, 450
risk ranking, 200–204
  See also risk matrix
risk-based RCM. *See* RCM3
running-time planning, 401–402

S

SAE JA 1011, 58, 59, 519, 520
  See also operating context
SAE JA 1012, 58, 519
safe-life limits, 264–265, 306
safety, 18
  behavior-based, 143–144
  and the net P-F interval, 279
  and one-time changes, 338–339
  and proactive maintenance, 211–213
  as an RCM achievement, 55, 503–505
  and reliability, 10
  secondary functions, 91
  See also physical risks
safety hazards, 66
safety instrumented functions (SIFs), 248
safety integrity level (SIL), 248
safety legislation, 213–214
safety officers, 111, 113
safety standards, 65–66
  See also standardization
scheduled discard tasks, 41, 441–442
  defined, 262
  effectiveness of, 267–269
  frequency of, 263–265
  technical feasibility of, 266–267
scheduled maintenance, 42
  no scheduled maintenance, 42, 351–352, 369

scheduled restoration, 41, 261–263, 305–306,
    441–442
  defined, 261
  effectiveness of, 267–269
  frequency of, 263–265
  technical feasibility of, 265–266
scheduled rework tasks. *See* scheduled restoration
Schwab, Klaus, 14
Second Industrial Revolution, 11
Second Machine Age. *See* Fourth Industrial
    Revolution
secondary damage, 156–157, 225
senior technicians, 480
services, failure of, 174
shift arrangements, 66–67
Shigeo Shingo, 335
significant assets, 465–467
simultaneous learning, 447
  See also review groups
skill-building, 479–480
skills availability, 69
S-N curves, 414–415
software, 376–378
SOPs. *See* standard operating procedures
spare part optimization, 42, 353
spares, 68–69
  and materials, 501
SPC. *See* statistical process control
staff turnover, 513
standard operating procedures, 391–392
standardization, 20–21, 57
  See also international standards; performance
    standards
standby systems, 46
start-up failures, 419–421
statistical process control, 284–285
statistically random failures, 272
statutory requirements, secondary functions,
    97–98
stocking policies, 68–69
strategy selection process, auditing, 385–386
stress
  and deterioration, 256–260
  peak, 270–271
  variable stress failure, 269–272
structural integrity, secondary functions, 92
superfluous functions, 98–99
supervisors, 480

T

task forces, 463–464
task selection process, 307–312
task times, excluding from protective devices'
    availability, 319–320
tasks
  combination of, 41
  combining, 240
  on-condition tasks, 40–41
  failure-finding, 41–42
  functional checks, 41–42

  one-time changes, 42, 54
  scheduled discard tasks, 41
  scheduled restoration, 41
teamwork, 57, 512
tearoom tick syndrome, 400
technical history records, 161, 433–435
technical validity, 474
Technological Revolution, 11
templating, 21, 467–468
  See also standardization
Third Industrial Revolution, 12–13
Third-Generation Maintenance, 311
time management, 457–459
tolerable risk, 37, 205, 209
  default actions for tolerable risk decisions, 42–43
  defined, 253
  economic risks (operational), 350
  and non-operational consequences, 222–225
  physical risks (safety and the environment),
    350–351
  RCM strategy selection process for, 44–45
  reducing, 443
  risk assessment, *198*
  risk management strategies, 191–194
Toyota Production System, 335
triggers, 140

U

ultimate-level-switches, 231
  See also protective devices
unavailability, 319–320
  See also availability
unit of measure, 371
uptime, 51
  See also availability
users, as sources of information about equipment,
    161–162
utilization, 63–64

V

variable stress failure, 269–272
vibration switches, 230–231
  See also protective devices

W

walk-around checks, 352–353
warning period, 277
  See also P-F intervals
Weibull distributions, 413–414, 415, 417
  shifted, 417–418
work in progress, 67–68
Work Management System (WMS), 46, 48
workflow, 459
work-related deaths, 8–10

Z

zero-based analysis, 151, 467